VETERINARY SCIENCES

VETERINARY SCIENCES

Editor

Reto Schreck

scitus academics

Veterinary Sciences

Edited by **Reto Schreck**

This Book is Printed on Acid-Free Paper. ♾

ISBN: 9781681179339
Library of Congress Control Number: 2017948963

SCITUS Academics LLC,
616, Corporate Way, Suite 2, 4766,
Valley Cottage, NY 10989

www.scitusacademics.com

Notice

Printed in United States of America.

Preface

Animals have played a crucial role in the lives of humans for thousands of years, and ensuring that our furry friends are happy and healthy has become a firmly established field of expertise. Veterinary science deals with the health and wellbeing of animals encompasses everything from preventative care to psychological analysis and complex surgical procedures. By its nature, veterinary science is comparative and gives rise to the basic science disciplines of comparative anatomy, comparative physiology, comparative pathology, and so forth. Veterinary research is directed at understanding and managing the determinants of health and disease at the level of both individuals and populations in order to inform veterinary practitioners, policy makers, animal owners and managers, and other health professionals who share or have a stake in veterinary medicine's purpose. The science of veterinary medicine continues to be rooted in comparative medicine which provides a powerful and historically successful approach to understanding biological phenomena that underlie health and disease. The practice of veterinary medicine continues to be informed by the values and needs of societies as they relate to animals.

This book covers all aspects of veterinary science that helps the theoretical and practical understanding of animal science and medicine relevant to any area of the health and management of domestic animals, poultry, aquatic animals, and wildlife. It covers specific aspects of applied veterinary medicine in pets and in livestock. This book is intended to students, researchers, and scientist exploring the advanced and latest research developments in the field of veterinary sciences and related academic disciplines. The book will be an ideal for veterinary practitioners, policy makers, veterinary managers, and other health professionals who share or have a stake in veterinary.

Table of Contents

CHAPTER 1

Genomic Tools and Animal Health

Ricardo Zanella

School of Agronomy and Veterinary Medicine, College of Veterinary Medicine, University of Passo Fundo, Passo Fundo RS 99052, Brazil

ABSTRACT

Animals have been selected to improve their productivity in order to increase the profitability to the producer. In this scenario, not much attention was given to health traits. As a consequence of that, selection was made for animals with higher production and a shortened productive life. In addition to that, the intense production system used in livestock has forced animals to be exposed to higher pathogen loads, therefore predisposing them to infections. Infectious diseases are known to be caused by micro-organisms that are able to infect and colonize the host, affecting their physiological functions and causing problems in their production and on animal welfare. Even with the best management practices, diseases are still the most important cause of economic losses in the animal industry. In this review article we have addressed the new tools that could be used to select animals to better cope with diseases and pathogens.

Keywords: animal health; GWAS; genomic selection; infectious diseases

1. INTRODUCTION

Animal selection has historically been based on production traits, and not much attention was given to selection for health traits. This resulted in modern cows with high milk yields and reduced fertility, and shortened productive lives, caused by increased susceptibility to postpartum metabolic disorders. However, even in the selection for animals with higher production, it was possible to observe individual differences to challenges in the host response to diseases [1]. Selection has been used to control or manage infectious diseases, such as anthelmintics for parasite control, ascaricides for tick control and antibiotics to control bacterial diseases, and this have created major resistance problems. As a consequence of that, breeders and agricultural industries are now facing an intense pressure to select animals for enhanced disease resistance and tolerance with better welfare conditions [2].

In the last decades, important attention was given to acute infections caused by emerging diseases, especially those that infect humans from other host species. Those are called zoonotic diseases, and are caused by different pathogenic agents. The circulation of zoonotic agents between humans, animals, and the environment is a concern for animal and public health. According to the Institute of Medicine [3], more than 65% of emerging infectious diseases were caused by zoonotic agents, resulting in annually indirect economic losses of over $200 billion [4]. In this context it is expected that infectious disease will cause major economic losses in the livestock sector worldwide.

Many diseases will affect the animals' productivity. Generally, it will occur in the early stages of infection (subclinical), when there is a lack of evidence of infection, which, if untreated, can progress to severe clinical disorders. The overall consequences of diseases on animal production are great. Generally, infected animals have lower performance rates, costs associated with animal treatment, increased mortality rates and possibly reproductive failure [5,6]. Complete knowledge of the disease mechanism is needed for the decision on the protocol and treatment to be used.

Vaccination and good management practices have been widely used to control and to avoid pathogens spreading in populations, providing variable levels of protection. However, the mechanism involving immune protection is not fully understood; therefore, not all the animals respond to the treatment, leading to the discovery of new approaches to address those failures [7,8]. It is common to observe in a vaccinated population some differences in the immune response between animals [7]. Also, differences in the severity of infection of animals exposed to similar doses of pathogens were found, suggesting a possible role of genetics involved with tolerance and resistance to infection [9]. Resistance can be defined as the host's ability to resist an infection or to control the parasite lifecycle, and tolerance is the host's ability to tolerate infection and show little or no effects of disease [9,10].

Up to this moment, the consequences of breeding animals for disease resistance or tolerance in relation to production traits are not known. It is known that selection for disease traits needs to be conducted within the herd, as opposed of production traits which can be done individually. This is because when working with infectious diseases, animals may infect each other directly or indirectly. Therefore, the health status between animals within the same herd is not independent of the other animals, since it depends on the host genetic makeup and also on the epidemiology of the disease.

Many factors have been identified to influence animal health; therefore, it is very complicated to evaluate and to study the host genetics and pathogen interaction. The most difficult concept in the experimental design is how to define and to establish the correct phenotype. How can we assume that an animal is resistant to a disease if we do not know the pathogen load that the animals have been exposed to or if the animal was even exposed? In this context, assumptions need to be made to evaluate animal exposure to pathogens, and it is expected that animals sharing the same environment are possibly experiencing the same amount and type of pathogens. The sensitivity and specificity of the diagnostic tools used are also very important to identify the epidemiologic status of animals in a herd. Different tests can result in a conflicting phenotypic classification, and therefore it can result in a broad range of outcomes with spurious results. Therefore, the success of genomic selection for health traits starts with correct phenotype collection and the accuracy of the diagnostic test. Animals can be infected without presenting clinical signs, or may be considered a false negative. The accuracy of the determination of the infectious status of the animal is directly related to the sensitivity and specificity of the diagnostic test [11], and different results can be obtained when different methodologies are applied. However, a true confirmation is obtained when the pathogen is isolated and sequenced from the organ or systems being affected by a specific disease. In addition to that, pathogen colonization will result in changes in the immune system of the animal, possibly altering the gene expression profile of tissues being infected or even having a global effect. Those changes in the host will have a direct or indirect effect on the disease phenotype.

In this review we have addressed and explored some of the challenges related to animal health, and some of the new tools that could be used to select animals to better cope with diseases and pathogens.

2. INFECTION-DISEASE TRANSMISSION

Several risk factors have been identified to influence whether an animal that was exposed to an agent will become infected or not. In addition to that, they can determine whether the infection will result in, sooner or later, clinical disease, depending on the host immune response. Some of those risk factors can be associated with animal and pathogen genetic composition, the nutritional state of the host at the time of infection, and the environmental conditions. Those factors can also be associated with the spread of the pathogen among other animals [1]. However, in all the cases of infection, the necessary condition is circulating levels of a specific pathogen in a population.

The transmission of pathogens from a reservoir may be continuous or sporadic; however, once the infection is in the host population it may follow different pathways. In a population, the presence of a reservoir can directly transmit the pathogen to the host (Figure 1, Arrow A) in a sporadic or in a continuous way [1]. Once the host population becomes infected, it can be their own continuous source of infection (Figure 1, Arrow C), or result in a cross-contamination to the reservoir (Figure 1, Arrow B). The selection being imposed on the pathogens in the host may differ from the one being imposed in the reservoir; therefore, the genetic makeup in the pathogens on different populations may vary, resulting in different levels of disease severity and infectivity. The initial step in controlling the spread of diseases is the identification and control of those risk factors.

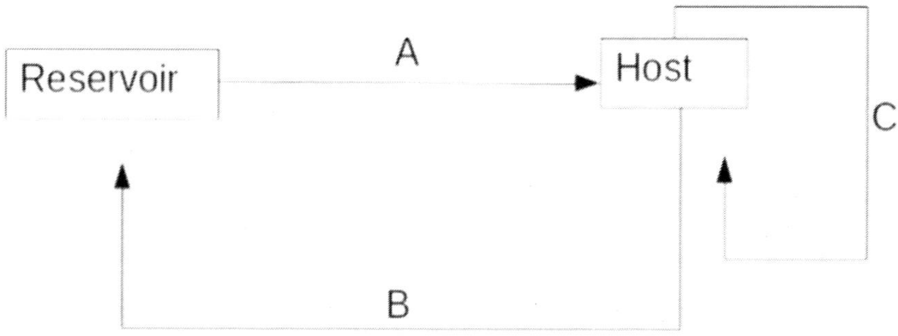

Figure 1. Model of disease transmission among or within populations.

3. SELECTION FOR RESISTANCE OR TOLERANCE

To control the spread of diseases, we first need to control the infection (resistance), and second it is necessary to reduce the severity of the disease (tolerance). The main objective to control the infection is to reduce the number of infected animals (prevalence) in a herd or population, reducing the infectious pressure among the existing population. There are two basic genetic approaches that can be used to select animals to fight a pathogenic infection (resistance and tolerance).

The first method is to select for animals that are resistant or less susceptible to infection; furthermore, susceptible animals can be subdivided as tolerant and less tolerant (Figure 2). Resistance was defined as the ability of an organism to prevent the infection when exposed to a pathogen [12]. Gonda and colleagues in 2007 [13] identified a QTL (Quantitative Trait Loci) on BTA20 affecting susceptibility to *Mycobacterium paratuberculosis* (*MAP*) infection in US Holstein cows. Using a similar approach for phenotypic classification, Settles and colleagues [14] tested animals for the presence of *MAP* with a different diagnostic method with better sensitivity and specificity. Using a genome-wide association analysis (GWAS), a chromosomal region on BTA3 associated with *MAP* tissue infection in a group of Holstein cows was identified. Investigation of this region on BTA3 using 42 single nucleotides to explore a 235 kb region identified a region of 10.6 kb, possibly harboring the causative mutation involved in *MAP* infection [15]. Further investigation using a gene set enrichment analysis identified the gene Endothelin-2 (*END2*) located near the SNP (Single Nucleotide Polymorphism) previously identified by Settles and colleagues [15] to be involved with *MAP* infection in cattle [16]. *EDN2* encodes a member of the protein family of secretory vasoconstrictive peptides and works as a ligand for the endothelin receptors, initiating an intracellular signaling cascade, which can modulate *MAP* tissue infection [9]. Therefore, this is a great positional and functional candidate gene involved in *MAP* infection.

To explore the role of *EDN2* and *MAP* infection, nuclear extracts from freshly harvested ileo-cecal lymph nodes from Holstein cows were used to characterize differences in the binding affinities of the sequences harboring the significant SNPs on BTA3. Luciferase reporter assays were performed to confirm the transcriptional activity of *EDN2*, demonstrating mobility shifts, indicating that one or more variants in the promoter of *EDN2* may result in susceptibility to *MAP* tissue infection in Holstein cattle [17].

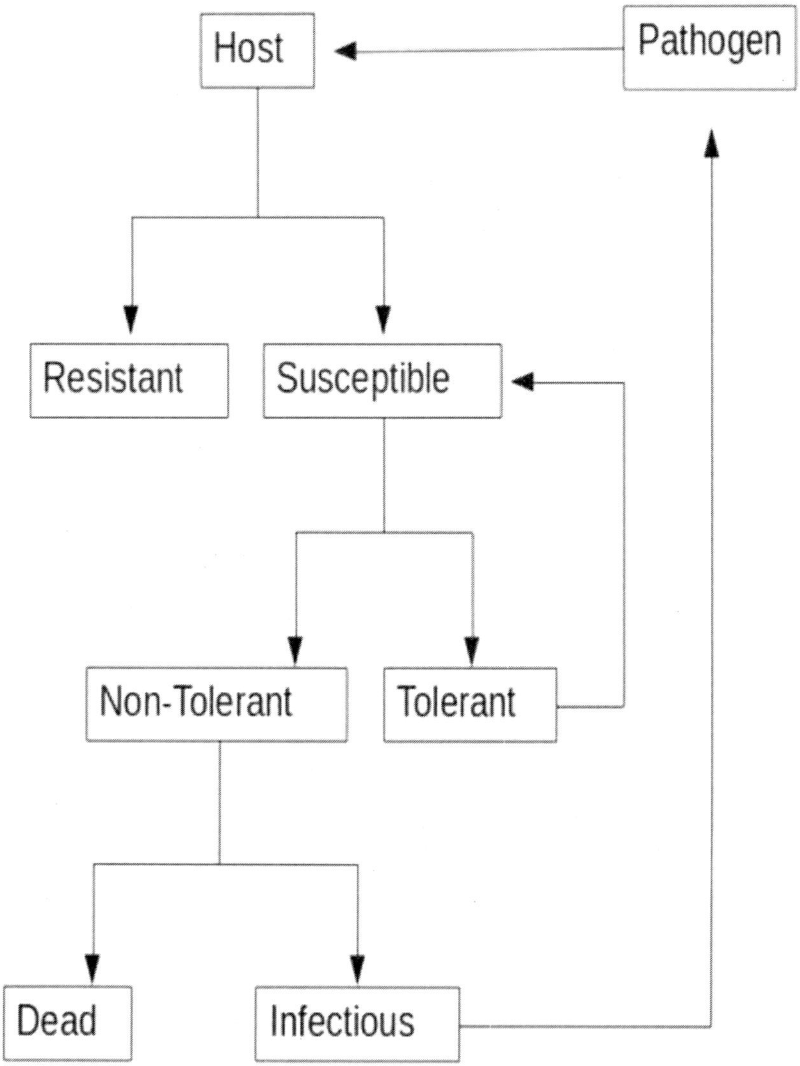

Figure 2. Schematic diagram of host × pathogen interaction-disease infection mode.

The second approach that can be used is to select for animals which are genetically tolerant to the pathogenic infection. Raberg and colleagues [12] presented empirical evidence for genetic variation leading to tolerance in animals using a mouse model to study malaria. Their definition of tolerance was based on individuals who respond to a pathogenic infection in such a way that limits the damage (fitness) to the host without affecting the pathogen burden. Tolerance is generally described as the relationship between the levels of infection intensity and fitness. Therefore, in a tolerant animal, the amount of infection has little or no effect on the fitness of the individual [10,15]. Production performance tends to be reduced in animals with higher pathogen loads, and this is illustrated in Figure 3, which shows a representation of the performance and level of infection.

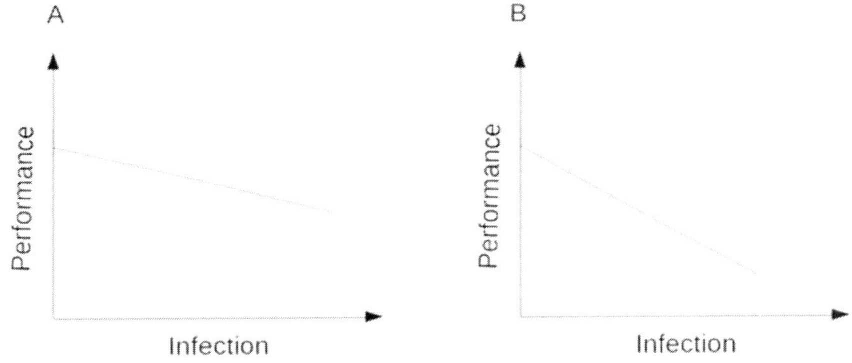

Figure 3. Susceptible animals can be tolerant or non-tolerant; a higher infection load associated with lower losses in the performance is an indication of higher tolerance (**A**) where increased losses in performance caused by increased infection load is an indication of a less tolerant animal (**B**).

Genetic tolerance has been evaluated in cattle that were infected with *MAP* and its genetic association was tested using a GWAS, and a region on BTA15 encoding the gene *GNA12* was associated. For this, tolerance was measured as the amount of *MAP* fecal shedding relative to the levels of *MAP* tissue infection [10,15]. High levels of fecal shedding on Johne's positive animals are generally associated with disease severity, since they are negatively correlated with milk production, fat content and protein content, lower productive life and an increased number of days between pregnancies, as well as high disease transmission rates to the herd [13,18,19,20]. Therefore, levels of *MAP* fecal shedding could be used as an accurate measurement of animal fitness.

Differences in the genetic background between animals also need to be taken into account when evaluating association with genetic markers, to reduce the possibility of identifying erroneous markers involved with the phenotype being studied. This test can be conducted using the pair-wise identical by state test (IBS) computed between all the animals in the study and plotted using a multi-dimensional scaling plot MDS for better visualization.

4. SELECTION FOR BETTER IMMUNE RESPONSE

Another approach that could be used is to select for animals that will have a better response when challenged with pathogens. Vaccination is considered the most powerful tool used to control and to diminish the parasite burden, reducing the clinical signs and the transmission of the pathogen within animals [21]. However, animal response to vaccination can vary within individuals, with some individuals not responding at all or having different levels of protection [22].

In the livestock sector, most of the economic losses have been associated with infection with pathogens. To overcome those losses, several strategies have been proposed and implemented with unsatisfactory results. For example, influenza virus is still considered the most prevalent respiratory pathogen circulating among animals and humans, causing a major impact to the economy and the population's health. Therefore, there is a need to develop new methodologies to increase the efficacy of the current methods.

Here, we present results of an association study of genetic markers with the immune response to influenza vaccination in a swine herd [7]. Animals were vaccinated with an inactivated H1N1pdm virus and tested for the association of genetic markers with immune response efficacy to influenza vaccination (IAV). The animal immune response was measured based on the presence or absence of antibodies to IAV (qualitative measurement), and also based on the levels of antibodies produced by the animals (quantitative).

Using the response to vaccination as a qualitative phenotype, a region on *SSC12* was identified to be associated with the immune response. When we tested the efficacy of the response to the vaccination as a quantitative phenotype, four additional associated chromosomes were identified: *SSC0*, *SSC1*, *SSC7* and *SSC15* [7]. These results suggest the complexity of the immune traits when used to test their interaction with host genetics.

However, even in vaccinated animals we can have responders that will become infected, and those animals can recover from infection but will present losses in production. Those losses are generally associated with lower production or even reproductivity failure (Figure 4).

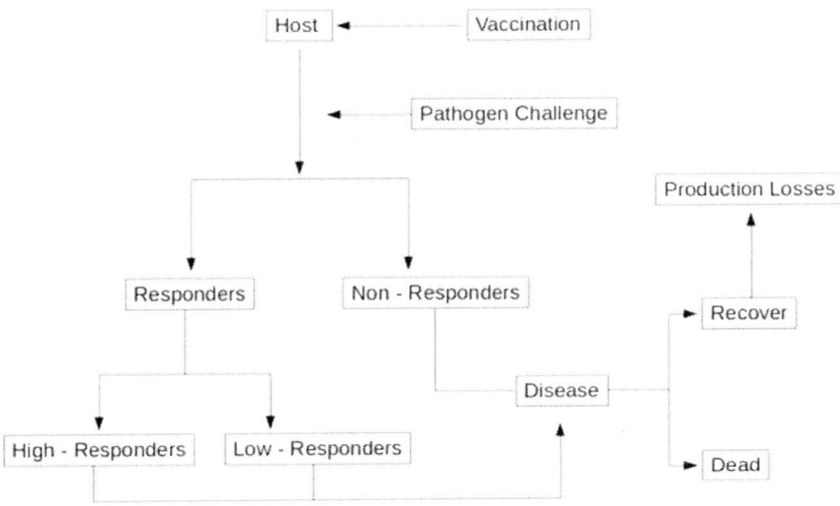

Figure 4. Representation of animal response to vaccination.

5. NEW APPROACHES TO OLD PROBLEMS

The great advantage of high-throughput sequencing is the ability to generate a large amount of genomic information to address complex problems. In addition to that, the use of genomic sequencing for the identification of specific pathogens associated with diseases has been facilitated, allowing for the identification of genomic variations associated with the pathogenicity of specific agents and the sensitivity to specific treatment drugs.

In dairy cows, clinical mastitis is considered a complex health disorder which will impact the farm's profitability and the animal welfare. The heritability associated with this disorder can range from 0.01 to 0.42, indicating the involvement of genetic components with the appearance of this disorder [23]. Using whole genomic association studies (GWAS), genetic markers were identified on chromosomes 2, 14 and 20 as associated with susceptibility to this disorder, indicating a polygenic effect [23]. Some of the genes located near the associated markers were involved with immune response activation and mammary gland metabolism. Despite those findings, it was proposed that susceptibility to clinical mastitis is regulated by different genes related to immune response which are pathogen-specific; therefore, animal selection for resistance to a specific pathogen might not work for different pathogens.

Recent studies have applied the RNA sequencing technique to unsolved biological problems, especially for the identification of genes involved with illness [24,25,26]. This new methodology uses the whole transcriptome sequencing of a specific tissue at a specific time point. This might be a useful tool to investigate the genetic mechanisms involved with the host infection and with the host immune response to an infection. Beside the advantage of using a small number of animals, this methodology will possibly be able to give the whole picture of what is happening transcriptomically when the animal is facing a challenge, especially within an infection.

The host-pathogen interaction has been described in a study using the RNA-Seq approach, which investigated both viral and host gene expression in a mouse model after H1N1 infection. An important gene involved with viral replication and host defense has been identified in this model. The use of this methodology could be applied to elucidate the host-pathogen interactions, increasing the knowledge of genes involved with host susceptibility or survival during infections [27].

As an example, animals with different genotypes (AA, AG, GG) for a specific marker on exon 5 of *EDN2* (the gene involved with resistance to *MAP* infection) had different levels of its expression in the lymph nodes. This indicates a possible suppression of the gene expression in animals with a specific genotype when exposed to specific challenges [28]. However, this study was conducted evaluating the transcription profile using only one gene in one tissue, without accounting for the effect of other genes. The use of high-throughput sequencing offers us an opportunity for the identification of global transcriptomic changes during specific challenges in different tissues at the same time.

6. CONCLUSIONS

Genetic selection for animals that are resistant, or tolerant or for animals that have a better immune response to vaccination can help diminish the impact of diseases in the livestock sector. However, it is not known what the best strategy is and what the long-term consequences are when selecting for those traits. However, if selection for those traits can reduce human exposure to infectious pathogens, it offers a potential to reduce the health risks in a population.

ACKNOWLEDGMENTS

The data used in this paper was originated from projects funded by: Washington State University Center for Reproductive Biology, Berry College, BJT grant 373167/2012-1 and 447470/2014-0 from the National Council of Scientific and Technological Development (CNPq) and ARD/PPP/FAPERGS/CNPq 08/2014. No funding was provided for publication fees.

CONFLICTS OF INTEREST

The author declares no conflict of interest.

REFERENCES

1. Bishop, S.C.; Stear, M.J. Genetic and epidemiological relationships between productivity and disease resistance: Gastrointe stinal parasite infection in growing lambs. Anim. Sci. 1999, 69, 515–524.
2. Bishop, S. Disease Genetics: From Concepts to Utilisation. In Presented at the Short Course at Iowa State University, Ames, IA, USA, 2–6 June 2014.
3. National Research Council (US) Committee on Achieving Sustainable Global Capacity for Surveillance and Response to Emerging Diseases of Zoonotic Origin. Sustaining Global Surveillance and Response to Emerging Zoonotic Diseases; Keusch, G.T., Pappaioanou, M., Gonzalez, M.C., Scott, K.A., Tsai, P., Eds.; National Academies Press: Washington, DC, USA, 2009.
4. World Bank. Volume One: Towards A One Health Approach for Controlling Zoonotic Diseases. World Bank: Washington, DC, USA, 2010. Available online: http:// documents. worldbank.org/curated/en/214701468338937565/Volume-one-towards- a-one-health-approach-for-controlling-zoonotic-diseases (accessed on 10 August 2016).
5. Snowder, G.D.; van Vleck, L.D.; Cundiff, L.V.; Bennett, G.L. Bovine respiratory disease in feedlot cattle: Environmental, genetic, and economic factors. J. Anim. Sci. 2006, 84, 1999–2008.
6. Garcia, M.D.; Thallman, R.M.; Wheeler, T.L.; Shackelford, S.D.; Casas, E. Effect of bovine respiratory disease and overall pathogenic disease incidence on carcass traits. J. Anim. Sci. 2010, 88, 491–496.
7. Zanella, R.; Gava, D.; Peixoto, J.O.; Schaefer, R.; Ciacci-Zanella, J.R.; Biondo, N.; da Silva, M.V.; Cantao, M.E.; Ledur, M.C. Unravelling the genetic components involved in the immune response of pigs vaccinated against influenza virus. Virus Res. 2015, 210, 327–336.
8. Davies, G.; Genini, S.; Bishop, S.C.; Guiffra, E. An assessment of opportunities to dissect host genetic variation in resistance to infectious diseases in livestock. Animal 2009, 3, 415–436.
9. McSpadden, K.; Caires, K.; Zanella, R. The effect of mycobacterium avium subspecies paratuberculosis exposure on animal health. Acta Sci. Vet. 2013, 41, 1095.

10. Zanella, R.; Settles, M.L.; McKay, S.D.; Schnabel, R.; Taylor, J.; Whitlock, R.H.; Schukken, Y.; Van Kessel, J.S.; Smith, J.M.; Neibergs, H.L. Identification of loci associated with tolerance to Johne's disease in Holstein cows. Anim. Genet.2011, 42, 28–38.
11. Fulton, R.W.; Confer, A.W. Laboratory test descriptions for bovine respiratory disease diagnosis and their strengths and weaknesses: Gold standards for diagnosis, do they exist? Can. Vet. J. 2012, 53, 754–761.
12. Raberg, L.; Sim, D.; Read, A.F. Disentangling genetic variation for resistance and tolerance to infectious diseases in animals. Science 2007, 318, 812–814.
13. Gonda, M.G.; Kirkpatrick, B.W.; Shook, G.E.; Collins, M.T. Identification of a QTL on BTA20 affecting susceptibility to Mycobacterium avium ssp. paratuberculosis infection in U.S. Holsteins. Anim. Genet. 2007, 38, 389–396.
14. Neibergs, H.L.; Settles, M.L.; Whitlock, R.H.; Taylor, J.F. GSEA-SNP identifies genes associated with Johne's disease in cattle. Mamm. Genome 2010, 21, 419–425.
15. Settles, M.; Zanella, R.; McKay, S.D.; Schnabel, R.D.; Taylor, J.F.; Whitlock, R.; Schukken, Y.; Van Kessel, J.S.; Smith, J.M.; Neibergs, H. A whole genome association analysis identifies loci associated with Mycobacterium avium subsp. paratuberculosis infection status in U.S. Holstein cattle. Anim. Genet. 2009, 40, 655–662.
16. Zanella, R.; Whitlock, R.H.; Neibergs, H.L. Short Communication: Refinement of genetic regions associated with Mycobacterium avium subspecies paratuberculosis tissue infection and to tolerance to Johne's disease. J. Dairy Sci.2011, 94, 4230–4236.
17. Neibergs, H.L.; Wang, Z.; Zanella, R.; Scraggs, E.R.; Van Tassell, C.; White, S.; Sieverkropp, K.D.; Neupane, M.; Knie, B.; Neibergs, H.M.; et al. Deciphering the role of EDN2 in Mycobacterium avium spp paratuberculosis tissue infection. In Proceedings of the Plant and Animal Genome XXI Conference, San Diego, CA, USA, 11–16 January 2013.
18. Benedictus, G.; Dijkhuizen, A.A.; Stelwagen, J. Economic losses due to paratuberculosis in dairy cattle. Vet. Rec. 1987, 121, 142–146.
19. Mcnab, W.B.; Meek, A.H.; Duncan, J.R.; Brooks, B.W.; Van Dreumel, A.A. An evaluation of selected screening tests for bovine paratuberculosis. Can. J. Vet. Res. 1991, 55, 252–259.
20. Nordlund, K.V.; Goodger, W.J.; Pelletier, J.; Collins, M.T. Associations between subclinical paratuberculosis and milk production, milk components, and somatic cell counts in dairy herds. J. Am. Vet. Med. Assoc. 1996, 208, 1872–1876.
21. Kimman, T.G.; Vanderbriel, R.J.; Hoebee, B. Genetic variation in the response to vaccination. Public Health Genom.2007, 10, 201–217.
22. Poland, G.A.; Ovsyannikova, I.G.; Kennedy, R.B.; Lambert, N.D.; Kirkland, J.L. A systems biology approach to the effect of aging, immunosenescence and vaccine response. Curr. Opin. Immunol. 2014, 29, 62–68.
23. Tiezzi, F.; Parker-Gaddis, K.L.; Cole, J.B.; Clay, J.S.; Maltecca, C. A Genome-wide association study for clinical mastitis in first parity U.S. holstein cows using single-step approach and genomic matrix re-weighting procedure. PLoS ONE 2015, 10, e0114919.
24. Wang, E.T.; Sandberg, R.; Luo, S.; Khrebtukova, I.; Zhang, L.; Mayr, C.; Kingsmore, S.F.; Schroth, G.P.; Burge, C.B. Alternative isoform regulation in human tissue transcriptomes. Nature 2008, 456, 470–476.
25. Wilhelm, B.T.; Briau, M.; Austin, P.; Faubert, A.; Boucher, G.; Chagnon, P.; Hope, K.; Girard, S.; Mayotte, N.; Landry, J.R.; et al. RNA-seq analysis of two closely related leukemia clones that differ in their self-renewal capacity. Blood2011, 117, e27–e38.
26. Blekhman, R.; Marioni, J.C.; Zumbo, P.; Stephens, M.; Gilad, Y. Sex-specific and lineage-specific alternative splicing in primates. Genome Res. 2010, 20, 180–189.
27. Wilk, E.; Pandey, A.K.; Leist, S.R.; Hatesuer, B.; Preusse, M.; Pommerenke, C. RNAseq expression analysis of resistant and susceptible mice after influenza A virus infection identifies novel genes associated with virus replication and important for host resistance to infection. BMC Genom. 2015, 16, 655.

28. Zanella, R. Identification of Chromosomal Regions Associated with Infectious Diseases in Cattle. Ph.D. Thesis, Washington State University, Pullman, WA, USA, 2011.

CHAPTER 2

Intestinal Organoids—Current and Future Applications

Andre M. C. Meneses [1,2,*], Kerstin Schneeberger [2], Hedwig S. Kruitwagen [2], Louis C. Penning [2], Frank G. van Steenbeek [2], Iwan A. Burgener [2] and Bart Spee [2]

[1]*Institute of Animal Health and Production, Universidade Federal Rural da Amazônia, Avenida Presidente Tancredo Neves 66077-830, Brazil*

[2]*Department of Clinical Sciences of Companion Animals, Faculty of Veterinary Medicine, Utrecht University, Yalelaan 104, Utrecht 3584 CM, The Netherlands*

ABSTRACT

Recent technical advances in the stem cell field have enabled the in vitro generation of complex structures resembling whole organs termed organoids. Most of these approaches employ culture systems that allow stem cell-derived or tissue progenitor cells to self-organize into three-dimensional (3D)-structures. Since organoids can be grown from different species (human, mouse, cat, dog), organs (intestine, kidney, brain, liver), and from patient-derived induced pluripotent stem cells, they create significant prospects for modelling development and diseases, for toxicology and drug discovery studies, and in the field of regenerative medicine. Here, we report on intestinal stem cells, organoid culture, organoid disease modeling, transplantation, specifically covering the current and future uses of this exciting new insight model to the field of veterinary medicine.

Keywords: intestinal organoids; dog; practical applications

1. INTRODUCTION

The intestinal tract is one of the most architecturally and functionally complex organs of the body [1] and the intestinal epithelium is an incredibly large mucosal surface with an extreme self-renewing capacity [2]. The canine intestine for medium sized dogs is about 4.5 m in length and is subdivided into five functional domains along the proximal-to-distal axis: the duodenum, jejunum, and ileum are segments of the small intestine, and the cecum and colon make up the large intestine [3].

In various pathological disorders or after medical treatments, such as chemotherapy and radiation therapy, the homeostasis between cell proliferation and cell death can become substantially disturbed, resulting in loss of epithelial integrity and local inflammation. The subsequent invasion of intestinal bacteria and the associated stimulation of the immune system further accelerate the course of the disease [4].

Recently a novel method that allows long-term culture of isolated intestinal crypts or intestinal stem cells was presented. Supplemented with the appropriate growth factor cocktail (epidermal growth factor, Noggin, R-spondin-1) and cultured in a three-dimensional extracellular matrix, these intestinal stem cells are capable of developing into organoids and enteroids, displaying many important functions of the normal intestinal epithelium (mini-guts) [5,6]. This culture model has been proven to serve as a powerful system to investigate regulatory and pathological mechanisms of the intestinal epithelium on a molecular level [4].

This organoid system has been extensively used for studying the regulation of intestinal stem cell self-renewal, growth, and differentiation. Additionally, in vitro expanded organoids may be used for gastrointestinal stem cell therapy in preclinical animal models, and in studies on colorectal tumor stem cells, among others [7].

2. INTESTINAL STEM CELLS

The small intestinal epithelium has enormous capacity for self-renewal, replacing itself every three to five days. The cellular basis for this regenerative potential has long been accepted to reside in multipotent intestinal stem cells (ISCs) [8]. ISCs reside near the bottom of the intestinal crypt. Their rapidly dividing, transit-amplifying (TA) daughter cells occupy the remainder of the crypts and flow onto the flanks of the villi, where they differentiate, absorb nutrients, and eventually die at the villus tips. The differentiated cell types include absorptive enterocytes, multiple secretory cells (Paneth cells, goblet cells, enteroendocrine cells, and tuft cells), and the M cells of Peyer's patches [9].

Absorptive enterocytes are the predominant cell type in the intestinal epithelium and play a central role in the digestion and absorption of luminal nutrients. Goblet cells are the second most predominant cell type and produce mucin to lubricate the luminal surface. This mucin layer traps bacterial flora before they attach to the epithelial surface. Enteroendocrine cells are a rare cell type and are scattered along the cryptvillus axis. Upon sensing neuroendocrine signals, enteroendocrine cells produce various hormones to control digestive enzyme secretion, metabolism, and bowel movements. In contrast to these three differentiated cell types, Paneth cells are long-lived (two to three months) differentiated cells that migrate downward and dwell at the crypt bottom; there they sterilize the crypt lumen through the production of antibacterial peptides. The colonic epithelium is devoid of Paneth cells but harbors Paneth-like cells, named deep crypt secretory cells or crypt base goblet cells [10].

Two types of ISC were identified. The first, termed +4 cells, was defined on the basis of DNA label retention, proliferation, and radiation injury response, and the second, known as crypt base columnar (CBC) cells, was defined using cytological lineage tracing. More recently, CBC and +4 cells have been defined using molecular markers and transgenic lineage-tracing techniques. CBC cells are the proliferative engines that drive cellular production in the crypts: they divide daily, with frequent turnover. +4 cells have been redefined as reserve or quiescent ISCs (qISCs): they divide infrequently under homeostatic conditions but can be induced to produce new CBC cells in response to injury or other stimuli [11].

One of the reasons why the intestinal field has failed to reach a consensus regarding the identity and location of intestinal stem cells in the past four decades has been the lack of well-validated markers for this cell population. Such markers are essential for driving advances in our understanding of intestinal stem cell biology, tissue homeostasis, repair, and cancer [12]. There are several cells markers to the CBC stem cells and for the +4 cells [13]. Although their specificity is controversial, some of them will be presented below and are graphically summarized in Figure 1.

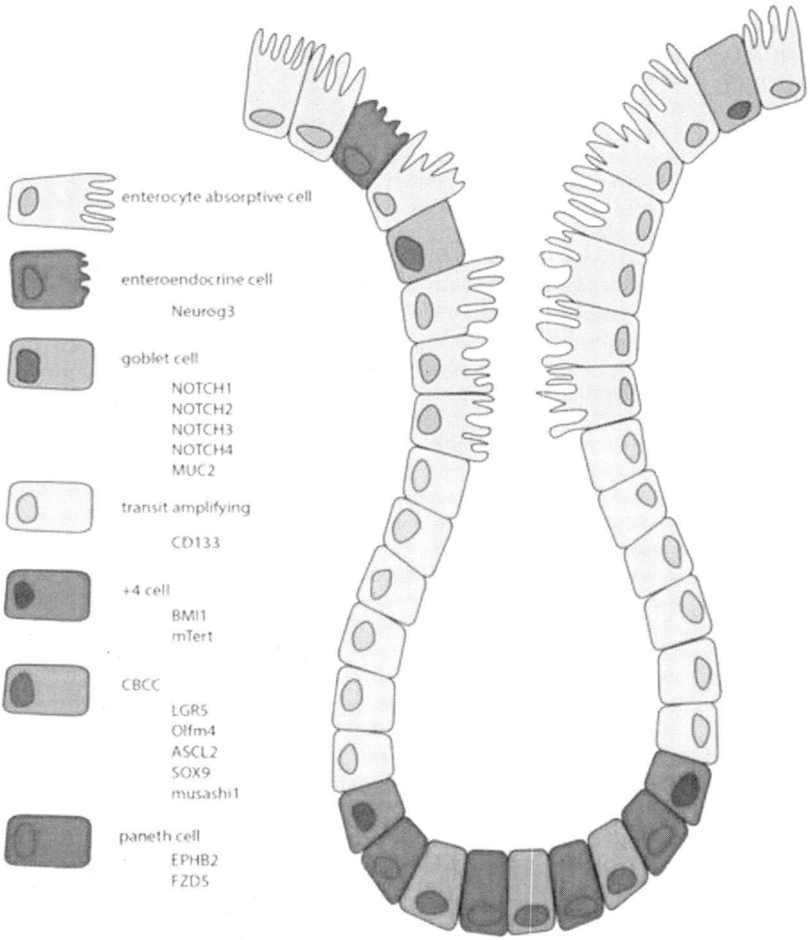

Figure 1. Schematic representation of the intestinal epithelium with all of the different cell types. **Neurog3**, Neurogenin3; **MUC2**, Mucin2; **CD133**, Prominin1; **BMI1**, B lymphoma Mo-MLV insertion region 1 homolog; mTert, Mouse telomerase reverse transcriptase; **CBCC**, crypt base columnar cells; **LGR5**, leucine-rich repeat-containing G-protein coupled receptor 5; **Olfm4**, Olfactomedin-4; **ASCL2**, homolog of the *Drosophila* Achaete-scute complex genes; **SOX9**, gene on chromosome 17q23 that encodes a member of the SOX (SRY-related HMG-box) family of transcription factors; **EPHB2**, gene on chromosome 1p36.1-p35 that encodes a member of the ephrin-B receptor subfamily of receptor tyrosine kinases; **FZD5**, Frizzled.

3. MARKERS OF THE DIFFERENT CELL TYPES PRESENT IN THE INTESTINAL EPITHELIUM

Leucine-rich repeat-containing G-protein coupled receptor 5/G-protein coupled receptor 49 (Lgr5/GPR49) was originally identified as a Wingless-related integration site (Wnt) target gene encoding an orphan G-protein-coupled receptor in colorectal cancer cell-lines [13]. Investigation of its expression pattern in normal intestine revealed it to be a highly specific marker for the proliferating CBC cells [14]. The consideration that Lgr5 is a robust marker of intestinal stem cells was further substantiated by ex vivo culture assays, where single Lgr5$^+$ cells were shown to form self-renewing epithelial organoids exhibiting a structure and composition highly reminiscent of crypt/villus epithelial units in vivo [6]. Importantly, cells that expressed low or no Lgr5 were incapable of forming these organoids. Taken together, these studies provide compelling evidence for Lgr5 being a marker of proliferating adult stem cells residing at the crypt base throughout the intestinal tract [13].

Bmi1 (B lymphoma Mo-MLV insertion region 1 homolog) is a member of the Polycomb group of transcriptional repressors, the function of which in the intestine is unknown. Bmi1 is an essential regulator of hematopoietic, neural, and lung epithelial stem cells. In the intestine, Bmi1 was initially detected in label-retaining stem cells located at the +4 position from the bottom of the crypt. This label retaining feature might indicate a quiescent nature or the capacity to asymmetrically segregate DNA strands. However, it was demonstrated that Bmi1 expression is unrestricted throughout the crypt compartment, similar to that of other proposed stem cell markers, such as Hopx (Homeodomain-only protein) and Tert (Telomerase reverse transcriptase). Furthermore, Bmi1 is a downstream effector of Notch in the ISC and progenitor compartment and Bmi1 is involved in ISC self-renewal [14].

CD133 (Prominin1 in mouse) was originally reported as a novel pentaspan glycoprotein expressed on neural and hematopoietic stem cells. More recently, it has been described as a marker of epithelial stem/progenitor populations in human kidney tubules and the prostate. However, a detailed analysis of reporter mice expressing lacZ under the control of the endogenous Prom1 promoter documented widespread expression in many adult epithelia, including the colon, casting doubt on its authenticity as a broad adult stem cell marker [13,15].

Telomerase is a ribonucleoprotein complex that helps maintain the telomeric ends of chromosomes, normally shortened with each cell division. Because loss of telomeric DNA beyond a critical threshold induces senescence in most somatic cells, maintenance or induction of telomerase activity provides a means of preventing cellular senescence that may be relevant for the self-renewal of tissue stem cells. Consistent with this hypothesis, loss of telomerase has been shown to result in intestinal villus atrophy, suggesting a functional requirement for telomerase activity and/or telomere maintenance in ISC function. mTert expression marks a slowly cycling ISC population distinct from Lgr5+ cells. mTert+ cells contribute to all differentiated intestinal cell types as well as the Lgr5+ cell population, persist long-term, are resistant to injury, and contribute to the regenerative response following tissue injury. Thus, a slowly cycling stem cell exists within the intestine alongside, and perhaps upstream of, the Lgr5+ population [16].

Olfactomedin-4 (Olfm4) has emerged from a gene signature for these Lgr5+ stem cells as a robust marker for murine small intestinal stem cells. Studies have shown that OLFM4 is highly expressed in crypt base columnar cells in human small intestine and colon. Moreover, it is expressed in cells within adenocarcinomas of the colon and can serve as a useful marker for Lgr5-type stem cells in human small intestine and colon [17]. Oflm4 is a secreted molecule that is specifically and robustly expressed on the stem cells of the small intestine, but is absent from the colon in mice. In contrast to Lgr5, expression of Oflm4 is not under the control of the Wnt pathway [13].

Ascl2 is a homolog of the *Drosophila* Achaete-scute complex genes and is a Wnt target gene. The Ascl2 gene encodes a basic helix-loop-helix (bHLH) transcription factor with an unusually restricted expression pattern (i.e., its expression is predominantly detected in extraembryonic tissues and in intestinal epithelium) [18]. Ascl2 has been proposed to be a master regulatory gene for Lgr5+ ISCs, since transgenic overexpression in mice induces hyperplastic and ectopic crypts in the villus compartment, while its deletion results in loss of Lgr5+ ISCs [19].

SOX transcription factors have the capacity to modulate stem/progenitor cell proliferation and differentiation in a dose-dependent manner. SOX9 is expressed in the small intestine epithelial stem cell zone and has been reported to be a downstream target of Wnt signaling by acting in a feedback loop to repress Wnt signaling, thus keeping proliferation under tight regulatory control. Distinct levels of SOX9 in the intestinal epithelium are associated with both proliferative and postmitotic cell types and, furthermore, these variable SOX9 levels likely play an important role in both the control of proliferative capacity of stem/progenitor populations and also the maturation of enteroendocrine cells, where low SOX9 expression supports proliferative capacity, and high SOX9 expression suppresses proliferation [20].

The receptor tyrosine kinase EphB2, is expressed in a decreasing gradient from the crypt base toward the differentiated cell compartment. EphB activity is required to establish the position of the different cell types in the crypts and the EphB2 expression gradient is required to compartmentalize cell populations in different territories of the healthy epithelium [21].

Wnt/b-catenin signaling is crucial for normal stem cell function in the intestinal epithelium. More specifically, Wnt3 signaling, provided by flanking Paneth cells, is necessary for the maintenance and function of CBC stem cells. In the absence of Wnt3, Wnt2b can compensate. The weak short range Wnt signal is augmented by R-spondin signaling through Lgr receptors. R-spondins are incorporated into a complex that contains Lrp (low-density lipoprotein receptor related proteins), Lgr, and Fzd (Frizzled); this complex facilitates Fzd-coupled Wnt/b-catenin signaling [22]. Of the ten mammalian Fzds, only Fzd7 is frequently upregulated in stem cell populations and cancers from diverse tissues [23]. Notably, FZD7 plays a non-redundant role in maintaining pluripotency of human embryonic stem cells [24] and might play a similar role in the intestinal stem cells. It was demonstrated that intestinal stem cells that do not express Fzd7 have an inherent defect in Wnt/b-catenin signaling, which compromises stem cell function under conditions of stress [22].

Neurogenin 3 (NEUROG3) has been investigated as a candidate gene responsible for congenital loss of intestinal enteroendocrine cells in humans because of its known role in enteroendocrine cell development in mouse [25]. The intestinal enteroendocrine cells consist of at least 15 different cell types classified essentially on the basis of their hormonal content with a specific geographical distribution. NEUROG3 is expressed in the fetal intestinal epithelium as well as immature cells located at the proliferative compartment of the crypts in the adult small intestine [26].

Notch is clearly active in ISCs and stimulates proliferation while blocking secretory cell differentiation. Furthermore, Notch is a direct activator of the ISC gene Olfm4 and the bHLH transcriptional repressor Hes1, which in turn blocks expression of the driver of secretory cell differentiation, Math1. The co-operation between Notch and Wnt in the intestine results in the amplification of Wnt-driven proliferation and Wnt-driven tumor formation is potentiated by constitutive Notch activation [27]. Notch is likely to target distinct stem and progenitor cell populations to regulate different aspects of intestinal homeostasis, although specific cellular targets have not been definitively identified. Crucial components of the Notch signaling pathway, including the Notch1 and Notch2 receptors, the ligands jagged 1, Dll1 and Dll4, and the Notch target genes hairy and enhancer of split 1 (Hes1), Hes5 and Hes6, have been localized to the proliferative zone of the intestinal crypts. Importantly, lineage tracing from cells undergoing active Notch signaling identified long-lived progenitors that gave rise to all the mature epithelial cell types, suggesting that Notch signaling was active in a stem cell. More specifically, Notch regulation of the CBC stem cell was suggested by the enrichment of Notch1 receptor mRNA in this cell type [28].

MUC2 encodes a large (>5000 amino acid residue) gel-forming mucin abundantly expressed by intestinal goblet cells and certain other mucus-secreting cells in normal, diseased, and neoplastic tissues. The 58-flanking region of the MUC2 gene has been isolated as well, and has been shown to have promoter activity in cultured cells. It is difficult to study the factors responsible for the regulation of this gene in cultured cells, however, in part because transfectable cell lines expressing the high levels of MUC2 found in intestinal goblet cells are nonexistent [29].

The RNA-binding protein Musashi 1 (Msi1), a regulator of asymmetric cell division, is also involved in stem cell maintenance. Particularly, in neural stem cells Msi1 is able to maintain stemness properties through Notch pathway activation. Independent immunoistochemical and in situ hybridization analyses demonstrated that Msi1 is expressed in the CBC cells immediately above the Paneth cells. Moreover, Msi1 overexpression in the intestine increases both Wnt and Notch pathways, and induces the upregulation of Lgr5 and Bmi1. Interestingly, although Msi1 is expressed in putative ISCS, in knockout mice lacking this marker, no defects in the development of the intestine are detected. Taken together, these observations demonstrated that Msi1 is not a specific ISCS marker, but is expressed in both ISCS and in their early progeny [7].

4. ORGANOID CULTURE

Currently, the two primary ways to generate human mini intestines include (a) isolation of intestinal crypts, which contain human adult stem cells, from donors or (b) use of human embryonic or inducible pluripotent stem cells (iPSCs). Intestinal crypts, containing human adult stem cells, can be isolated either from surgically resected tissue or from biopsies, embedded in Matrigel, and cultured as ex vivo self-perpetuating three-dimensional primary cultures in growth factor enriched media [30].

Initially, these cultures produce polarized three-dimensional spheroid-like structures with the apical domain facing inside newly formed lumens and basolateral surfaces in contact with the Matrigel and external media. These are referred to as enteroids (derived from small intestine) or colonoids (derived from colon) and contain only epithelial cells types derived from the crypt-based stem cells. Human three-dimensional intestinal tissue cultures, termed organoids, can also be generated in vitro from iPSCs [30]. Intestinal organoids are generated from iPSCs that are differentiated to form mid- and hindgut tissue. From the mid- or hindgut epithelial cell layers, three-dimensional spheroid buds are further cultured in Matrigel along with pro-intestinal growth factors [25].

Organoids have been successfully generated from many regions of the mouse gastrointestinal tract, ranging from the tongue through to the colon Recently, human iPSC-derived functional cholangiocyte organoids have also been generated by the modulation of the Activin A and Notch signaling pathways. A notable difference between organoids derived from primary tissue and ESCs/iPSCs is the presence of cell types other than the intended lineage in the latter. This is because the factors used for directed differentiation of ESCs/iPSCs are not completely efficient in driving all the cells towards the lineage of choice, thus many ectodermal and endodermal organoids, such as those of the intestine, stomach, and kidney, have reported the limited presence of mesenchymal cell type [31].

Although a variety of culture systems have been described, only recently have long-term culture systems become available that maintain basic crypt physiology, based on the growth factor requirements observed in vivo. A Matrigel based culture system was established that allows the formation of ever-expanding organoids, or "mini-guts", in vitro from a single Lgr5+ stem cell [6].

An essential component of these cultures is the Wnt agonist Rspondin1, the ligand of Lgr5. The other constituents are epidermal growth factor (EGF) and the bone morphogenetic protein (BMP) inhibitor Noggin [2]. For colon crypt culture, Wnt ligand is an additional factor required to maintain Lgr5-CBC cells, because the epithelium makes little, if any, Wnt [9]. It has been suggested that some of these growth factors in small intestine are derived from the neighboring Paneth cells in vivo, while other cues are likely provided by intestinal subepithelial myofibroblasts (ISEMFs). Paneth cells are tightly nested between ISCs in the crypt base; they act as anti-microbial defensive cells, while also producing supportive growth factors for neighboring ISCs. It has been reported that Paneth cell-derived Wnt3 is necessary for in vitro intestinal epithelial cultures. However, ISC niche maintenance in vivo is preserved even with conditional epithelial Wnt3 deletion and Paneth cell ablation. This suggests the presence of redundant sources of Wnt signaling from surrounding non-epithelial cells within the niche [32].

The mini-guts faithfully recapitulate the central features of normal small intestinal epithelium. They consist of crypts (with resident Lgr5 cells, Paneth cells, and TA cells) that feed into a central lumen lined by mature epithelial cells of all villus lineages. Self-renewal kinetics resemble the in vivo situation: cells are born in the crypts proliferate, differentiate, and are shed into the central lumen about five days later [2]. The isolated crypts require Matrigel (BD Biosciences, San Jose, CA, USA), a 3D laminin- and collagen-rich matrix that mimics the basal lamina. Single crypts can be readily isolated from mouse or human intestine by ethylenediaminetetraacetic acid (EDTA)-based Ca2+/Mg2+ chelation [9], as well as in dogs (data not shown).

In vitro-generated organoids occur as cysts with a central lumen flanked by a simple, highly polarized villus epithelium. Multiple cryptlike structures project outward. The basal side of the cells is oriented toward the outside, touching the matrigel, whereas enterocyte brush borders form the luminal surface. Secretion by Paneth and goblet cells occurs toward the lumen. The organoids can be passaged weekly at a 1:5 ratio for at least 1.5 years, with a phenotype and karyotype that remain unchanged. Mechanically disrupted organoids rapidly reseal. Self-renewal kinetics and cell-type composition closely resemble the in vivo situation in small intestinal epithelial derived organoids. Notably, a timer of unknown molecular nature remains active in the absence of the in vivo wear-and-tear: two to three days after terminal differentiation, the cells exfoliate into the lumen [9].

In a medium containing Wnt3A in addition to R-spondin-1 (Rspo1), the sharp Wnt gradient is lost and epithelial mini-guts become symmetric, round cysts, consisting of a homogeneous population of stem and progenitor cells. Epithelial mini-guts grown from adenomatous polyposis coli (APC)—mutant adenoma cells display the same symmetric shape, which is not surprising, because APC loss leads to constitutive Wnt pathway activation. These observations imply that the typical crypt-villus architecture is suppressed under conditions of homogeneous (rather than focal) Wnt signaling [33]. The recently published immobilized Wnt protein can be used to improve intestinal stem cell niche architecture [34].

The addition of Rspo1, Noggin, and epidermal growth factor (EGF), which are all essential to small intestine culture, did not maintain the growth of mouse colonic crypts. Therefore authors developed the following "TMDU (Tokyo Medical and Dental University) protocol": crypts were embedded in type I collagen in serum-free medium with Wnt3a, hepatocyte growth factor (HGF), and BSA, in addition to Rspo1, Noggin, and EGF. Although Noggin, EGF, and HGF were not essential for growth of the colonic crypts, each enhanced their growth [35].

When a single cultured stem cell is followed over time, a small symmetric cyst forms. The stochastic appearance of a Paneth cell constitutes the "symmetry-breaking" event: a bud forms around the cell; within two to three days this bud develops into a cryptlike structure with stem and Paneth cells. Every nondifferentiated cell that touches a Paneth cell is (or becomes) an Lgr5- CBC cell, presumably driven by potent Wnt and Notch signals from the Paneth cell [8,33]. The proliferative stem cell—Paneth niche pushes itself outward from the central cyst. This process is most likely driven by repulsive EphB-EphrinB interactions, as described in vivo [35].

Although mouse models and genetic analyses have greatly increased the general understanding of embryonic gastrointestinal (GI) development, mouse development does not always recapitulate human [33] development, particularly in development of the intestinal stem cell niche.

Organoids have been described in numerous tissues form humans and mice [36]; regarding companion animals, only one paper described canine liver organoids [37]. Since growth factor dependency for the various organ/species derived organoids varies greatly, the time needed to discover the optimal culture conditions can be cumbersome.

5. ORGANOID DISEASE MODELING AND TRANSPLANTATION

Gastrointestinal (GI) organoids have been used to model human genetic diseases, infections, inflammatory bowel diseases (IBD), and malignancies. They can be used too to conduct drug screens and identify personalized approaches for treating disease [38].

Also of interest is the direct use of intestinal organoids as therapy. Intestinal organoids have been transplanted into damaged colon for tissue repair [35], but only with limited engraftment success [39].

Enteroids are showing great promise as models to study the interaction between intestinal pathogens and the intestinal epithelium. To date, human enteroids have been shown to model human rotavirus (RV) infection, cholera toxin effect on transport, and several aspects of enterohemorrhagic *Escherichia coli*-related human diarrhea. RV replicates and produces infectious viruses in human enteroids, with viral replication increasing over 96 h [30].

The evolution of the enteroid/organoid models to include sub-epithelial myofibroblasts, smooth muscle cells, enteric nerves, immune cells, microbiome, and endothelial cells is expected to help facilitate our understanding of complex intestinal organ development. In addition, methods have been established to alter gene expression either by overexpression or shRNA-mediated knockdown using lipophilic transfection reagents and virus mediated transduction [40,41].

These new models may provide valuable insight into developing methodologies for intestinal regeneration that lead to therapeutic options including repair of the damaged epithelium that occurs in ulcerative colitis, microvillus inclusion disease, or intestinal transplantation to treat short bowel syndrome [42]. Despite these evolutions in therapeutic possibilities, the in vivo transplantation in humans remains a challenge [43].

In spite of the successful rodent studies [44], successful generation of neomucosa using an intestinal cell-cluster in transplantation was first described in dogs in 2009. The authors showed that the isolation and transplantation of 3D cultured cells can be used to generate neomucosa in a dog model [45]. However the molecular characterization of the presumed "organoids" was not rigorously described.

The intestinal organoids can be utilized not only to investigate the crucial aspects of gastrointestinal biology, but their usage could also facilitate functional studies, diagnoses, and personalized treatments of various disorders including cancer [46]. Deregulation of Wnt signaling is associated with onset of cancer, most notably carcinoma of the colon and rectum [46]. Other studies have employed CRISPR/Cas9-mediated gene editing of healthy organoids to directly evaluate candidate gene function in tissue physiology and carcinogenesis [47].

Despite the strengths that organoids offer over traditional cell lines, their use is restricted. For example, although organoids are complex and often retain multiple cell lineages, they lack many of the cellular inputs present in an in vivo system. Complexity can be increased through coculture with additional cell types, but organoids should still be viewed as a reductionist system [35].

Recently the first generation of 3D human gastric organoids was reported through directed in vitro differentiation of hESCs and hiPSCs. The authors recapitulated in vivo human stomach development by inducing first the PSCs to form definitive endoderm spheroids, which were subsequently stimulated by retinoic acid to generate the foregut. To permit 3D growth and further differentiation, spheroids were transferred into a retinoic acid-supplemented Matrigel culture, leading to the formation of domains with gastric mucous and endocrine cells, and LGR5-expressing SCs. Notably, to study the etiology of *Helicobacter pylori*-mediated disease, they injected the bacterium directly into the lumen of the organoids, and observed that the virulence factor CagA bound and activated the c-Met (receptor for HGF), inducing epithelial proliferation. This epithelial pathophysiological response makes the organoids a promising model for human gastric disease [48].

Transplanted human intestinal organoids have displayed mature intestine characteristics such as peptide uptake, but therapeutic application would require additional functions, such as the ability to perform the peristaltic movements essential for food bolus transition along the digestive tract [49].

6. FUTURE OF ORGANOIDS IN VETERINARY MEDICINE

Despite the great advances in the field of intestinal organoids, the great challenge is, as described by Fatehullah et al. [48], to rapidly and safely translate the knowledge of mouse stem cells into the human arena. Efforts to exploit intestinal stem cells for regenerative medicine have accelerated over the past decade, driven mostly by the identification of cell-surface markers facilitating the isolation of pure stem cell populations and the development of near-physiological culture methods supporting stem cell expansion and epithelial growth. To keep momentum and to ensure safety issues in the mouse-to-human translation, dogs are ideally suited as experimental animals and clinical patients, having GI- and liver-pathologies highly similar to humans.

With their robust genomic integrity, it will be feasible to use organoids as a tool for regenerative medicine and biomedical research, and in the near future create tissues for autologous transplantation. Besides that, the development of "cancer organoids" creates significant prospects for personalizing and optimizing current treatments [49].

Studies involving molecularly characterized intestinal organoids in dogs are lacking. Because of that, a task-force at the Faculty of Veterinary Medicine of the Utrecht University (Utrecht, the Netherlands) was created to fill this gap. Recent studies have shown that maintenance of canine intestinal organoids are workable (data not published) and that practical applications can be tested, as for the development and study of viral infections such canine parvovirus and coronavirus. This system can be also used for other species like porcines, equines, ruminants, and felines, but studies in these species are punctual and scarce, and advances and technique adaptations will be necessary to use this method for them.

In the future, organoid technologies will undoubtedly provide a methodological window, allowing us to understand development in all species, including companion animals, and disease in depth, and will enabling new therapeutic options and a better understanding of the pathophysiological processes of various diseases.

ACKNOWLEDGMENTS

Special thanks for CNPq, Brazil for the postdoctoral scholarship to Andre M. C. Meneses.

AUTHOR CONTRIBUTIONS

Andre M. C. Meneses wrote the paper. Kerstin Schneeberger analyzed and corrected the paper. Hedwig S. Kruitwagen analyzed and corrected the paper. Louis C. Penning analyzed and corrected the paper. Frank G. van Steenbeek conceived the picture, analyzed and corrected the paper. Iwan A. Burgener analyzed and corrected the paper. Bart Spee analyzed and corrected the paper.

CONFLICTS OF INTEREST

The authors declare no conflict of interest.

REFERENCES

1. Wells, J.M.; Spence, J.R. How to make an intestine. Development 2014, 141, 752–760.
2. Clevers, H. The intestinal crypt, a prototype stem cell compartment. Cell 2013, 154, 274–284.
3. Kararli, T. Comparison of the gastrointestinal anatomy, physiology, and biochemistry of humans and commonly used laboratory-animals. Biopharm. Drug Dispos. 1995, 16, 351–380.
4. Grabinger, T.; Luks, L.; Kostadinova, F.; Zimberlin, C.; Medema, J.P.; Leist, M.; Brunner, T. Ex vivo culture of intestinal crypt organoids as a model system for assessing cell death induction in intestinal epithelial cells and enteropathy. Cell Death Dis. 2014, 5, e1228.
5. Barker, N.; van Oudenaarden, A.; Clevers, H. Identifying the stem cell of the intestinal crypt: Strategies and pitfalls. Cell Stem Cell 2012, 11, 452–460.

6. Sato, T.; Vries, R.G.; Snippert, H.J.; van de Wetering, M.; Barker, N.; Stange, D.E.; van Es, J.H.; Abo, A.; Kujala, P.; Peters, P.J.; et al. Single Lgr5 stem cells build crypt-villus structures in vitro without a mesenchymal niche. Nature2009, 459, 262–266.
7. Tesori, V.; Puglisi, M.A.; Lattanzi, W.; Gasbarrini, G.B.; Gasbarrini, A. Update on small intestinal stem cells. World J. Gastroenterol. 2013, 19, 4671–4678.
8. Carlone, D.L.; Breault, D.T. Tales from the crypt: The expanding role of slow cycling intestinal stem cells. Cell Stem Cell 2012, 10, 2–4.
9. Sato, T.; Clevers, H. Growing self-organizing mini-guts from a single intestinal stem Cell: Mechanism and applications. Science 2013, 340, 1190.
10. Date, S.; Sato, T. Mini-gut organoids: Reconstitution of the stem cell niche. Ann. Rev. Cell Dev. Biol. 2015, 31, 269–289.
11. Goodell, M.A.; Nguyen, H.; Shroyer, N. Somatic stem cell heterogeneity: Diversity in the blood, skin and intestinal stem cell compartments. Nat. Rev. Mol. Cell Biol. 2015, 16, 299–309.
12. Lin, S.A.; Barker, N. Gastrointestinal stem cells in self-renewal and cancer. J. Gastroenterol. 2011, 46, 1039–1055.
13. Leushacke, M.; Barker, N. Ex vivo culture of the intestinal epithelium: Strategies and applications. Gut 2014, 63, 1345–1354.
14. López-Arribillaga, E.; Rodilla, V.; Pellegrinet, L.; Guiu, J.; Iglesias, M.; Roman, A.C.; Gutarra, S.; González, S.; Muñoz-Cánoves, P.; Fernández-Salguero, P.; et al. Bmi1 regulates murine intestinal stem cell proliferation and self-renewal downstream of Notch. Development 2015, 142, 41–50.
15. Zhu, L.; Gibson, P.; Currle, D.S.; Tong, Y.; Richardson, R.J.; Bayazitov, I.T.; Poppleton, H.; Zakharenko, S.; Ellison, D.W.; Gilbertson, R.J. Prominin 1 marks intestinal stem cells that are susceptible to neoplastic transformation. Nature2009, 457, 603–607.
16. Montgomerya, R.K.; Carloneb, D.L.; Richmonda, C.A.; Farillab, L.; Kranendonkc, M.E.G.; Hendersonc, D.E.; Baffour-Awuaha, N.Y.; Ambruzsc, D.M.; Foglic, L.K.; Algraa, S.; et al. Mouse telomerase reverse transcriptase (mTert) expression marks slowly cycling intestinal stem cells. PNAS 2011, 108, 179–184.
17. Van der Flier, L.; Haegebarth, A.; Stange, D.E.; van de Wetering, M.; Clevers, H. OLFM4 is a robust marker for stem cells in human intestine and marks a subset of colorectal cancer cells. J. Gastroenterol. 2009, 137, 15–17.
18. Van der Flier, L.G.; Van Gijn, M.E.; Hatzis, P.; Kujala, P.; Haegebarth, A.; Stange, D.E.; Begthel, H.; Van den Born, M.; Guryev, V.; Oving, I.; et al. Transcription factor achaete scute-like 2 controls intestinal stem cell fate. Cell 2009, 136, 903–912.
19. Yan, K.S.; Kuo, C.J. Ascl2 reinforces intestinal stem cell identity. Cell Stem Cell 2015, 16, 105–106.
20. Formeister, E.J.; Sionas, A.L.; Lorance, D.K.; Barkley, C.L.; Lee, G.H.; Magness, S.T. Distinct SOX9 levels differentially mark stem/progenitor populations and enteroendocrine cells of the small intestine epithelium. Am. J. Physiol. Gastrointest. Liver Physiol. 2009, 296, G1108–G1118.
21. Merlos-Suarez, A.; Barriga, F.M.; Jung, P.; Iglesias, M.; Cespedes, M.V.; Rossell, D.; Sevillano, M.; Hernando-Momblona, X.; Silva-Diz, V.; Munoz, P.; et al. The intestinal stem cell signature identifies colorectal cancer stem cells and predicts disease relapse. Cell Stem Cell 2011, 8, 511–524.
22. Flanagan, D.J.; Phesse, T.J.; Barker, N.; Schwab, R.H.M.; Amin, N.; Malaterre, J.; Stange, D.E.; Nowell, C.J.; Currie, S.A.; Saw, J.T.S.; et al. Frizzled7 functions as a wnt receptor in intestinal epithelial lgr5+ stem cells. Stem Cell Rep.2015, 4, 759–767.
23. Vincan, E.; Barker, N. The upstream components of the WNT signalling pathway in the dynamic EMT and MET associated with colorectal cancer progression. Clin. Exp. Metastasis 2008, 25, 657–663.
24. Fernandeza, A.; Hugginsa, I.J.; Pernaa, L.; Brafmana, D.; Lub, D.; Yaob, S.; Gaasterlandc, T.; Carsonb, D.A.; Willert, K. The WNT receptor FZD7 is required for maintenance of the pluripotent state in human embryonic stem cells. PNAS2014, 111, 1409–1414.

25. Spence, J.R.; Mayhew, C.N.; Rankin, S.A.; Kuhar, M.; Vallance, J.E.; Tolle, K.; Hoskins, E.E.; Kalinichenko, V.V.; Wells, S.I.; Zorn, A.M.; et al. Directed differentiation of human pluripotent stem cells into intestinal tissue in vitro. Nature2011, 470, 105–109.
26. Jenny, M.; Uhl, C.; Roche, C.; Duluc, I.; Guillermin, V.; Guillemot, F.; Jensen, J.; Kedinger, M.; Gradhowl, G. Neurogenin3 is differentially required for endocrine cell fate specification in the intestinal and gastric epithelium. EMBO J. 2002, 21, 6338–6347.
27. Germanna, M.; Xua, H.; Malaterrea, J.; Sampurnoa, S.; Huyghed, M.; Cheasleya, D.; Fred, S.; Ramsay, R.G. Tripartite interactions between WNT signaling, Notch and Myb for stem/progenitor cell functions during intestinal tumorigenesis. Stem Cell Res. 2014, 13, 355–366.
28. VanDussen, K.L.; Carulli, A.J.; Keeley, T.M.; Patel, S.R.; Puthoff, B.J.; Magness, S.T.; Tran, I.T.; Maillard, I.; Siebel, C.; Kolterud, A.; et al. Notch signaling modulates proliferation and differentiation of intestinal crypt base columnar stem cells. Development 2012, 139, 488–497.
29. Gum, J.R., Jr.; Hicks, J.W.; Gillespie, A.M.; Carlson, E.J.; Komuves, L.; Karnik, S.; Hong, J.C.; Epstein, C.J.; Kim, Y.S. Goblet cell-specific expression mediated by the MUC2 mucin gene promoter in the intestine of transgenic mice. Am. J. Physiol. Gastrointest. Liver Physiol. 1999, 276, G666–G676.
30. Zachos, N.C.; Kovbasnjuk, O.; Foulke-Abel, J.; In, J.; Blutt, S.E.; de Jonge, H.R.; Estes, M.K.; Donowitz, M. Human enteroids/colonoids and intestinal organoids functionally recapitulate normal intestinal physiology and pathophysiology. J. Biol. Chem. 2016, 291, 3759–3766.
31. Fatehullah, A.; Tan, S.H.; Barker, N. Organoids as an in vitro model of human development and disease. Nat. Cell Biol. 2016, 18, 246–254.
32. Lei, N.Y.; Jabaji, Z.; Wang, J.; Joshi, V.S.; Brinkley, G.J.; Khalil, H.; Wang, F.; Jaroszewicz, A.; Pellegrini, M.; Li, L.; et al. Intestinal subepithelial myofibroblasts support the growth of intestinal epithelial stem cells. PLoS ONE 2014, 9, e84651.
33. Sato, T.; van Es, J.H.; Snippert, H.J.; Stange, D.E.; Vries, R.G.; van den Born, M.; Barker, N.; Shroyer, N.F.; van de Wetering, M.; Clevers, H. Paneth cells constitute the niche for Lgr5 stem cells in intestinal crypts. Nature 2011, 469, 415–418.
34. Lowndes, M.; Rotherham, M.; Price, J.C.; El Haj, A.J.; Habib, S.J. Immobilized WNT proteins act as a stem cell niche for tissue engineering. Stem Cell Rep. 2016, 7, 126–137.
35. Dedhia, P.H.; Bertaux-Skeirik, N.; Zavros, Y.; Spence, J.R. Organoid models of human gastrointestinal development and disease. Gastroenterology 2016, 150, 1098–1112.
36. Barker, N. Adult intestinal stem cell: Critical drivers of epithelial homesostasis and regeneration. Mol. Cell Biol. 2014, 15, 19–33.
37. Nantasanti, S.; Spee, B.; Kruitwagen, H.S.; Chen, C.; Geijsen, N.; Oosterhoff, L.A.; van Wolferen, M.E.; Pelaez, N.; Fieten, H.; Wubbolts, R.W.; et al. Disease modeling and gene therapy of copper storage disease in canine hepatic organoids. Stem Cell Rep. 2015, 5, 895–907.
38. Yui, S.; Nakamura, T.; Sato, T.; Nemoto, Y.; Mizutani, T.; Zheng, X.; Ichinose, S.; Nagaishi, T.; Okamoto, R.; Tsuchiya, K.; et al. Functional engraftment of colon epithelium expanded in vitro from a single adult Lgr5+ stem cell. Nat. Med.2012, 18, 618–623.
39. Yin, X.; Mead, B.E.; Safaee, H.; Langer, R.; Karp, J.M.; Levy, O. Engineering stem cell organoids. Cell Stem Cell 2016, 18, 25–38.
40. Schwank, G.; Andersson-Rolf, A.; Koo, B.K.; Sasaki, N.; Clevers, H. Generation of BAC transgenic epithelial organoids. PLoS ONE 2013, 8, e76871.
41. Koo, B.K.; Stange, D.E.; Sato, T.; Karthaus, W.; Farin, H.F.; Huch, M.; van Es, J.H.; Clevers, H. Controlled gene expression in primary LG5 organoid cultures. Nat. Methods 2011, 9, 81–83.
42. Spurrier, R.G.; Grikscheit, T.C. Tissue engineering the small intestine. Clin. Gastroenterol. Hepatol. 2013, 11, 354–358.

43. Van Rijn, J.M.; Schneeberger, K.; Wiegerinck, C.L.; Nieuwenhuis, E.E.S.; Middendorp, S. Novel approaches: Tissue engineering and stem cells—In vitro modelling of the gut. Best Pract. Res. Clin. Gastroenterol. 2016, 30, 281–293.
44. Tait, I.S.; Flint, N.; Campbell, F.C.; Evans, G.S. Generation of neomucosa in vivo by transplantation of dissociated rat postnatal small intestinal epithelium. Differentiation 1994, 56, 91–100.
45. Agopian, V.G.; Chen, D.C.; Avansino, J.R.; Stelzner, M. Intestinal stem cell organoid transplantation generates neomucosa in dogs. J. Gastrointest. Surg. 2009, 13, 971–982.
46. Krausova, M.; Korinek, V. WNT signaling in adult intestinal stem cells and cancer. Cell. Signal. 2014, 26, 570–579.
47. Polakis, P. WNT Signaling in Cancer. Cold Spring Harb. Perspect. Biol. 2012, 4, a008052.
48. Xinaris, C.; Brizi, V.; Remuzzi, G. Organoid models and applications in biomedical research. Nephron 2015, 130, 191–199.
49. Watson, C.L.; Mahe, M.M.; Múnera, J.; Howell, J.C.; Sundaran, N.; Poling, H.M.; Schweitzer, J.I.; Vallance, J.E.; Mayhew, C.N.; Sun, Y.; et al. An in vivo model of human small intestine using pluripotent stem cells. Nat. Med. 2014, 20, 1310–1314.

CHAPTER 3

Exfoliative Endometrial Cytology in Embryo Donor Cows—Comparison of Sampling Localizations for the Diagnosis of Subclinical Endometritis

Janna Egberts [1,2,*], Jan Detterer [1], Arno Park [1] and Sabine Meinecke-Tillmann [2]

[1]*AI- and ET-Center Georgsheil, VOST, Am Bahndamm 4, Südbrookmerland 26624, Germany*

[2]*Institute for Reproductive Biology, University of Veterinary Medicine Hannover, Bünteweg 2, Hannover 30559, Germany*

ABSTRACT

Subclinical endometritis has a major effect on the reproductive performance of dairy cows and also on the success of embryo collection. Thus it is important to minimize the number of false-negative diagnoses. In order to evaluate the question of whether or not a single cytobrush sample is representative of the whole endometrium, 53 German Holstein embryo donor cows in the northwest of Germany were examined via the cytobrush method at three different localizations of the uterus: the uterine body about 0.5 cm cranial of the cervical canal and both uterine horns about 1.5 cm cranial of the bifurcation. Although the mean percentage of polymorphonuclear neutrophils at the three locations is not significantly different ($p = 0.64$), the individual variations lead to the conclusion that more than one sample of the endometrium should be taken into account when diagnosing subclinical endometritis in embryo donor cows.

Keywords: exfoliative cytology; cytobrush; subclinical endometritis; dairy cows; embryo donor cows; PMN

1. INTRODUCTION

Subclinical endometritis (SE) is defined as a postpartum inflammation of the endometrium in the absence of any signs of clinical endometritis such as purulent vaginal discharge [1]. In order to diagnose SE, the proportion of polymorphonuclear neutrophils (PMN) in endometrial samples is evaluated.

Endometrial cytology has been proven to be a reliable method to detect SE, whereas bacteriological testing is not recommended since SE seems to be more of a problem of the recovery of the uterus postpartum than an infection [1,2,3,4,5,6,7]. Due to the impact of estrus cyclity, including the infiltration of PMN, which is lower on superficial compared to deeper layers of the endometrium, uterine biopsies are more sensitive to physiological changes than endometrial cytology [8,9]. Hence it is recognized that there is low correlation between endometrial biopsies and exfoliative cytology for diagnosing uterine disease [5].

Although there is no gold standard for the diagnosis of SE, cytobrushing has been used in the majority of studies and has been considered to be reliable, quick and easy [2,4,5,6,9,10,11] in contrast to low-volume flush [3,12]. Samples are taken from the most superficial layers of the endometrium, which makes cytobrushing insensitive to changes during the estrus cycle [5], which results in a reduction of false-positive outcomes. However, so far, and without regard to the possibility of localized events, cytobrush samples are mainly taken from a single intrauterine localization, based on the assumption that the diagnostic conditions would not differ throughout the whole endometrium, which has been reported for endometrial biopsies [13].

SE in cattle affects the reproductive performance as far as increasing the interval between calving and conception by about 30 days and decreasing the pregnancy rate by about 16% [1,3,4,5]. This results in major economic losses. An influence on the success of embryo collection has also been reported, where more unfertilized oocytes and fewer transferable embryos of poorer quality are collected from cows suffering from SE compared to healthy donor cows [10,11].

To allow for the proper management of those animals and to improve the success of embryo collection, it is important that the number of false-negative diagnoses is minimized.

The aim of this study was to answer the question of whether a cytological smear preparation taken from a single intrauterine localization is representative of the whole endometrium.

2. MATERIALS AND METHODS

Fifty-three German Holstein cows in their first to tenth lactation (4.3 ± 1.8 lactations/animal; means ± SD) were investigated between D 64 and 1018 postpartum (D 215 ± 165 p.p.; means ± SD). The animals (client-owned: informed client-consent) were presented as embryo donors in a commercial embryo transfer program. Before superovulation, all participants were clinically examined during interoestrus (D 6–13 of the oestrus cycle). With regard to standard transrectal palpation and ultrasound scanning (Easi-Scan®, BCF™ Technology, broadband straight linear rectal probe, 7.5 MHz) of the reproductive tract as well as vaginoscopic inspection of the vagina and external os of the cervix showed no signs of clinical illness.

Exfoliative cytology samples were obtained with commercial cytobrushes (Gynobrush®, Heinz Herenz Medizinalbedarf GmbH, Hamburg, Germany), which had been attached to a universal A.I. gun without mandrin (Kombicolor, IMV, L'Aigle, France). These extended cytobrushes had a total length of 54 cm. In order to expose only the head of the cytobrush to the desired location of the endometrium, sterile casing aluminum pipes (47 cm long, outer diameter: 6 mm, inner diameter: 4 mm) with rounded tips were used to pass the cervix and to reach the desired intrauterine locations under transrectal control. To protect the sample from vaginal contamination and to ensure a clean sampling system to pass the cervix the whole system was placed in an A.I. sanitary sheath (IMV, L'Aigle, France), which was only perforated directly in front of the external os of the cervix.

Exfoliative cytology samples were taken by two experienced examiners from three different uterine localizations per animal, including both uterine horns about 1.5 cm cranial of the bifurcation and the uterine body about 0.5 cm cranial of the internal os of the cervix (standardized conditions). These localizations were chosen referring to the regions of interest in an ultrasonic echotexture study of the bovine uterus [14]. Sampling started in the uterine body (UB), followed by the right (RUH) and left uterine horn (LUH), each time with a fresh sampling system. The obtained cellular material was spread out on microscope slides, air dried, fixed, stained with DiffQuick® and covered with a HistoKitt® fixed cover glass. A total of 300 cells/slide were differentiated microscopically (light microscope, 400× magnification, Olympus, BH-2, Tokyo, Japan) by a single skilled investigator and the percentage of PMN was assessed [6]. The threshold level of PMN indicating SE was set at 5% [6,15]. A total of 159 microscopic slides were investigated.

All results are displayed with means ± SD. Statistical analysis was done using One Way Analysis of Variance on Ranks with InStat3 (GraphPad Software, Inc., La Jolla, California, USA).

3. RESULTS

The mean percentage of PMN found throughout all differentiated smears was 2.4% ± 1.7%. Although statistically not significant ($p = 0.64$), the mean percentage of PMN found in the uterine body (2.2% ± 1.8%) was generally lower than in the right (2.3% ± 1.5%) and left uterine horns (2.5% ± 1.7%).

In 13 smears the PMN threshold of 5% was exceeded, and 18.9% (10/53) of the cows were diagnosed with SE. Intra-individual differences became obvious, yet in each one of these SE-positive animals, only one or two of the samples exceeded the threshold level for SE. In two individuals SE was verified in both uterine horns (individual results: UB: 4% and 1.4% PMN; RUH: 6.4% and 5% PMN; LUH: 5.7% and 5% PMN), the third surpassed the threshold in the uterine body as well as in the right uterine horn (individual results: UB: 7.7% PMN; RUH: 5.4% PMN; LUH: 1.4% PMN) and the other seven animals were diagnosed with SE only in the right (individual results: UB: 3.7% and 3% PMN; RUH: 5.4% and 5.4% PMN; LUH: 4.4% and 4% PMN) or left uterine horn (individual results: UB: 0.7%, 2.7% and 2.0% PMN; RUH: 4%, 3% and 3.4% PMN; LUH: 5%, 6.7% and 5% PMN) or the uterine body (individual results: UB: 5% and 6.7% PMN; RUH: 0.4% and 3% PMN; LUH: 4.7% and 4.4% PMN).

4. CONCLUSIONS

The results demonstrate that it might be advisable, for safe diagnosis of subclinical endometritis in embryo donor cows or cows well past the postpartum period, to take at least two cytological samples from different intrauterine localizations, preferably the uterine horns. The cow should be seen as positive for SE if at least one sample exceeds the threshold level of 5% PMN. A single exfoliative cytology does not seem to represent the entire endometrium. These findings might also be an argument to investigate this problem further using the low-volume flush technique, where a larger region of the endometrium is sampled.

ACKNOWLEDGMENTS

The authors would like to thank the participating farmers, their personnel and all the other involved people for their support and help.

AUTHOR CONTRIBUTIONS

Janna Egberts performed the experiments and statistical analysis. Jan Detterer supervised the experiments in the field, Jan Detterer and Arno Park were involved in cytobrush sampling. Janna Egberts, Jan Detterer and Sabine Meinecke-Tillmann participated in the interpretation of the results. Janna Egberts wrote the manuscript. Sabine Meinecke-Tillmann and Jan Detterer conceived and designed the experiments and provided editorial advice. Sabine Meinecke-Tillmann supervised the study.

CONFLICTS OF INTEREST

The authors declare no conflict of interest.

REFERENCES

1. Sheldon, I.M.; Lewis, G.S.; LeBlanc, S.; Gilbert, R.O. Defining postpartum uterine disease in cattle. Theriogenology2006, 65, 1516–1530.

2. Kasimanickam, R.; Duffield, T.F.; Foster, R.A.; Gartley, C.J.; Leslie, K.E.; Walton, J.S.; Johnson, W.H. Endometrial cytology and ultrasonography for the detection of subclinical endometritis in postpartum dairy cows. Theriogenology2004, 62, 9–23.
3. Gilbert, R.O.; Shin, S.T.; Guard, C.L.; Erb, H.N.; Frajblat, M. Prevalence of endometritis and its effects on reproductive performance of dairy cows. Theriogenology 2005, 64, 1879–1888.
4. Barański, W.; Podhalicz-Dzięgielewska, M.; Zduńczyk, S.; Janowski, T. The diagnosis and prevalence of subclinical endometritis in cows evaluated by different cytologic thresholds. Theriogenology 2012, 78, 1939–1947.
5. Madoz, L.V.; Giuliodori, M.J.; Migliorisi, A.L.; Jaureguiberry, M.; de la Sota, R.L. Endometrial cytology, biopsy and bacteriology for the diagnosis of subclinical endometritis in grazing dairy cows. J. Dairy Sci. 2014, 97, 195–201.
6. Prunner, I.; Wagener, K.; Pothmann, H.; Ehling-Schulz, M.; Drillich, M. Risk factors for uterine diseases on small- and medium-sized dairy farms determined by clinical, bacteriological, and cytological examinations. Theriogenology 2014, 82, 857–865.
7. Sens, A.; Heuwieser, W. Presence of Escherichia coli, Trueperella pyogenes, α-hemolytic streptococci, and coagulase-negative staphylococci and prevalence of subclinical endometritis. J. Dairy Sci. 2013, 96, 6347–6354.
8. Chapwanya, A.; Meade, K.G.; Narciandi, F.; Stanley, P.; Mee, J.F.; Doherty, M.L.; Callanan, J.J.; O'Farrelly, C. Endometrial biopsy: A valuable clinical and research toolin bovine reproduction. Theriogenology 2010, 73, 988–994.
9. Madoz, L.V.; Giuliodori, A.L.; Jaureguiberry, M.; Plöntzke, J.; Drillich, M.; de la Sota, R.L. The relationship between endometrial cytology during estrous cycle and cutoff points for the diagnosis of subclinical endometritis in grazing dairy cows. J. Dairy Sci. 2013, 96, 4333–4339.
10. Carvalho, P.D.; Souza, A.H.; Sartori, R.; Hackbart, R.S.; Dresch, A.R.; Vieira, L.M.; Baruselli, P.S.; Guenther, J.N.; Fricke, P.M.; Shaver, R.D.; et al. Effects of deep-horn AI on fertilization and embryo production in superovulated cows and heifers. Theriogenology 2013, 80, 1074–1081.
11. Fernandez-Sanchez, F.I.; Barrio-Lopez, M.; Quintela-Arias, L.A.; Becerra-Gonzales, J.J.; Peña-Martinez, A.I.; Martinez-Bello, D.; Garcia-Herradon, P.; Perez-Marín, C.C. Use of endometrial cytology and metabolic profiles for selection of embryo donor cows. Span. J. Agric. Res. 2014, 12, 664–671.
12. Kasimanickam, R.; Duffield, T.F.; Foster, R.A.; Gartley, C.J.; Leslie, K.E.; Walton, J.S.; Johnson, W.H. A comparison of the Cytobrush and uterine lavage techniques to evaluate endometrial cytology in clinically normal postpartum dairy cows. Can. Vet. J. 2005, 46, 255–259.
13. Bonnett, B.N.; Miller, R.B.; Martin, S.W.; Etherington, W.G.; Buckrell, B.C. Endometrial biopsy in Holstein-Friesioan dairy cows II. Correlations between Histological Criteria. Can. J. Vet. Res. 1991, 55, 162–167.
14. Schmauder, S.; Weber, F.; Kiossis, E.; Bollwein, H. Cyclic changes in endometrial echotexture of cows using a computer-assisted program for the analysis of first- and second-order grey level statistics of B-Mode ultrasound images. Anim. Reprod. Sci. 2007, 106, 153–161.
15. LeBlanc, S.J. Reproductive tract inflammatory disease in postpartum dairy cows. Anim. Cons. 2014, 8, 54–63.

CHAPTER 4

The Comparative Diagnostic Features of Canine and Human Lymphoma

Davis M. Seelig [1,*] ,Anne C. Avery [2] ,E. J. Ehrhart [2,] andMichael A. Linden [3]

[1]*Department of Veterinary Clinical Sciences, College of Veterinary Medicine, University of Minnesota, St. Paul, MN 55455, USA*

[2]*Department of Microbiology, Immunology, and Pathology and the Flint Animal Cancer Center, College of Veterinary Medicine and Biomedical Sciences, Colorado State University, Fort Collins, CO 80523, USA*

[3]*Department of Laboratory Medicine and Pathology, University of Minnesota, Minneapolis, MN 55455, USA*

ABSTRACT

The non-Hodgkin lymphomas (NHLs) are a heterogeneous family of lymphoid malignancies that are among the most common neoplasms of both dogs and humans. Owing to shared molecular, signaling, incidence, and pathologic features, there is a strong framework supporting the utilization of canine lymphoma as a comparative, large animal model of human NHL. In alignment with the biologic similarities, the current approach towards the diagnosis and classification of canine lymphoma is based upon the human World Health Organization guidelines. While this approach has contributed to an increasing appreciation of the potential biological scope of canine lymphoma, it has also become apparent that the most appropriate diagnostic philosophy must be multimodal, namely by requiring knowledge of microscopic, immunophenotypic, and clinical features before establishing a final disease diagnosis. This review seeks to illustrate the comparative similarities and differences in the diagnosis of canine lymphoma through the presentation of the microscopic and immunophenotypic features of its most common forms.

Keywords: lymphoma; comparative oncology; canine; non-Hodgkin lymphoma; comparative pathology

1. INTRODUCTION

The non-Hodgkin lymphomas (NHL) are a heterogeneous family of related, yet very distinct, lymphoid malignancies that are among the most common neoplasms in both human and dog. In humans, NHL is the 7th most common cancer type, representing 4.3% of all new cancer cases with an annual incidence of 19.7 per 100,000 [1]. Although the true incidence of canine lymphoma is somewhat uncertain as most dogs with the disease are not reported, most studies agree that lymphoma is the most common hematopoietic neoplasm of the dog and among the three most common canine neoplasms [2,3,4,5,6]. The largest, and most widely cited studies, report that lymphoma represents approximately 6% of all canine malignancies and 90% of all hematopoietic neoplasia with an annual incidence up to 24 cases per 100,000 dogs [4,5,7]. In addition to similar incidence rates, there is compelling evidence that the incidence of lymphoma is increasing in both species. The most recent data from a pet insurance study reports that the age-adjusted incidence of canine lymphoma has increased to 107 cases per 100,000 dogs at risk [8] whereas the incidence of human NHL has increased in the past four decades in both the USA and other countries, including a nearly doubling of the incidence rate between 1970 and 1990 [9]. In addition to their shared incidence, there is a robust and ever-expanding literature describing the shared biologic, genetic, and, in accordance with this review, their shared pathology. In both species diffuse large B-cell lymphoma (DLBCL) is the most common subtype (Figure 1), but this is where the similarities end [10,11,12,13].

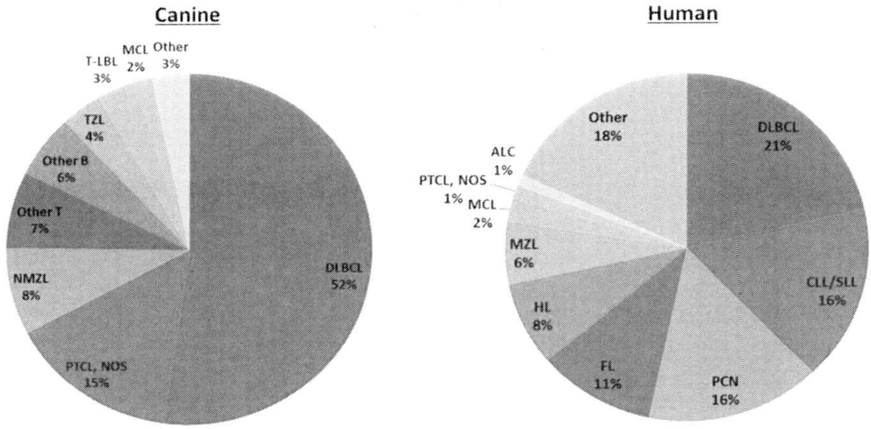

Figure 1. Incidence of canine (**A**) and human (**B**) lymphoma. Canine data reflect summary data from three publications whereas human data originate from one. DLBCL; diffuse large B-cell lymphoma, PTCL, NOS; peripheral T-cell lymphoma, not otherwise specified, NMZL; nodal marginal zone lymphoma, TZL; T-zone lymphoma, T-LBL; T-cell lymphoblastic lymphoma, MCL; mantle cell lymphoma, CLL/SLL; chronic lymphocytic leukemia/small lymphocytic lymphoma, PCN; plasma cell neoplasia, FL; follicular lymphoma, HL; Hodgkin's lymphoma, and ALC; anaplastic large cell lymphoma.

In the dog, the next most common subtypes, namely peripheral T-cell lymphoma, not otherwise specified (PTCL, NOS), nodal marginal zone lymphoma (NMZL), and T-zone lymphoma (TZL), the latter of which is considered a morphologic variant of PTCL, NOS in humans, are much less common in humans. In light of the rapidly evolving field of canine diagnostic hematopathology, the expanding access to advanced immunophenotyping options, and an increasing utilization of a standardized diagnostic algorithm (see below), it is likely that many of the published large canine studies that form the basis of current incidence data underestimate certain forms of lymphoma, including those with novel histologic architectures (e.g., nodular forms of lymphoma).

2. CLASSIFICATION OF CANINE AND HUMAN LYMPHOMA

According to the most recent World Health Organization (WHO) diagnostic guidelines, there are over 60 forms of human NHL. While it is beyond the scope of this review to summarize the evolution of human NHL classification, the inextricable link between diagnosis and classification and recent work validating a canine version of these guidelines necessitates a brief summary of its philosophical underpinnings. The process of lymphoma classification seeks to segregate individual disease subtypes according to reproducible and clinically relevant guidelines that facilitate efficient communication and enable relevant treatment and research. According to the WHO, lymphoma subtypes are defined according to their cytomorphology, immunophenotype, genetic, molecular, and clinical features [14]. The driving philosophies of this system are two-fold, namely: (1) that neoplastic lymphocytes are malignant counterparts of benign cells at varying stages of differentiation; and (2) that morphologically-indistinguishable neoplastic cells could be characterized as different lineages according to their genetic, molecular, and immunophenotypic characteristics. Although its scope was limited by a lack of species-specific genetic and/or immunophenotypic information, recent work has confirmed the applicability of the 2008 WHO guidelines to canine lymphoma [15]. Using a canine-adapted WHO system, a series of 18 international veterinary pathologists were able to obtain an overall accuracy of 83% with inter-observer variability of 65% when provided glass slides and immunophenotype information for 300 cases of canine lymphoma. This accuracy was increased to 87% when the case series was limited to the six most common forms of disease [15].

Although the clinical implications of classification has been recognized for many years in human hematopathology, the increasing recognition of distinct biological behavioral differences amongst canine lymphoma subtypes is more recent. A number of reports have identified distinct subtypes of canine lymphoma with a slow rate of biologic progression, so-called “indolent” lymphomas. These include marginal zone lymphoma, mantle cell lymphoma, and T-zone lymphoma. In contrast to subtype agnostic survival times of 10–12 months for canine B-cell lymphoma and 6 months for canine T-cell lymphoma, subtype specific survival times of 21 and 33 months for indolent forms of canine B-cell and T-cell lymphoma, respectively have been reported [16].

2.1. Basic Histologic Approach

The current standard for the diagnosis and classification of human lymphoma, which also serves as the basis for the current canine recommendations, is the 2008 REAL-WHO guidelines [17,18]. On histologic samples, these guidelines emphasize three features, namely: (1) tumor architecture; (2) tumor cell morphology (*i.e.*, size, shape, and nuclear/cytoplasmic features); and (3) mitotic activity. These features, in concert with knowledge of immunophenotype, clinical disease burden, and, in the case of human NHL, genetic features, are used to establish a final disease diagnosis. According to the current guidelines, there nearly 50 discrete subtypes of human NHL (Table 1). Using the human WHO as a template, 20 canine diseases have been described (Table 1).

In practice, application of the WHO histologic algorithm requires 3–5 μm thick, hematoxylin and eosin (HE) stained tissue sections. It is important to note that the histologic criteria used to diagnose and classify canine lymphoma are entirely based upon the human WHO classification system and, as such, a high degree of morphologic overlap between human and canine variants of the same subtype is to be anticipated. Tissue sections from wholly excised lymph nodes are considered to be the gold-standard for the diagnosis and classification of lymphoma [18,19]. The diagnostic utility of non-excisional lymph node biopsy samples has yet to be studied in veterinary medicine.

Using the 2008 WHO guidelines as a template, canine nodal lymphoma is initially divided according to tumor architecture as either a "diffuse" or "nodular/follicular" disease (Figure 2). In the case of the former, all or nearly all of the lymph node architecture is effaced by the neoplastic infiltrate whereas in the latter, the neoplastic infiltrate is organized in fashion that either mimics, or spares some portion, of lymph node tissue. Cytologically, tumor cells are classified as either small, intermediate, or large. Neoplastic cell size is determined based upon nuclear size relative to a known standard, traditionally either red blood cell (RBC) or a small lymphocyte. In human hematopathology, the gold standard for assessing cell size is an endothelial cell nucleus, although a small lymphocyte may be used. Based upon this comparison, neoplastic cells are classified as small (nuclei approximately 1–1.25 × the diameter of a red blood cell or smaller than an endothelial cell nucleus), intermediate (nuclei approximately 1.25–2 times the diameter of a red blood cell or equal to the size of an endothelial cell nucleus) or large (nuclei ≥ 2.0 × the diameter of a red blood cell or larger than an endothelial cell nucleus). Finally, using additional cytoplasmic (volume and staining intensity) and nuclear (shape, chromatin pattern, and nucleolar features) features in concert with knowledge of immunophenotype, a final histologic diagnosis can be established (Figure 2).

Table 1. Canine and human non-Hodgkin lymphomas according to the 2008 WHO guidelines.

B-Cell Neoplasms	*T-Cell Neoplasms*
B-lymphoblastic leukemia/lymphoma	**T-lymphoblastic leukemia/lymphoma**
Chronic lymphocytic leukemia/small lymphocytic lymphoma	T-cell prolymphocytic leukemia
B-cell prolymphocytic leukemia	**T-cell large granular lymphocytic leukemia**
Splenic marginal zone lymphoma	Chronic lymphoproliferative disorder of NK cells
Hairy cell leukemia	Aggressive NK-cell leukemia
Splenic lymphoma/leukemia, unclassifiable	EBV-positive T-cell disease of childhood
Lymphoplasmacytic lymphoma	Hydroa vacciniforme-like lymphoma
Heavy chain diseases	Adult T-cell leukemia/lymphoma
Plasma cell myeloma	Extranodal NK/T-cell lymphoma, nasal type
Solitary plasmacytoma of bone	**Enteropathy-associated T-cell lymphoma** *
Extraosseous plasmacytoma	**Hepatosplenic T-cell lymphoma**
MALT lymphoma	Subcutaneous panniculitis-like T-cell lymphoma
Nodal marginal zone lymphoma	**Mycosis fungoides**
Follicular lymphoma	**Sézary syndrome**
Primary cutaneous follicle centre lymphoma	Cutaneous $CD30^+$ T-cell disorders
Mantle cell lymphoma	**Peripheral T-cell lymphoma, NOS**
Diffuse large B-cell lymphoma	**T-Zone lymphoma** **
Lymphomatoid granulomatosis	Angioimmunoblastic T-cell lymphoma
B-Cell Neoplasms	*T-Cell Neoplasms*
Primary mediastinal large B-cell lymphoma	Anaplastic large cell lymphoma, ALK-positive
Intravascular large B-cell lymphoma	Anaplastic large cell lymphoma, ALK-negative
ALK-positive large B-cell lymphoma	
Plasmablastic lymphoma	
Large B-cell lymphoma in Castleman disease	
Primary effusion lymphoma	
Burkitt lymphoma	
B-cell lymphoma, unclassifiable	

Non-bolded entities represent malignancies in humans only whereas bolded entities reflect malignancies reported in both humans and canines. * A neoplasm with an EATL-like morphology has been described in dogs, but an association with celiac disease has not been documented. ** In humans, TZL is considered a morphologic variant of PTCL, NOS.

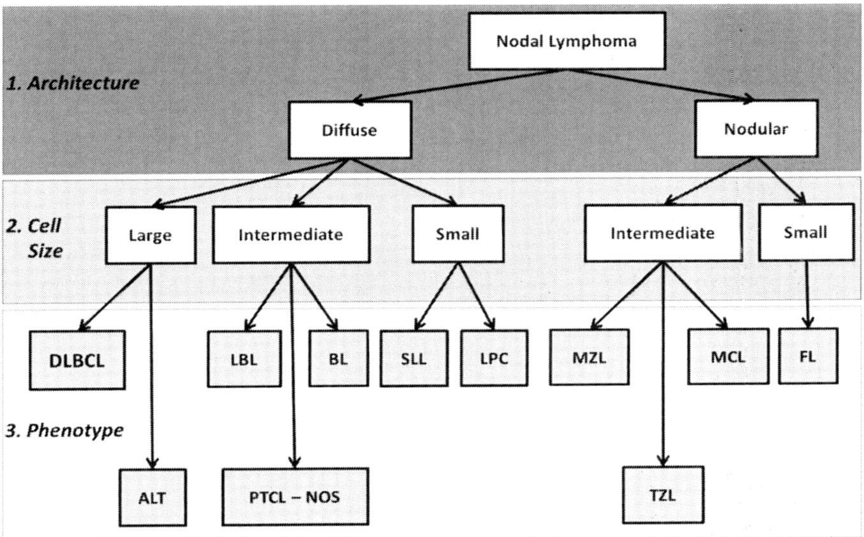

Figure 2. The histologic approach towards the classification of canine nodal lymphoma. Using excisional lymph node sections, lymphoma is initially divided into diffuse (effacing) or nodular (non-effacing) forms of the disease. Next, using a red blood cell or a small lymphocyte as a guideline, the neoplastic population is divided into large, small, and intermediate forms of the disease. Finally, using knowledge of additional cellular and nuclear features, including mitotic rate, and immunophenotype, a final diagnosis is established. DLBCL; diffuse large B-cell lymphoma, PTCL, NOS; peripheral T-cell lymphoma, not otherwise specified, NMZL; nodal marginal zone lymphoma, TZL; T-zone lymphoma, T-LBL; T-cell lymphoblastic lymphoma, MCL; mantle cell lymphoma, CLL/SLL; chronic lymphocytic leukemia/small lymphocytic lymphoma, PCN; plasma cell neoplasia, FL; follicular lymphoma, HL; Hodgkins lymphoma, and ALC; anaplastic large cell lymphoma.

Knowledge of neoplastic cell immunophenotype is an absolute requirement for the complete diagnosis and classification of lymphoma in either species. A number of tools, including flow cytometry, immunocytochemistry, and lymphoid clonality assays, can provide immunophenotype, but Immunohistochemistry (IHC) is considered the gold standard [20]. The importance of IHC in the characterization of canine lymphoma was illustrated by Flood-Knapik *et al.* which reported a reclassification of 20% of indolent lymphomas cases with the addition of IHC data [16].

2.2. FNAC in the Diagnosis of Lymphoma

In light of their shared pathology, the philosophical approach towards the diagnosis and classification of lymphoma in both man and dog is similar. Although there are technologic constraints and economic realities that do not allow the direct duplication of human NHL diagnostic algorithms to the dog, both human and veterinary pathologists generally adopt a stepwise approach. When faced with an enlarged peripheral lymph node, the chief objective of both is to refine the provisional list of differentials which includes reactive hyperplasia, lymphadenitis, lymphoma, and metastatic disease. Fine needle aspiration cytology (FNAC) may be suited for these initial investigations owing to its simplicity, low-cost, safety, and minimally-invasive sampling.

As the vast majority (>80%) of the cases of canine nodal lymphomas present with diffuse effacement of one or more peripheral lymph nodes by large neoplastic cells, fine needle aspiration cytology (FNAC) is a widely accepted as a highly sensitive and specific first-line lymphoma diagnostic. In humans, FNAC has a >90% sensitivity and specificity in differentiating a malignant process from reactive hyperplasia [21,22]. In both species, the results of the FNAC are often used to trigger a number of second tier studies, including flow cytometry, immunocytochemistry, biopsy with immunohistochemistry, clonality assays, and cytogenetic analysis [17]. While FNAC has numerous advantages, it is generally unsuitable as a solitary diagnostic for certain subtypes of lymphoma, including nodular and small-cell variants [23,24]. In general, while FNAC plays an important role in the management of human patients, a tissue-based diagnosis by needle core or excisional biopsy (with immunophenotyping), is essentially required prior to initiating (chemo) therapy.

2.3. Flow Cytometry in the Diagnosis and Classification of Lymphoproliferative Disorders

When used in concert with FNAC, flow cytometry immunophenotyping offers many advantages. Namely, it requires only a small volume of sample, it can evaluate multiple antigens on one cell, it can give quantitative results, and it can detect rare and phenotypically aberranT-cells. As an instrument of diagnostic hematopathology, flow cytometry (FC) chiefly detects and subclassifies lymphoid neoplasms through the identification of tumor-specific immunophenotypic signatures, including atypical light scatter (chiefly size) and antigen expression profiles. Atypical antigen expression is reflected by either monoclonal light chain expression (in the case of human B-cell neoplasms) or in the detection of cells with an antigen profile incongruous with normal lymphoid cells.

Monoclonal, or restricted light chain expression, is the most useful surrogate marker of B-cell clonality in humans and, in practice, is generally defined by kappa:lambda ratios of >10 or lambda:kappa ratios of >5 [25]. In contrast, given that both neoplastic and reactive B-cell populations are dominated by lambda-expressing cells, a similar approach is not feasible in dogs [26]. As such, in the evaluation of canine B-cell lymphomas, FC is chiefly used to provide adjunctive immunophenotyping and prognostic information, rather than act as a primary diagnostic tool [27]. Although a similar surrogate marker of monoclonality does not exist for T-cells, immunophenotypic aberrancy, as defined by abnormal increases or decreases (including complete loss of surface antigen expression) in normal T-cell antigens or unexpected co-expression profiles, is often used to diagnose T-cell neoplasia [28]. Using subtype-specific antigen expression profiles (Table 2), the combination of FNAC and FC has a sensitivity and specificity of 97% and 94%, respectively for the diagnosis and classification of human NHL in a series of 72 cases with histologic follow-up [28].

In canine hematopathology, although the diagnostic philosophy with respect to FC is similar, the limited species-specific reagents and the novel biological features of canine lymphoma, require an alternate approach. In contrast to human hematopathology, in which FC panels using 15–25+ antibodies (cocktailed in more than one tube) are recommended, most canine FC laboratories use panels with 10–12 antibodies [24,29]. Moreover, as the vast majority of both normal and malignant lymphoid populations express the lambda variant, light chain restriction is not a practical surrogate marker of clonality in canine lymphoproliferative diseases [26,30]. While no study has yet examined the performance of FC as a primary diagnostic tool for canine lymphoma, a number of recent studies have confirmed its utility as a tool for advanced classification of certain forms of canine lymphoma, for lineage assignment, and for the provision of prognostic information.

The sections that follow will, using an evidence-based approach, review the diagnostic approach for the most common forms of canine lymphoma. Incidence data for specific subtypes of canine lymphoma classified according to the updated WHO algorithm stems from three studies, which comprise a total of 1675 cases [11,12,13]. Using this data, this comparative diagnostic review will emphasize the most common forms of canine nodal lymphoma, namely: (1) Precursor disease; (2) DLBCL; (3) Marginal Zone Lymphoma; (4) Mantle Cell Lymphoma; (5) Peripheral T-cell Lymphoma NOS; and (6) T-zone lymphoma.

Table 2. Comparative histologic and immunophenotypic characteristics of common forms of Canine and human non-Hodgkin lymphoma.

Disease	Canine		Human	
	Histology	Phenotype	Histology	Phenotype
Precursor Neoplasms	Architecture: Diffuse	*T-cell:* CD45$^+$, CD34$^{+/-}$, CD5$^{+/-}$, CD3$^{+/-}$, CD4$^{+/-}$, CD8$^-$	Architecture: Diffuse	*T-cell:* CD34$^+$, TdT$^+$, cytoplasmic CD3$^+$, CD5$^+$
	Cytology: Intermediate sized cells, round nuclei, scant cytoplasm, high mitotic rate.	*B-cell:* CD45$^+$, CD18$^+$, CD34$^{+/-}$, CD79a$^+$, CD21$^{+/-}$	Cytology: Small to large cells, varied cytoplasmic volume, high mitotic rate.	*B-cell:* CD34$^+$, TdT$^+$, CD19$^+$, CD79a$^+$, sIg
DLBCL	Architecture: Diffuse Cytology: Large cells with round nuclei. One (central, immunoblast) or multiple (peripheral, centroblasts) nucleoli. High mitotic rate and "starry sky" appearance.	CD20$^+$, CD21$^+$, CD45$^+$, CD79a$^+$, Pax5$^+$, MHC II$^+$	Architecture: Diffuse Cytology: Large cells with round nuclei. Both immunoblasts and centroblasts. High mitotic rate and "starry sky" appearance in a subset.	sIg$^+$, CD45$^+$, CD20$^+$, CD22$^+$, CD79a$^+$, Pax5$^+$, CD10$^{+/-}$
Mantle Cell Lymphoma	Architecture: Nodular Cytology: Small to intermediate-sized cells; scant cytoplasm, round nuclei with dense chromatin, and inconspicuous nucleoli. Varied mitotic rate.	CD20$^+$, CD21$^+$, CD45$^+$, CD79a$^+$, MHC II$^+$	Architecture: Nodular or diffuse Cytology: Very heterogeneous, small to large cells, round to irregular nuclei; varied nucleoli and mitotic rate	sIg$^+$, CD19$^+$, CD20$^+$, CD22$^+$, CD79a$^+$, CD79b$^+$, CD5$^+$, cyclin D1$^+$, CD23$^-$, CD10$^-$
Splenic Marginal Zone Lymphoma	Architecture: Nodular Cytology: Intermediate-sized cells, abundant pale cytoplasm, irregular nuclei with peripheralized chromatin and, a single nucleolus; rare mitotic figures.	CD20$^+$, CD21$^+$, CD45$^+$, CD79a$^+$, MHC II$^+$	Architecture: Nodular Cytology: Biphasic-Small cells with scant cytoplasm and round nuclei, intermediate-sized cells with abundant cytoplasm and irregular nuclei; rare mitotic figures.	sIg$^+$, CD20$^+$, IgD$^+$, BCL2$^+$, BCL6$^-$, CD5$^-$, CD10
Nodal Marginal Zone Lymphoma	Architecture: Nodular Cytology: Mixed. Mostly intermediate-sized cells with pale cytoplasm, irregular nuclei, peripheral chromatin, and one nucleolus.	CD20$^+$, CD21$^+$, CD45$^+$, CD79a$^+$, MHC II$^+$	Architecture: Nodular to diffuse. Cytology: Intermediate-sized cells with pale cytoplasm and irregular nuclei (centrocyte-like) and large cells with abundant pale cytoplasm and irregular nuclei (monocytoid).	sIg$^+$, CD19$^+$, CD20$^+$, CD22$^+$, CD79a$^+$, CD5$^-$, CD10$^-$, BCL6$^-$, and Cyclin D1$^-$
Follicular Lymphoma	Architecture: Nodular Cytology: Mixed. Mostly small cells with clear cytoplasm, pale chromatin, and inconspicuous nucleoli (centrocytes) with fewer large cells with dark blue cytoplasm, vesicular nuclei, and 1–3 nucleoli (centroblasts).	CD20$^+$, CD21$^+$, CD45$^+$, CD79a$^+$, MHC II$^+$	Architecture: Nodular Cytology: Similar to the dog.	sIg$^+$, CD45$^+$, CD19$^+$, CD20$^+$, CD22$^+$, CD79a$^+$, Pax5$^+$, CD10$^+$, BCL6$^+$, BCL2$^+$
PTCL-NOS	Architecture: Diffuse Cytology: Small to large (heterogeneous) with irregular nuclei, variable chromatin, prominent nucleoli, and varied mitotic activity.	CD3$^+$, CD79a$^-$, CD21$^-$, CD45$^+$, CD5$^+$, CD4$^{+/-}$, CD8$^{+/-}$	Architecture: Diffuse Cytology: Small to large (heterogeneous) with irregular nuclei, variable chromatin, prominent nucleoli, and varied mitotic activity	CD3$^+$, CD4$^+$ /> CD8$^{+/-}$, CD7$^{+/-}$, CD5$^{+/-}$, CD2$^+$, CD30$^{+/-}$
TZL	Architecture: Nodular Cytology: Small to intermediate-sized cells, moderate amount of pale cytoplasm, oval to elliptical nuclei with sharp, shallow indentations, nucleoli and mitotic figures are sparse.	CD45$^-$, CD3$^+$, CD5$^+$, CD21$^+$, CD4$^{+/-}$, CD8$^{+/-}$	*Histologic Variant of PTCL, NOS* Architecture: Nodular Cytology: Small to intermediate-sized cells; moderate amount of pale cytoplasm; oval to elliptical nuclei with sharp, shallow indentations; nucleoli and mitotic figures are sparse	CD2$^+$, CD3$^+$, CD5$^+$, CD4$^+$

3. PRECURSOR NEOPLASMS

Lymphoblastic lymphoma (LBL) and acute lymphoblastic leukemias (ALL) are precursor cell neoplasms derived from immature lymphoid cells of either B (B-LBL/ALL) or T-cell (T-LBL/ALL) lineage. They are believed to arise from lineage-committed precursors in either the bone marrow (B-LBL/ALL) or thymus (T-LBL/ALL). The distinction between LBL and ALL of either phenotype is based upon the tumor tissue burden rather than any unique cellular features. Namely, if the disease is dominated by blood or bone marrow involvement (*i.e.*, >20%–25% neoplastic cells), the diagnosis is ALL. Alternately, if tumor burden predominantly involves non-blood/non-marrow tissue compartments the diagnosis is LBL. Practically, the distinction between ALL/LBL is thought to be somewhat arbitrary as the biologic behavior and prognosis of LBLs and ALLs of a particular phenotype are considered to be equivalent.

The precursor neoplasms comprise approximately 2% of all human NHL and, in total, the B-cell form of the diseases predominates over the T-cell form. However, when segregated into LBL and ALL types, the incidence of the B- and T-cell forms of the disease differs [31]. With hALL, the B-cell form of the disease (hB-ALL) predominates, representing 75%–80% of all acute leukemias, whereas in contrast, hLBL is far more often a T-cell disease, with an approximate ratio of 9:1 [32]. Although they are overall uncommon neoplasms, the precursor diseases are reported to be slightly more frequent in the dog than in humans, with LBL representing 3%–9% of canine lymphomas [11,12,13]. Similar to humans, canine T-LBL (64%–100%) is reported to be much more common than canine B-LBL (0%–34%), however it is important to note that the largest effort to classify canine lymphoma using human criteria, precursor B-cell lymphomas were not present in the 600 cases studied [11,12,13]. One report suggests that cB-ALL is much more common than cT-ALL [33].

The diagnostic approach to the precursor diseases in both species is similar. In both human and canine variants, LBL is a diffuse disease comprised of a uniform population of neoplastic cells (referred to as lymphoblasts) that occasionally infiltrate the perinodal tissue (Figure 3) [34]. The cells are small-to-intermediate in size with round to convoluted nuclei containing variable (dispersed to condensed) chromatin and faint to indistinct nucleoli. Cytoplasmic volume is scant and, often, LBL has a high mitotic rate and a starry sky appearance [11,12,13,34,35,36]. Similarly, the diagnosis of either hALL or cALL necessitates the identification of >20%–30% blasts in blood or bone marrow [37,38]. In both species, the two forms of precursor disease (*i.e.*, LBL *vs.* ALL) and the two phenotypes (*i.e.*, T *vs.* B-cell) cannot be distinguished using light microscopy.

Immunophenotypically, detection of the stem cell antigen CD34 and terminal deoxynucleotidyl transferase (TdT) are hallmarks in the diagnosis of hALL/LBL [34,36]. Human B-ALL/LBL is nearly always positive for B-cell markers CD10, CD19, cytoplasmic CD79a, and CD22 with dim to absent CD45 expression, whereas T-ALL/LBL consistently express T-cell markers cytoplasmic CD3, CD5, and CD4 and CD8 [32,36]. The expression of TdT and the lack of surface IG are useful in distinguishing precursor B-cell disease from mature B-cell neoplasms, whereas the expression of TdT, CD34, and the presence of cytoplasmic CD3 without surface CD3 can be used to distinguish mature T-cell neoplasms from precursor T-cell disease [36]. The expression of CD45 in contrast, the immunophenotypic profile of the canine precursor diseases, owing to their rarity and the lack of a standardized diagnostic approach is murkier.

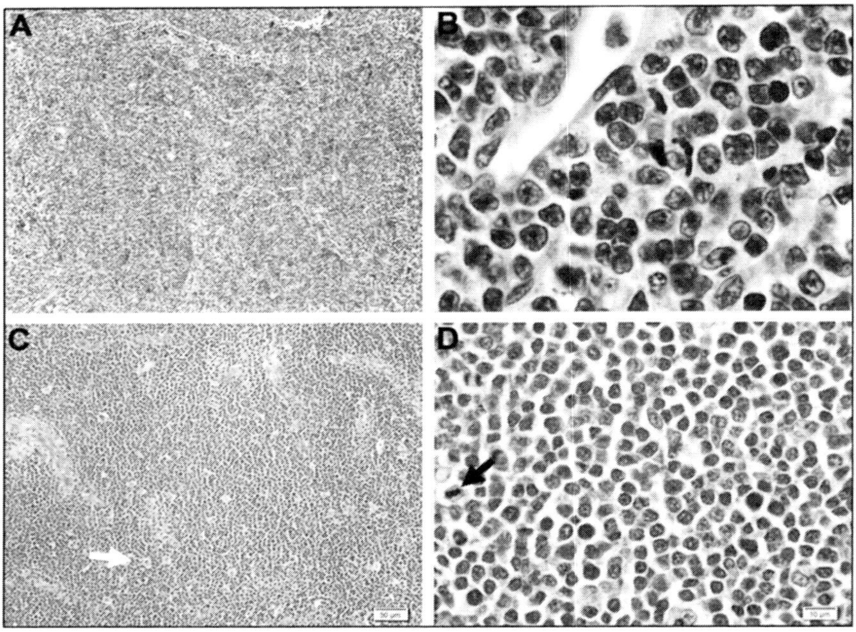

Figure 3. The histologic features of human and canine precursor disease (LBL/ALL). In both species (**A**, **B**, human and **C**, **D**, canine) the architecture is diffuse (**A** and **C**). Morphologically, the neoplastic cells in both species are small-to-intermediate in size with round to convoluted nuclei containing dispersed to clumped chromatin and faint to indistinct nucleoli. A starry sky appearance (white arrow; (**C**) and high mitotic rate (black arrow); (**D**) may be seen in both.

According to published reports, the canine precursor diseases, like their human counterparts, express either B or T-cell antigens (e.g., CD79a, CD21, CD3, and/or CD5) [33,38,39,40,41,42,43,44], but in contrast, demonstrate varied CD34 expression (*i.e.*, rare in cLBL, but common in cALL) [11,12,13,43,45]. To date, immunostaining for TdT has not proven useful in the diagnosis of canine precursor disease, although it is unclear of these negative findings reflect true biologic differences between the canine and human forms of the disease or if they represent antibody shortcomings [46]. However, it is important to emphasize that these reports utilize highly variable diagnostic techniques. For example, most reports of cLBL utilize archived, formalin-fixed, paraffin-embedded tissues, which are unsuitable for canine anti-CD34 IHC, whereas cALL studies primarily use flow cytometry, where detection of CD34 is more robust. In practice, most canine precursor neoplasms (predominantly acute leukemias) are phenotyped by flow cytometry rather than histology. While it is generally accepted that CD34 expression is diagnostic for precursor leukemia, there are no consensus criteria for differentiating true precursor leukemias from other forms of lymphoproliferative disease when CD34 expression is lacking [47]. Moreover, there is limited understanding as to the role of lineage-specific markers for the immunophenotyping of canine precursor disease. For example, although CD21 expression as detected by flow cytometry is variably used in the literature to diagnose cB-ALL, in other species CD21 is not expressed until B-cell reach maturity, and is not expressed on human B-ALL/LBL [48]. And, although CD79a is more commonly used to ascribe a B-cell lineage to human precursor disease, and may be useful in cALL [33], it has not yet been established that CD79a is specific for B-cell ALL/LBL in dogs [49].

4. T-CELL NEOPLASMS

As a percentage of all NHL, the mature T-cell lymphomas are less common than their B-cell counterparts in both species, although they are much more common in the dog than human representing 30%–40% and 6%–10% of all lymphomas by species, respectively [10,11,13,50]. In the dog, the most common forms appear to be T-zone lymphoma and PTC-NOS whereas in humans, the most common forms are PTCL-NOS and anaplastic large cell lymphoma [10].

4.1. Peripheral T-Cell, Not Otherwise Specified (PTCL, NOS)

While designated as a specific subtype of PTCL in the current WHO classification, PTCL, NOS is a heterogeneous group of nodal and extranodal mature T-cell neoplasms that demonstrate features which set them apart from other forms of PTCL. In both species, the category of PTC, NOS is a "wastebasket" or "catchall" category. Based upon their morphological, genetic, and immunophenotypic heterogeneity, it is accepted that PTCL, NOS contains many different unique entities, however no reliable criteria have been established that allow for consistent identification of any one specific entity. In both human and canine, peripheral T-cell lymphoma not-otherwise specified (PTCL, NOS) is the most common T-cell lymphoproliferative disease, representing approximately 30% and 39%–65% of all human and canine cases, respectively [11,13,51]. However, as a percentage of all forms of lymphoma, PTCL, NOS is uncommon in humans, representing less than 1% of all NHL, but comparatively much more common in dogs, at 15% of all lymphomas.

The cellular origins of both human PTCL, NOS (hPTCL, NOS) and cPTCL, NOS are likely heterogeneous, but are presumed to be derived from post-thymic T-cells in various stages of transformation [51]. Gene expression profiling (GEP) studies suggest that hPTCL, NOS can be subdivided into at least three cell-of-origin subgroups, namely cytotoxic, helper, and follicular helper PTCL, NOS [52,53]. Initial works by the International Peripheral T-cell Lymphoma Project suggests that molecular subtyping may have clinical implications as the cytotoxic tumors appear to behave significantly worse [54]. Clinically, patients with hPTCL, NOS most often present with disseminated disease characterized by generalized lymphadenopathy and extranodal involvement, with skin, GI tract, and bone marrow being the most commonly affected sites [17,55,56]. Similarly, canine PTCL, NOS (cPTCL, NOS) has a varied pattern of tissue involvement, including nodal and extranodal (subcutaneous, hepatosplenic, mediastinal, and intestinal) manifestations [39].

The diagnosis of PTCL, NOS in either species, because it lacks any single specific morphologic, genetic, or immunophenotypic signature, is considered to be a process of exclusion. It is noteworthy that this pathologic heterogeneity has resulted in poor (*i.e.*, 75%) diagnostic agreement amongst human hematopathologists in the diagnosis of PTCL, NOS [57]. Similar studies have not been conducted in veterinary hematopathology. Histologically, both human and canine PTCL, NOS are usually diffuse diseases although rare follicular variants of hPTCL, NOS are recognized (Figure 4) [58]. The cytomorphology of the neoplastic cells in both species is highly variable and can include varying proportions, either monomorphic or polymorphic, of small, medium, or large cells. Reflective of this, a previous canine-adapted version of the Kiel classification scheme divided PTCL, NOS into 6, morphologically-unique subtypes, namely pleomorphic small-cell, small clear-cell, pleomorphic mixed, pleomorphic large-cell, immunoblastic, and plasmacytoid lymphomas [11]. However in both species, most cases consist of medium-to-large sized cells with irregular nuclei, a variable chromatin pattern, prominent nucleoli, and variable mitotic activity [15,39,40,59].

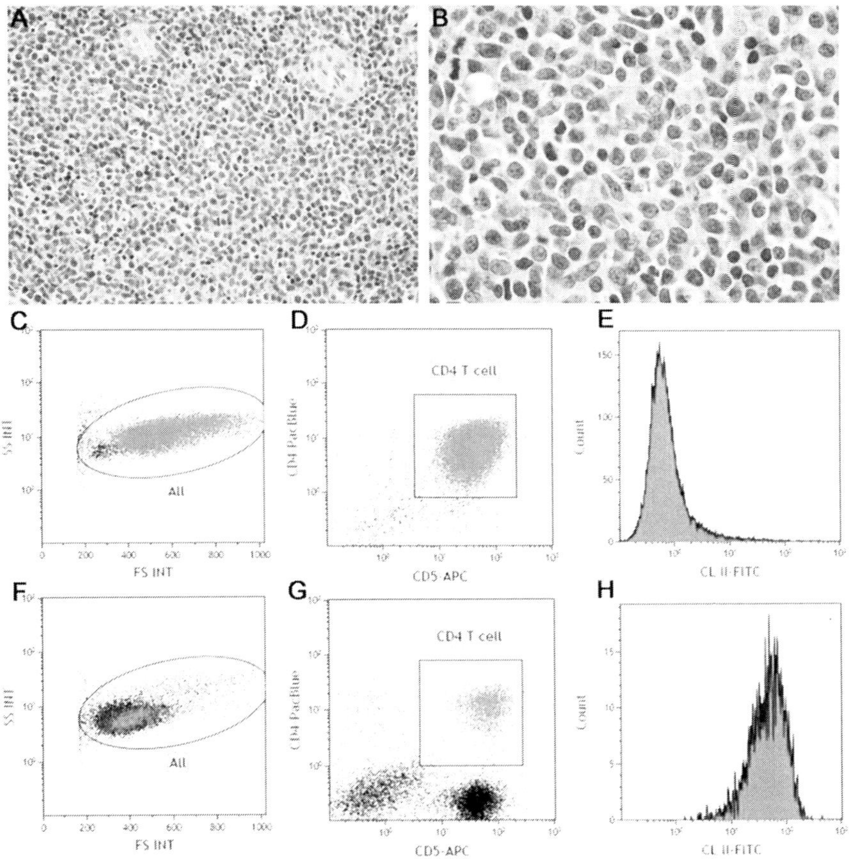

Figure 4. The histologic and flow cytometric features of canine peripheral T-cell lymphoma, not otherwise specified (PTCL, NOS). Architecturally, the neoplastic infiltrate is diffuse (**A**) and the infiltrating cells are heterogeneous, with cell size varying from small to large with pleomorphic, oval to cleaved nuclei (**B**). Flow cytometry of the neoplastic cells (colored green, **C–E**) in reported cases of PTCL, NOS reveals cells that are large by light scatter (**C**), $CD5^+$, $CD4^+$ (**D**), and demonstrate decreased expression of MHC Class II (**E**). These findings contrast to the T-cells within a normal canine lymph node (green, **F–H**), which are small by light scatter (**F**), CD5+, split between $CD4^+$ and $CD4^-$ (**G**), and demonstrate higher levels of MHC Class II expression (**H**).

There is no consistent immunophenotypic signature for hPTCL, NOS and specific immunophenotypes do not consistently correspond to a clinically-meaningful entity [60]. However, neoplastic cells in both human and canine PTCL, NOS are typically positive for mature T-cell markers, including CD3 and CD5 [39,40]. Phenotypic aberrancy in hPTCL, NOS, as evident by loss of either CD5 and CD7 or CD4/CD8 dual positivity or negativity are also common [59,60,61]. Nodal cases of hPTCL, NOS are most commonly CD4+/CD8− [56]. Additional commonly expressed markers in hPTCL, NOS include TIA-1 (T-cell intracellular antigen-1), CXCL13, PD-1, CD52, and CD30 [62,63,64], and many of these markers help to support a helper *vs.* cytotoxic differentiation. Although the extent of the immunophenotype of cPTCL, NOS has not been described, like its human counterpart, the nodal variant is most commonly reported as a CD45+, CD3+, CD4+, CD8−, MHC II low, CD21 negative disease, with frequent loss of CD5 expression [39,40,41].

4.2. T-Zone Lymphoma

T-zone lymphoma (TZL) is a variant of both human and canine T-cell lymphoma characterized by a unique histologic and cytologic appearance. Although no publications directly address its incidence, TZL appears to account for less than 2% of human PTCLs [65]. Although the true incidence of cTZL has not been defined it appears to be much more common in veterinary medicine, accounting for 16%–62% of all canine indolent lymphomas. In the current WHO guidelines, human TZL (hTZL) is classified as a *morphologic variant* of PTCL-NOS rather than a *distinct subtype* of NHL [56]. In contrast, due to its clinical, histologic, and immunophenotypic novelty, canine TZL (cTZL) is widely accepted to be distinct subtype of canine lymphoma.

Although the diagnosis of TZL in both species necessitates tissue evaluation and recognition of its novel histologic and cytologic features, Seelig *et al.* report that its unique immunophenotypic features may allow the diagnosis of cTZL to be made by flow cytometry alone [66]. According to the original Kiel classification scheme, hTZL is non-leukemic type of lymphoma containing all components of the T-regions of lymphoid tissue and can be considered as the diametric counterpart to follicular lymphoma [67]. Architecturally, in both canine and human, TZL is a non-effacing disease in which there is retention of B-cell follicles, including germinal centers, in the face of dramatic inter-follicular expansion owing to proliferating neoplastic T-cells (Figure 5) [67,68]. Expansion of the interfollicular space by the neoplastic population results in displacement and compression of the residual follicles against a thinned perinodal capsule [66]. Owing to its non-effacing nature, TZL can be classified as a nodular disease. Cytologically, the neoplastic cells in similar in both canine and human TZL. In both species, they are small to medium-sized with a moderate amount of lightly stained cytoplasm and oval to elliptic nuclei with sharp, shallow indentations, finely granular chromatin, and inapparent nucleoli [67,69]. In both species, small numbers of large cells may be noted. In cTZL, mitotic figures are generally rare, but the human form they are described as either uncommon or very frequent [67,68]. In cases in which mitotic figures are evident, their significance is uncertain. Anecdotally, the neoplastic cells in cTZL are often described as having cytoplasmic tails, which are described on cytology samples, as either "hand mirrors" or "uropods" [70].

In contrast to cTZL, human patients with TZL are described as seldom showing peripheral blood involvement [67]. In the dog, peripheral blood lymphocytosis is described in 48%–64% of cases and neoplastic cells can be detected in the peripheral blood in 100% of cases in which peripheral blood was available for flow cytometric analysis, including many cases in which the absolute lymphocyte count was normal [16,66,69]. Bone marrow involvement in both hTZL and cTZL appears uncommon and peripheral blood cytopenias are very uncommon [69,71]. Limited hTZL-specific immunophenotypic information is available, but the tumor cells are described as CD2+, CD3+, CD5+, and CD4+ [72]. To date, no studies of hTZL using modern immunophenotyping techniques or reagents have been reported. The immunophenotype of cTZL is typical of a mature T-cell neoplasm through its expression of both CD3+ and CD5+ and the lack of for B-cell antigens CD20, Pax5, and CD79a. However, TZL is unique among canine lymphoma through its CD45− and CD21+ phenotype [66,69]. The expression of CD4 and CD8 is variable as single positive, dual positive, and dual negative variants have been described [66,69,73].

5. B-CELL NEOPLASMS

B-cell lymphomas arise during different steps of B-lymphocyte development and represent their malignant counterpart and, as such, pre-germinal and post-germinal center lymphomas can be identified [74]. In both man and dog, the B-cell phenotype is the most common form of lymphoma, representing 85%–90% and 62%–75% of all human and canine lymphomas, respectively [11,13,17].

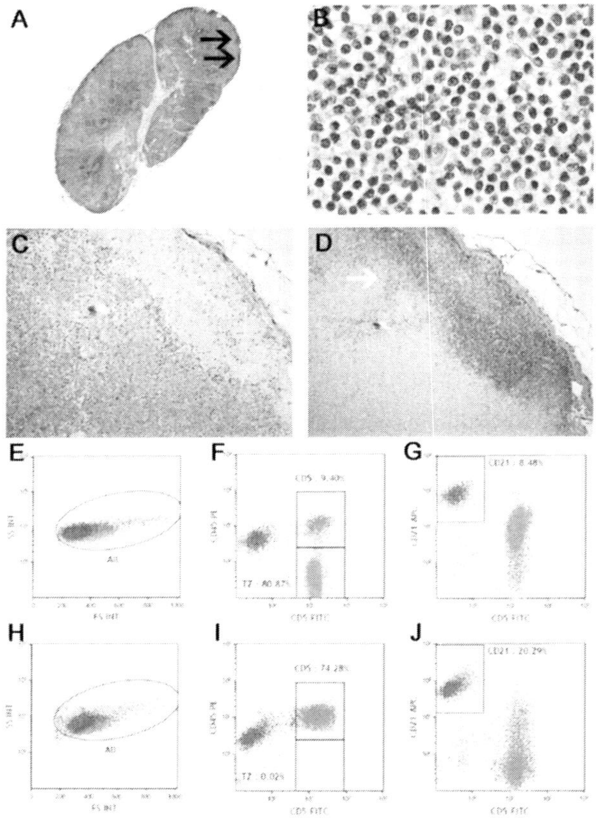

Figure 5. The histologic, immunohistochemical, and flow cytometric appearance of canine T-zone lymphoma (TZL). Affected nodal tissue is characterized by expansion and effacement of the T-cell areas of the cortex and medulla with resulting capsular displacement and compression of residual follicles (black arrows, **A**). The neoplastic cells are intermediately-sized and contain a moderate amount of pale blue cytoplasm and ovoid nuclei with sharp, shallow indentations, finely granular chromatin, and inapparent nucleoli (**B**). Anti-CD3 (red, **C**) and anti-Pax5 (red, **D**) immunohistochemistry demonstrates the T-cell phenotype of the neoplastic cells and further illustrates the compressed, B-cell rich residual follicles (white arrow, **D**). Flow cytometrically, in contrast to a normal canine peripheral lymph node (**H–J**), TZL (**E–G**) is characterized by the presence of an intermediate-sized population of CD5+ cells (colored blue) which are phenotypically aberrant owing to their lack of CD45 expression (**F** *vs.* **I**) and increased CD21 expression (**G** *vs.* **J**).

5.1. Diffuse Large B-Cell Lymphoma

In both dog and man, diffuse large B-cell lymphoma (DLBCL) is the most common mature lymphoid malignancy. In humans, DLBCL represents the most common form of adult NHL and the most common form of BCL, representing 30%–40% of all adult NHL and 37% of BCL [17,31]. Similarly, DLBCL is the most common form of lymphoma in the dog, although the overall incidence is higher, representing 47%–56% of all canine lymphomas and 73% of canine BCL [11,13]. Although the neoplastic cells share generally similar morphologic features between the two species, human DLBCL (hDLBCL) can be subdivided according to genetic, immunophenotypic, and molecular studies into four subgroups: (1) DLBCL—not otherwise specified (NOS); (2) DLBCL with a predominant extranodal location; (3) Large cell lymphomas of terminally differentiated B-cells, and (4) B-cell neoplasms with intermediate features (aka borderline cases) [17]. At present, although the morphologic and immunophenotypic heterogeneity of cDLBCL has not been reported, there is evidence suggesting genetic variation is likely to exist in cDLBCL [75,76] and that such genetic variation correlates with biologic behavior [77].

In both species, DLBCL most commonly occurs as a *de novo* disease, although transformation from a less aggressive BCL, including MZL and FL (*i.e.*, Richter syndrome/Richter transfromation) is well-characterized in hDLBCL. [78]. Although similar events have been described in the dog, further work is needed to define its pathologic scope and disease-specific frequency [79]. Work by Frantz *et al.* by demonstrating molecular overlap between MZL and DLBCL, in concert with the morphologic overlap between these two entities, suggests that an unknown percentage of cases of cDLBCL might represent transformed cMZL [77]. Clinically, both canine and human DLBCL patients present with rapidly progressive, generally painless nodal or extranodal disease with clinical signs reflecting tumor burden and location [80]. Hematologically, approximately 10%–25% of hDLBCL cases present with peripheral blood involvement at the time of diagnosis diagnosis (more likely to be detected with sensitive techniques such as flow cytometry) and, although the frequency of blood involvement in dogs is less certain, one study found that in 50% of 14 cases, neoplastic B-cells could be found in peripheral blood by flow cytometry [81,82,83].

The definitive diagnosis of both human and canine DLBCL requires histopathological examination of affected tissues [27,80]. Architecturally both hDLBCL and cDLBCL are diffuse diseases with coagulative necrosis and extension into the surrounding tissue (Figure 6) [15,84]. At low magnification, a "starry-sky" appearance resulting from the scattered

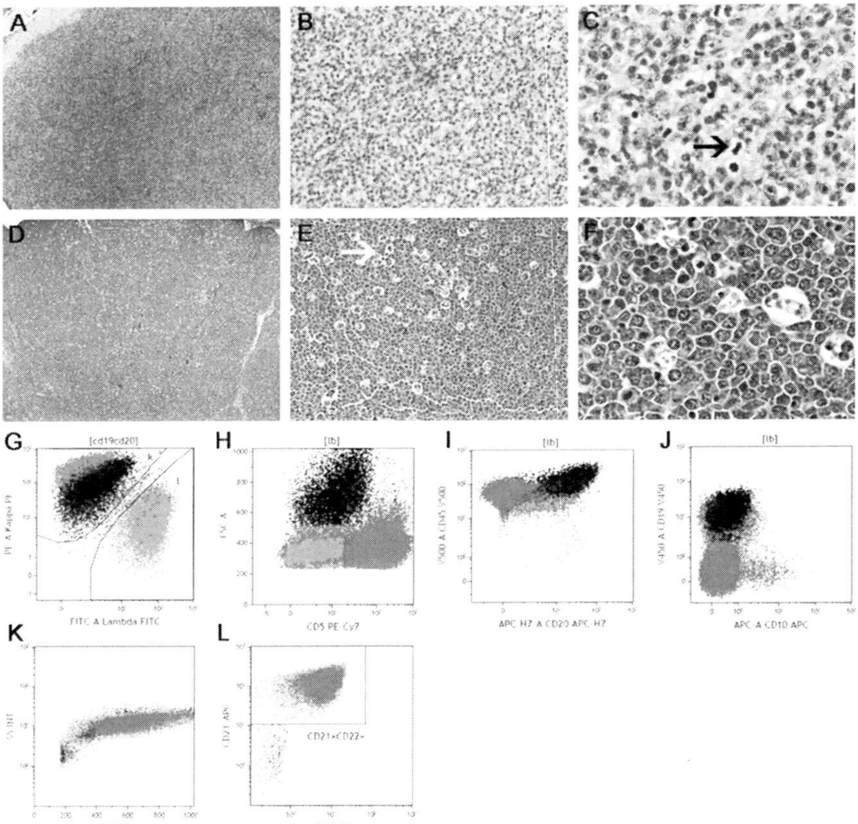

Figure 6. The histologic and flow cytometric features of human and canine diffuse large B-cell lymphoma (DLBCL). In both species, DLBCL is a diffuse disease (**A**–**C**, human and **D**–**F**, canine) and often demonstrates a starry-sky appearance due to phagocytic histiocytes (white arrow, **E**). Neoplastic cells are large with varying amounts of cytoplasm and vesicular nuclei with prominent nucleoli. The mitotic rate is often high (black arrow, **F**). Flow cytometrically, the neoplastic cells (colored black) in hDLBCL are light chain restricted (**G**), large by forward light scatter, CD5− (**H**), CD45+, CD20+ (**I**), CD19+, and CD10− (**J**). Similarly, the neoplastic cells in cDLBCL (colored red) are large by forward scatter (**K**) and are CD21+ and CD22+ (**L**).

representation of histiocytes with phagocytosed cellular debris is commonly observed in many cases [18,84]. The presence of remnant follicular structures may be apparent, including germinal centers and mantle zone cells. Cytomorphologically, a defining feature of DLBCL in both species is the size of the neoplastic cell. In hDLBCL, the cells are described as intermediate to large with nuclei at least twice the size of a small lymphocyte and usually larger than a tissue macrophage or endothelial cell nucleus [78]. In cDLBCL, the cells are similarly large. In both species, the number of mitotic figures is typically quite high and the amount of reactive T-lymphocytes and/or histiocytes is variable [78].

According to their nuclear morphology, neoplastic cells in both human and canine DLBCL are commonly characterized as resembling centroblasts (CB), immunoblasts (IB), or a form intermediate to the two. Centroblasts are medium to large-sized cells with round to oval nuclei with vesicular chromatin, multiple small nucleoli opposed to the nuclear membrane, and a thin rim of amphophilic cytoplasm. Immunoblasts are similarly sized with round to oval vesicular nuclei, a single, large centrally-located nucleolus, and a broad rim of basophilic cytoplasm. Based upon the percentage of either cell type, DLBCL can be cytologically subclassified as either immunoblastic (*i.e.*, > 90% immunoblasts) or centroblastic (*i.e.*, >90% centroblasts) [18,80]. In hDLBCL, the clinical relevance of morphologic subclassification is inconclusive as some studies have reported a worse outcome for immunoblastic hDLBCL, whereas others have not [85,86]. To date, no studies have established any clinical significance to the stratification of canine lymphoma into IB or CB subtypes. There is no characteristic immunophenotype for either human or canine DLBCL as both express pan-B markers, including CD79a, Pax5, and CD20, and CD21 [27,84].

According to gene expression profiling (GEP), hDLBCL can be further segregated into three molecular subtypes, namely germinal center B-cell-like (GCB-DLBCL), activated B-cell-like (ABC-DLBCL), and primary mediastinal B-cell lymphoma (PMBL). Each of these subtypes has notable biological and therapeutic differences, illustrating the prognostic significance of such an approach [87,88,89]. Given the practical limitations of GEP as a primary diagnostic modality, IHC has been proposed as a surrogate. The most commonly used algorithm (Hans) segregates DLBCL into GCB-type and non-GCB types according to their expression of CD10, BCL6, and IRF4/MUM-1 [90]. Attempts to stratify canine DLBCL into GCB and ABC-like categories using gene expression data have been limited, but have shown that GEP can organize canine BCL into multiple subtypes with prognostic implications [75,77]. However, the limited attempts to apply the Hans algorithm to canine large cell lymphomas have not been successful [75].

5.2. Marginal Zone Lymphoma

The marginal zone lymphomas (MZLs) are a group of similarly named, but biologically unique forms of mature B-cell neoplasia that originate from post-follicular memory lymphocytes of the marginal zone, which is a unique antigen filtration region of the spleen and lymph node [91]. In both humans and dogs, three variants of MZL are described, namely: (1) nodal marginal zone lymphoma (NMZL), (2) splenic marginal zone lymphoma (SMZL) and (3) mucosal associated lymphoid tissue lymphoma (MALT lymphoma) [17]. In all subtypes of both hMZL and cMZL, the neoplastic cells are characterized by many common morphologic, histologic, and immunophenotypic characteristics [91]. Despite this, each has very unique diagnostic features as well as recurrent cytogenetics/translocations.

5.2.1. Splenic Marginal Zone Lymphoma

Splenic marginal zone lymphoma (SMZL) is an indolent, small-cell B-cell lymphoma characterized by a unique pattern of splenic infiltration and widespread bone marrow involvement (with characteristic intrasinusoidal distribution). It was formally included in the WHO classification system in 2001 and, despite its MZL nomenclature, it is a distinct from either NMZL and MALT lymphoma [92]. Splenic MZL is uncommon form of human B-cell lymphoma that accounts for 0.6% of cases of NHL and 20% of hMZLs [36]. It is a similarly uncommon in the dog accounting for 0.2%–1.0% of canine lymphoma and 2%–16% of cMZLs [11,13].

Clinically, hSMZL is characterized by generalized splenomegaly with concurrent peripheral blood and bone marrow involvement with minimal to absent lymphadenopathy [32,34,36,93]. In contrast, dogs with SMZL commonly present with either a solitary splenic mass in (79%) or multiple splenic nodules (24%) whereas generalized splenomegaly is uncommon [33,45,68]. Moreover, the frequency of peripheral blood involvement in cSMZL as documented by a peripheral blood lymphocytosis appears to be much lower (2.9% *vs.* 90%) than its human counterpart [16,33]. In contrast, hemoabdomen as a result of spontaneous splenic rupture is far more common in the cSMZL (22% of cSMZL) as compared to hSMZL, where it is very rarely described [33,44,45,68,94].

The diagnosis of hSMZL requires the identification of a mature B-cell neoplasm with a unique pattern of tissue involvement, cytology, and characteristic cytogenetic abnormalities. Architecturally, the tumor infiltrate is nodular and consists of small lymphocytes that surround and replace the splenic white pulp germinal centers, efface the follicle mantle zone, and merge with a peripheral zone of larger cells [35]. The architecture of canine SMZL is similar to its human counterpart and is characterized by a multi-nodular infiltrate in which the number and size of white pulp follicles, which often encircle fading remnants of germinal centers, is increased and begin to coalesce [11,15,45,68]. Cytomorphologically, hSMZL appears to be slightly distinct from cSMZL. In the former, infiltrated follicles maintain a biphasic appearance with a centrally located small lymphocyte population with round nuclei and scant cytoplasm, surrounded by a cuff of larger cells [35]. These larger cells are slightly larger than the small cells with slightly irregular nuclear contours and moderate to abundant pale cytoplasm [37]. While morphologically distinct, both the central core of small lymphocytes and the outer zone of larger cells have been shown to be part of the neoplastic clone [92]. In contrast, the neoplastic cells in cSMZL are reported to be more monomorphic [95]. They are medium-sized and contain an intermediate amount of lightly-stained pink cytoplasm. Nuclei are vesicular with peripheralized chromatin and a prominent single central nucleolus [15]. In both species, mitotic figures are rare [95] and neoplastic areas may be surrounded by areas of hemorrhage or EMH [45]. Immunophenotypically, neoplastic cells in hSMZL are CD20+, IgD+, BCL2+, CD3−, CD5−, CD10−, and BCL6− [51]. In the dog, neoplastic SMZL cells are described as CD79a and CD20 positive and CD3 negative [95,96].

5.2.2. Nodal Marginal Zone Lymphoma

Nodal marginal zone lymphoma (NMZL), like the other forms of MZL, is a primary B-cell neoplasm originating from marginal zone cells. Morphologically, NMZL is composed of tumor cells similar to those encountered in either extranodal or splenic subtypes of MZL, however the disease burden is isolated to lymph nodes. As such, diagnosing NMZL requires knowledge of disease distribution [97]. Human NMZL (hNMZL) is uncommon, accounting for less than 2% of all NHLs and approximately 10% of all MZLs [91]. In contrast, NMZL is a more common disease in dogs, accounting for 6%–10% of all forms of canine lymphoma and 75%–89% of all cMZLs [11,12,13]. Clinically, hNMZL is characterized by peripheral lymphadenopathy without evidence of either extranodal or splenic sites. In hNMZL, bone marrow involvement is apparent in 30%–45% of cases, although peripheral blood involvement is rather rare (approximately 10%) and cytopenias are infrequent. Clinically, canine NMZL (cNMZL) presents as single or generalized lymph node enlargement and no studies have evaluated bone marrow or peripheral blood involvement [13]. The diagnosis of both NMZL can be challenging in both species as it lacks a specific clinical, morphological, or immunophenotypic signature.

Further challenging the diagnosis of NMZL is its heterogeneous histology, which may be more dramatic in the human variant of the disease. Architecturally, hNMZL can present with a number of patterns, including nodular, perifollicular, interfollicular, diffuse, and inverse follicular variants (Figure 7) [91,98]. In each, there may be sporadic retention of normal nodal architecture, including residual germinal centers that may or may not be colonized by tumor cells [91]. In contrast, cNMZL appears to have a more uniform histology. Architecturally, cNMZL is reported to be nodular resulting from the expansion of perifollicular marginal zone lymphocytes surrounding fading remnants of germinal centers (Figure 7) [15]. Like its architecture, the cytologic appearance of hNMZL is also heterogeneous. Although the neoplastic cells in hNMZL are generally medium in size, they lack the monotony seen in seen in other forms of lymphoma. Most cases contain two types of cells, namely: (1) a dominant, small to medium-sized population containing pale cytoplasm and nuclear irregularities (*i.e.*, centrocyte-like); and (2) a minority large cell population with oval to reniform nuclei and abundant pale cytoplasm (*i.e.*, monocytoid). Canine NMZL is also a mixed cell disease, although the dominanT-cell type is centrocyte-like in its morphology, namely intermediate sized with abundant pale cytoplasm and irregular nuclei containing peripheralized chromatin and a single central nucleolus (Figure 7) [68].

While hNMZL is considered to an indolent form of B-cell lymphoma, transformation to large-cell disease (*i.e.*, DLBCL) is reported to occur in 7%–31% of patients [97,99,100,101,102,103]. However it is important to note that the histologic criteria for hNMZL transformation are not well-defined. Proposed hNMZL transformation criteria include blast percentage (transformation defined as the large, centroblast or immunoblast percentage exceeding 20% or 50% of all tumor cells) or the identification of "sheets" of such cells. The presence of sheets of large cells is the most agreed upon marker of transformation as no difference in survival can be detected using a 20% cutoff criteria, whereas but sheets of large cells may be associated a worse clinical outcome [97,104]. Similarly, transformation of cNMZL to cDLBCL is thought to occur [18] and, correlate with this, gene expression profiling has revealed that the two diseases cannot be distinguished [77]. Although as in human hematopathology, there are no defining histologic criteria for this process and differentiating cNMZL from cDLBCL can be very challenging. Features reported to be suggestive of cNMZL include: (1) the identification of nodular foci of proliferation; and (2) intermediate (rather than large) nuclear size [13]. One author has proposed grading cNMZL according to its stage of development as "early, mid or late" based upon the area of tissue involved, the degree of coalescence of neoplastic proliferation, and the number of cells in mitosis [68]. It is unclear if this grading offers any prognostic or therapeutic implications. In summary, the importance of large cells in both canine and human NMZL continues to be debated by pathologists and whether or when these cells reflect a either a morphologic variant of MZL or true transformation to DLBCL.

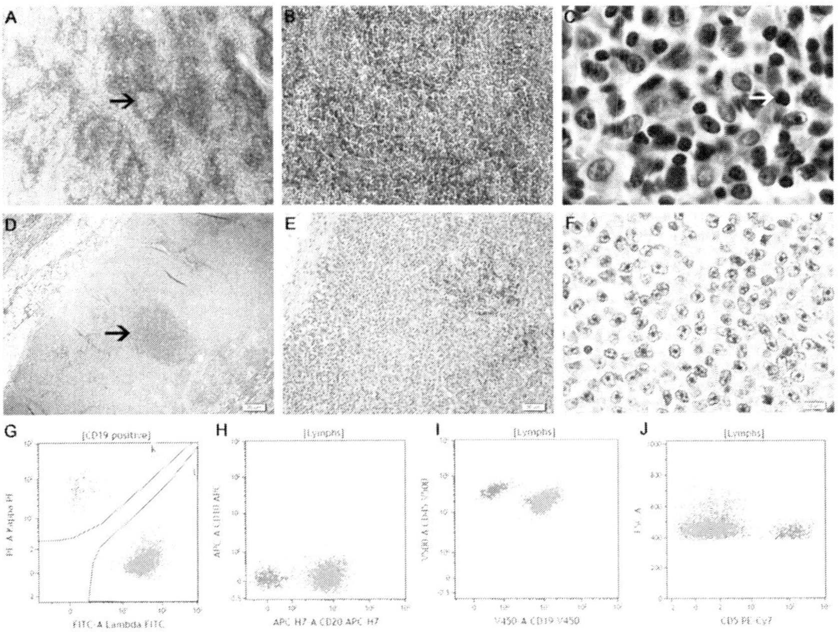

Figure 7. This histologic and flow cytometric appearance of human and canine nodal marginal zone lymphoma (**A–C**, human and **D–F**, canine). At low magnification both species demonstrate a nodular architectural pattern characterized by expansion of germinal centers and perifollicular areas (**A, D**) by a population of intermediate-sized neoplastic cells. In both, residual follicular structures are common (black arrow, **A** and **D**). At higher magnification, the neoplastic infiltrate in hMZL consists of transformed intermediate (white arrow, **C**) and large cells (gray arrow, **C**). In contrast, neoplastic cells in cMZL are predominately intermediate sized with abundant pale cytoplasm and irregular nuclei containing peripheralized chromatin and a single central nucleolus (**F**). Flow cytometrically, the neoplastic cells (colored blue, **G–J**) in hMZL are lambda light chain restricted, CD20+, CD45+, CD19+, CD5−, and CD10−.

The immunophenotype of hNMZL is identical to other hMZL variants, including expression of mature B-cell antigens (CD19, CD20, CD22, CD79a) and an absence of germinal center antigens (CD10, bcl-6, cyclin D1) [91]. The neoplastic cells in cNMZL demonstrate a mature B-cell phenotype as reflected by CD79a+, CD20+, and CD3−, and nearly all cases demonstrate clonal IgH rearrangements with a small fraction having both clonal IgH and TCRg rearrangements [16,68].

5.3. Mantle Cell Lymphoma

In both humans and dogs, mantle cell lymphoma (MCL) is an uncommon form of B-cell lymphoma, representing 3%–6% and 0.7%–1.8% of NHLs in the two species, respectively [11,13,105,106]. However, this is largely where the similarities end. Clinically, human MCL (hMCL) is an intermediate-grade disease characterized by generalized lymphadenopathy and frequent extranodal involvement, including spleen, bone marrow, liver, and GI tract infiltration [107]. In humans, peripheral blood involvement can be identified in up to 70% of patients on routine examination and this number increases to nearly 100% with flow cytometry [108,109,110]. Bone marrow involvement is reported in 50%–91% of patients and cytopenias, most notably anemia and thrombocytopenia, are seen in up to 40% of patients [108,111,112]. In contrast, canine MCL (cMCL) primarily affects the spleen (80% of the reported cases are of splenic origin) with only occasional visceral node involvement [11,13,16,68]. The vast majority of cases lack detectable peripheral node infiltration and peripheral blood or bone marrow involvement appears to be very uncommon. Biologically, cMCL is reported to be an indolent disease and a series of nine dogs with splenic MC had a 1 year survival rate of 88.9% and an overall survival period of 502 days [113].

The diagnosis of hMCL is based upon the identification of a mature B-cell neoplasm composed of small to intermediate-sized cells with a unique immunophenotypic and genetic profile [108]. Architecturally, hMCL is characterized by three possible growth patterns, two of which are nodular (the mantle zone and nodular forms) and one of which is diffuse (the diffuse form) (Figure 8) [108]. The mantle zone pattern results in expansion of the follicle mantle area by tumor cells surrounding a reactive germinal center whereas the nodular pattern consists of nodules of tumor cells that lack a remnant germinal center [108]. The diffuse form of the disease is characterized by sheets of neoplastic cells in which rare residual germinal centers may be seen. The architectural heterogeneity of hMCL has not been described in cMCL and, to date, it is described primarily as a nodular disease although diffuse areas may be

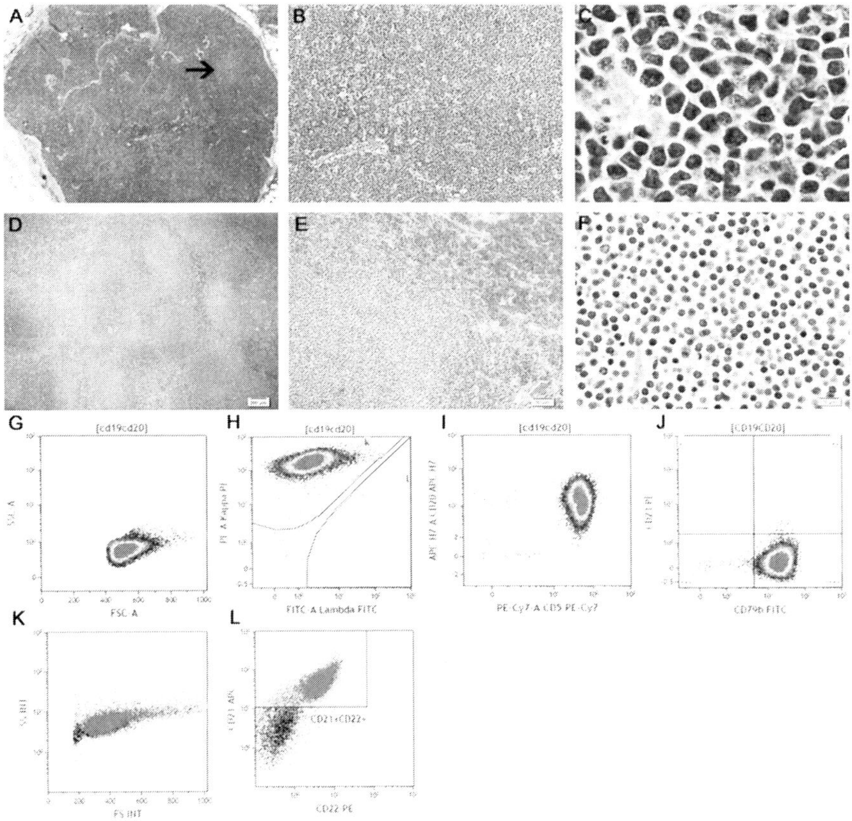

Figure 8. The histologic and flow cytometric features of human and canine mantle cell lymphoma (**A**–**C**, human and **D**–**F**, canine). Architecturally, hMCL is a heterogeneous disease, although the nodular (mantle zone) pattern is presented (**A**, **B**). The tumor infiltrate expands the mantle cell cuff and surrounds a reactive germinal center (black arrow, **A**). The neoplastic cells are small to intermediately sized, contain scant cytoplasm and demonstrate indented nuclear contours with evenly dispersed chromatin (**C**). Architecturally, cMCL is characterized by discrete to coalescing nodules of a monotonous population of neoplastic cells (**D**, **E**). Morphologically, these cells are small to intermediate in size with round nuclei, condensed chromatin, and inapparent nucleoli (**F**). By flow cytometry, the cells are intermediate-sized by forward scatter (**G**), light chain restricted (**H**), CD20+, CD5+ (**I**), CD79b+, and CD23− (**J**). The neoplastic cells in cMCL (colored red) are small to intermediate in size (**K**) and are CD21+ and CD22+ (**L**).

seen in later stages of the disease [18]. The nodules of neoplastic cells in cMCL, which may coalesce, consist of a monotonous population of cells that are often centered around end arterioles and/or a faded, hypocellular germinal center (Figure 8) [18]. The architecture of cMCL has been described as the histologic inverse of cMZL [18].

Morphologically, hMCL is similarly heterogeneous and four morphologies are described, namely classic (common or typical), small, blastoid, or pleomorphic variants [114]. In the classical form, the cells are monomorphic, small to medium in size, contain scant cytoplasm and demonstrate indented nuclear contours, evenly dispersed chromatin, scan cytoplasm, inapparent nucleoli, and few mitoses [17]. Rare large cells may be seen. The small cell form is dominated by small lymphocytes with round nuclei; the blastoid form is composed of medium-sized tumor cells with round nuclei, finely dispersed chromatin, and numerous mitoses, whereas the pleomorphic variant is comprised of large cells with irregular nuclear morphologies [108]. The classic and small-cell forms of the disease typically have a low mitotic index and cases with increased mitotic activity are associated with a worse prognosis [108]. In contrast to the other morphologic manifestations, the blastoid and pleomorphic variant have poor prognoses. Like its architecture, the cytomorphology of cMCL is more uniform than its human counterpart. The neoplastic cells are small to intermediate in size and contain a scant amount of cytoplasm and round to rarely indented nuclei containing dense chromatin, and inconspicuous nucleoli. The mitotic rate is very low to high [11,68].

In light of this histologic heterogeneity, immunophenotyping is mandatory for the diagnosis of hMCL [107]. Phenotypically, hMCL is a mature B-cell neoplasm characterized by positivity for the B-cell markers CD19, CD20, CD22, CD79a, CD79b, with surface Ig expression and lambda light-chain restriction in approximately 2/3rds of cases [107]. Unique among other forms of B-cell neoplasia is the near-ubiquitous expression of the T-cell associated antigen CD5 and the expression of cyclin D1 [107,108]. Moreover, most cases of hMCL are negative for CD10 and Bcl6 (which aids in excluding follicular lymphoma, DLBCL, and precursor disease). Immunophenotypically, cMCL is a mature B-cell neoplasm characterized by CD79a and CD20 expression [12,16].

5.4. Follicular Lymphoma

Follicular lymphoma (FL) is a mature B-cell neoplasm with similar pathologic features between dog and man, but very different incidences. Whereas FL is the second most common form of human lymphoma, representing 20% of NHL, canine FL (cFL) is quite rare and reported to represent less than 1% of canine lymphomas [11,15,31]. Clinically, most human patients with FL have generalized lymph node involvement, including peripheral, mediastinal, and retroperitoneal nodes [115,116]. Bone marrow involvement is present in a majority of patients at staging [116]. Given its rarity, the clinical features of cFL are scarcely reported, but as in human FL (hFL), multinodal disease including both peripheral and visceral nodes, appears common [68] . In both species, the magnitude of lymph node enlargement is mild to moderate and generally less than other forms of NHL [18,117].

In both species, FL is best diagnosed using excisional lymph node biopsy and FNA samples are not considered adequate [18,68,116]. Both hFL and cFL are nodular diseases characterized by complete effacement of the nodal architecture by neoplastic follicles (Figure 9). Owing to its unique architecture, distinguishing the neoplastic follicles in FL from the benign follicles in reactive hyperplasia can be challenging. However, neoplastic follicles lack two architectural features seen in their benign counterpart, namely: germinal center polarization and a well-defined and continuous mantle zone cuff. Germinal center polarization is a histologic hallmark of antigen-induced lymphoid hyperplasia and reflects the proliferation of distinct mature B-cell populations. Histologically, germinal center polarization manifests as segregation of the germinal center into two zones, namely an apical light zone, composed of small-to-medium sized cells that contain a relatively abundant amount of pale-staining cytoplasm (*i.e.*, centrocytes) and a deeper dark zone containing large cells that contain a narrow rim of dark-staining cytoplasm (*i.e.*, centroblasts).

Cytologically, although the neoplastic follicles in both human and canine FL lack polarization, they are composed of cells that are morphologically similar to those in normal germinal centers. Neoplastic *centrocytes* are small cells with clear cytoplasm, an irregular nuclear outline, pale and evenly dispersed chromatin, and inconspicuous or small nucleoli. Neoplastic *centroblasts* are large cells with a moderate amount of dark basophilic cytoplasm, round to oval to indented, vesicular nuclei with peripherally-dispersed chromatin and 1–3, peripherally-positioned nucleoli [11,68,118]. In both species, centrocytes are usually more common than centroblasts in the neoplastic follicles and mitotic activity is generally low [116]. In addition to follicle architecture, FL can be differentiated from lymphoid hyperplasia through the detection of intravascular invasion or a more numerous and more monotonous centrocyte population, the absence of an appreciable plasma cell population or starry-sky macrophages, irregular-shaped follicles with attenuated mantle zones, follicles arranged "back-to-back," and extension of cells beyond the capsule into perinodal tissues [116,118].

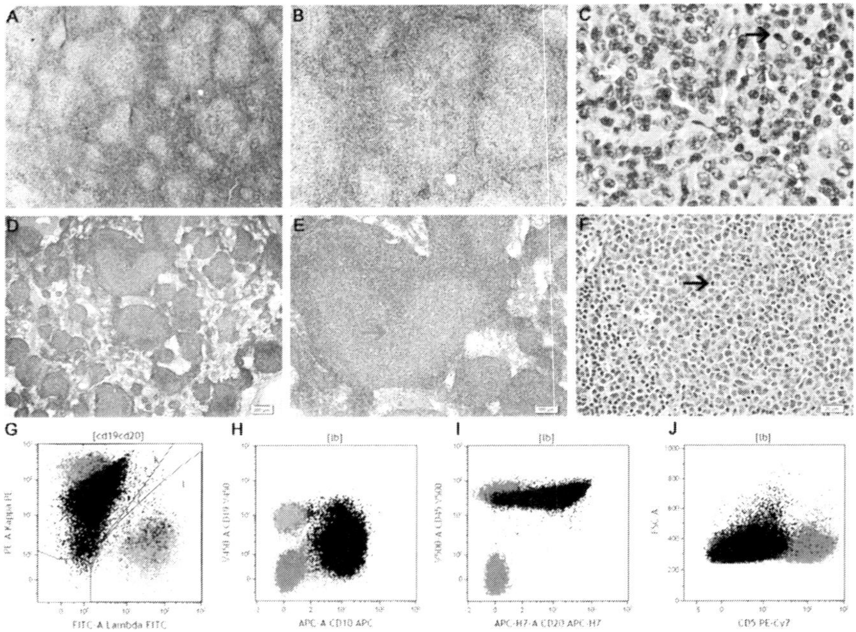

Figure 9. The histologic features of human and canine follicular lymphoma (**A–C**, human and **D–F**, canine). Architecturally, in both species, the disease is nodular (**A, C**) and the neoplastic follicles lack germinal center polarization (gray arrow, **B** and **E**). Morphologically, the tumor infiltrate in both species is bimorphic, composed of predominantly smaller neoplastic centrocytes (black arrow, **C** and **F**) and rare admixed larger neoplastic centroblasts (white arrow, **C** and **F**). By flow cytometry, the neoplastic cells in human FL (colored black) are light chain restricted (**G**), CD10+, variably CD19+ (**H**), CD20+, CD45+ (**I**), but CD5−, and are intermediate size by forward light scatter (**J**).

Human FL can be further graded by counting the number of centroblasts in 10–20, randomly-selected 400× microscopic high-power fields (hpf) into grade 1 (1–5 centroblasts), grade 2 (6–15 centroblasts), or grade 3 (>15 centroblasts) forms of the disease [17]. Grade 3 FL can be further split into "A" and "B" forms of the disease, with grade 3A demonstrating more than 15 centroblasts per hpf but with a persistent centrocyte population and persistent of a follicular dendritic cell meshwork whereas grade 3B forms have solid sheets of centroblasts and lack a follicular dendritic cell meshwork [116]. While grade generally correlates with disease aggressiveness, there is disagreement regarding the interobserver reproducibility and clinical significance of this system

[117,119]. Accordingly, in light of their shared clinical behavior, grades 1 and 2 hFL have recently been combined into a single "low grade" category, termed WHO grade 1–2 [116]. To date, likely owing to its rarity, a histologic grading scheme for cFL has not been systematically evaluated. Immunophenotypically, neoplastic cells in hFL express typical B-cell antigens (CD19, CD20, CD22, CD79a, and PAX5) and most cases express the germinal center-associated protein CD10 and BCL6 [116,118,120]. Similarly, cFL has been shown to express mature B-cell markers (CD79a) while demonstrating a lack of staining with mature T-cell markers (*i.e.*, CD3) [11,12,16,68].

6. CONCLUSIONS

In light of their complementary biology, including overlapping incidence, genetic, and clinical features, there is tremendous interest in utilizing the dog as an animal model for human NHL. However, implicit in this approach is a need to determine the degree to which canine and human lymphomas are diagnostically similar. While the high reproducibility, and subsequent increasing acceptance, of canine-adapted WHO guidelines may be a watershed moment in veterinary hematopathology, subsequent studies are needed to further refine these guidelines, confirm their clinical and diagnostic utility, and identify and validate novel reagents and techniques.

AUTHOR CONTRIBUTIONS

Davis M. Seelig, Anne C. Avery, E. J. Ehrhart, and Michael A. Linden drafted the manuscript. All authors read and approved the final manuscript.

CONFLICTS OF INTEREST

The authors declare no conflict of interest.

REFERENCES

1. SEER Cancer Statistics Review, 1975–2012. Howlader, N., Krapcho, M., Garshell, J., Miller, D., Altekruse, S.F., Kosary, C.L., Yu, M., Ruhl, J., Tatalovich, Z., Mariotto, A., et

al., Eds.; National Cancer Institute: Bethesda, MD, USA, 2015. National Cancer Institute: Bethesda, MD, USA, 2015.
2. Merlo, D.F.; Rossi, L.; Pellegrino, C.; Ceppi, M.; Cardellino, U.; Capurro, C.; Ratto, A.; Sambucco, P.L.; Sestito, V.; Tanara, G.; et al. Cancer incidence in pet dogs: Findings of the animal tumor registry of Genoa, Italy. J. Vet. Intern. Med. 2008, 22, 976–984.
3. Dorn, C.R.; Taylor, D.O.; Chaulk, L.E.; Hibbard, H.H. The prevalence of spontaneous neoplasms in a defined canine population. Am. J. Public Health Nations Health 1966, 56, 254–265.
4. Dorn, C.R.; Taylor, D.O.; Schneider, R.; Hibbard, H.H.; Klauber, M.R. Survey of animal neoplasms in Alameda and Contra Costa Counties, California. II. Cancer morbidity in dogs and cats from Alameda County. J. Natl. Cancer Inst.1968, 40, 307–318.
5. Dorn, C.R.; Taylor, D.O.; Schneider, R. The epidemiology of canine leukemia and lymphoma. Bibl. Haematol 1970, 403–415.
6. Teske, E. Malignant lymphomas in dogs: A review with reference to non-Hodgkin lymphoma in man. Tijdschr Diergeneeskd 1994, 119, 705–717.
7. Dorn, C.R.; Taylor, D.O.; Hibbard, H.H. Epizootiologic characteristics of canine and feline leukemia and lymphoma. Am. J. Vet. Res. 1967, 28, 993–1001.
8. Dobson, J.M.; Samuel, S.; Milstein, H.; Rogers, K.; Wood, J.L. Canine neoplasia in the UK: Estimates of incidence rates from a population of insured dogs. J. Small Anim. Pract. 2002, 43, 240–246.
9. Chiu, B.C.; Hou, N. Epidemiology and etiology of non-hodgkin lymphoma. Cancer Treat. Res. 2015, 165, 1–25.
10. Bassig, B.A.; Au, W.-Y.; Mang, O.; Ngan, R.; Morton, M.L.; Ip, D.K.M.; Hu, W.; Zheng, T.; Seow, W.J.; Xu, J.; et al.Subtype-specific incidence rates of lymphoid malignancies in Hong Kong compared to the United States, 2001–2010. Cancer Epidemiol. 2016, 42, 15–23.
11. Ponce, F.; Marchal, T.; Magnol, J.P.; Turinelli, V.; Ledieu, D.; Bonnefont, C.; Pastor, M.; Delignette, M.L.; Fournel-Fleury, C. A morphological study of 608 cases of canine malignant lymphoma in France with a focus on comparative similarities between canine and human lymphoma morphology. Vet. Pathol. 2010, 47, 414–433.
12. Vezzali, E.; Parodi, A.L.; Marcato, P.S.; Bettini, G. Histopathologic classification of 171 cases of canine and feline non-Hodgkin lymphoma according to the WHO. Vet. Comp. Oncol. 2010, 8, 38–49.
13. Valli, V.E.; Kass, P.H.; San Myint, M.; Scott, F. Canine lymphomas: Association of classification type, disease stage, tumor subtype, mitotic rate, and treatment with survival. Vet. Pathol. 2013, 50, 738–748.
14. Campo, E.; Swerdlow, S.H.; Harris, N.L.; Pileri, S.; Stein, H.; Jaffe, E.S. The 2008 WHO classification of lymphoid neoplasms and beyond: Evolving concepts and practical applications. Blood 2011, 117, 5019–5032.
15. Valli, V.E.; Myint, M.S.; Barthel, A.; Bienzle, D.; Caswell, J.; Colbatzky, F.; Durham, A.; Ehrhart, E.J.; Johnson, Y.; Jones, C.; et al. Classification of canine malignant lymphomas according to the World Health Organization criteria. Vet. Pathol. 2011, 48, 198–211.
16. Flood-Knapik, K.E.; Durham, A.C.; Gregor, T.P.; Sanchez, M.D.; Durney, M.E.; Sorenmo, K.U. Clinical, histopathological and immunohistochemical characterization of canine indolent lymphoma. Vet. Comp. Oncol. 2013, 11, 272–286.
17. Swerdlow, S.H.; Harris, N.L.; Jaffe, E.S.; Pileri, S.A.; Stein, H.; Thiele, J.; Vardiman, J.W. WHO classification of tumours of the haematopoietic and lymphoid tissues, 4th ed.; WHO Classification of Tumours; IARC: Lyon, France, 2008.
18. Valli, V.E. Veterinary comparative hematopathology, 1st ed.; Blackwell Pub: Ames, IA, USA, 2007.
19. Ghielmini, M. ESMO Guidelines consensus conference on malignant lymphoma 2011 part 1: Diffuse large B-cell lymphoma (DLBCL), follicular lymphoma (FL) and chronic lymphocytic leukemia (CLL). Ann. Oncol. 2013, 24, 561–576.
20. Thalheim, L.; Williams, L.E.; Borst, L.B.; Fogle, J.E.; Suter, S.E. Lymphoma immunophenotype of dogs determined by immunohistochemistry, flow cytometry,

and polymerase chain reaction for antigen receptor rearrangements. J. Vet. Intern. Med. 2013, 27, 1509–1516.

21. Oh, E.J.; Hong, S.W.; Jeong, H.J.; Yoon, S.O. The diagnostic approach to fine-needle aspiration of malignant lymphoma: using cytomorphology and immunocytochemistry with cell transfer method. Diagn Cytopathol. 2014, 42, 671–679.
22. Lioe, T.F.; Elliott, H.; Allen, D.C.; Spence, R.A. The role of fine needle aspiration cytology (FNAC) in the investigation of superficial lymphadenopathy; uses and limitations of the technique. Cytopathology 1999, 10, 291–297.
23. Meda, B.A.; Buss, D.H.; Woodruff, R.D.; Cappellari, J.O.; Rainer, R.O.; Powell, B.L.; Geisinger, K.R. Diagnosis and subclassification of primary and recurrent lymphoma. The usefulness and limitations of combined fine-needle aspiration cytomorphology and flow cytometry. Am. J. Clin. Pathol. 2000, 113, 688–699.
24. Bienzle, D.; Vernau, W. The diagnostic assessment of canine lymphoma: Implications for treatment. Clin. Lab. Med.2011, 31, 21–39.
25. Demurtas, A.; Accinelli, G.; Pacchioni, D.; Godio, L.; Novero, D.; Bussolati, G.; Palestro, G.; Papotti, M.; Stacchini, A. Utility of flow cytometry immunophenotyping in fine-needle aspirate cytologic diagnosis of non-Hodgkin lymphoma: A series of 252 cases and review of the literature. Appl Immunohistochem Mol. Morphol. 2010, 18, 311–322.
26. Sokolowska, J.; Micun, J.; Zabielska, K.; Malicka, E.; Lechowski, R. Immunohis tochemical study of expression of immunoglobulins in canine B-cell lymphomas. Pol. J. Vet. Sci. 2010, 13, 623–628.
27. Rao, S.; Lana, S.; Eickhoff, J.; Marcus, E.; Avery, P.R.; Morley, P.S.; Avery, A.C. Class II major histocompatibility complex expression and cell size independently predict survival in canine B-cell lymphoma. J. Vet. Intern. Med. 2011, 25, 1097–1105.
28. Barrena, S. Flow cytometry immunophenotyping of fine-needle aspiration specimens: Utility in the diagnosis and classification of non-Hodgkin lymphomas. Histopathology 2011, 58, 906–918.
29. Greig, B.; Oldaker, T.; Warzynski, M.; Wood, B. 2006 Bethesda International Consensus recommendations on the immunophenotypic analysis of hematolymphoid neoplasia by flow cytometry: recommendations for training and education to perform clinical flow cytometry. Cytometry. B Clin. Cytom. 2007, 72 Suppl 1, S23–S33.
30. Arun, S.S.; Breuer, W.; Hermanns, W. Immunohistochemical examination of light-chain expression (lambda/kappa ratio) in canine, feline, equine, bovine and porcine plasma cells. Zentralbl. Veterinarmed. A. 1996, 43, 573–576.
31. A clinical evaluation of the International Lymphoma Study Group classification of non-Hodgkin's lymphoma. The Non-Hodgkin's Lymphoma Classification Project. Blood 1997, 89, 3909–3918.
32. Bassan, R.; Maino, E.; Cortelazzo, S. Lymphoblastic lymphoma: An updated review on biology, diagnosis and treatment. Eur. J. Haematol. 2016, 96, 447–460.
33. Tasca, S.; Carli, E.; Caldin, M.; Menegazzo, L.; Furlanello, T.; Gallego, L.S. Hematologic abnormalities and flow cytometric immunophenotyping results in dogs with hematopoietic neoplasia: 210 Cases (2002–2006). Vet. Clin. Pathol. 2009, 38, 2–12.
34. Racke, F.K.; Borowitz, M.J. Precursor B- and T-Cell Neoplasms. In Hematopathology; Jaffe, E.S., Harris, N.L., Vardiman, J.W., Campo, E., Arber, D.A., Eds.; Elsevier Saunders: St. Louis, MO, USA, 2011; pp. 629–639.
35. You, M.J.; Medeiros, L.J.; Hsi, E.D. T-lymphoblastic leukemia/lymphoma. Am. J. Clin Pathol. 2015, 144, 411–422.
36. Cortelazzo, S.; Ponzoni, M.; Ferreri, A.J.; Hoelzer, D. Lymphoblastic lymphoma. Crit Rev. Oncol. Hematol. 2011, 79, 330–343.
37. Jain, N.C. Proposed criteria for classification of acute myeloid leukemia in dogs and cats. Vet. Clin. Pathol. 1991, 20, 63–82.

38. Stokol, T.; Schaefer, D.M.; Shuman, M.; Belcher, N.; Dong, L. Alkaline phosphatase is a useful cytochemical marker for the diagnosis of acute myelomonocytic and monocytic leukemia in the dog. Vet. Clin. Pathol. 2015, 44, 79–93.
39. Fournel-Fleury, C. Canine T-cell lymphomas: A morphological, immunological, and clinical study of 46 new cases. Vet. Pathol. 2002, 39, 92–109.
40. Avery, P.R.; Burton, J.; Bromberek, J.L.; Seelig, D.M.; Elmslie, R.; Correa, S.; Ehrhart, E.J.; Morley, P.S.; Avery, A.C. Flow cytometric characterization and clinical outcome of CD4+ T-cell lymphoma in dogs: 67 cases. J. Vet. Intern. Med.2014, 28, 538–546.
41. Lurie, D.M.; Milner, R.J.; Suter, S.E.; Vernau, W. Immunophenotypic and cytomorphologic subclassification of T-cell lymphoma in the boxer breed. Vet. Immunol Immunopathol 2008, 125, 102–110.
42. Gelain, M.E.; Mazzilli, M.; Riondato, F.; Marconato, L.; Comazzi, S. Aberrant phenotypes and quantitative antigen expression in different subtypes of canine lymphoma by flow cytometry. Veterinary Immunology & Immunopathology2008, 121, 179–188.
43. Williams, M.J.; Avery, A.C.; Lana, S.E.; Hillers, K.R.; Bachand, A.M.; Avery, P.R. Canine lymphoproliferative disease characterized by lymphocytosis: Immunophenotypic markers of prognosis. J. Vet. Intern. Med. 2008, 22, 596–601.
44. Adam, F.; Villiers, E.; Watson, S.; Coyne, K.; Blackwood, L. Clinical pathological and epidemiological assessment of morphologically and immunologically confirmed canine leukaemia. Vet. Comp. Oncol. 2009, 7, 181–195.
45. Vernau, W.; Moore, P.F. An immunophenotypic study of canine leukemias and preliminary assessment of clonality by polymerase chain reaction. Vet. Immunol Immunopathol 1999, 69, 145–164.
46. Ponce, F.; Magnol, J.P.; Blavier, A.; Bonnefont, C.; Ghernati, I.; Felman, P.; Buff, S.; Fournel-Fleury, C. Clinical, morphological, and immunological study of 13 cases of canine lymphoblastic lymphoma: Comparison with the human entity. Comp. Clin. Path. 2003, 12, 75–83.
47. Williams, M.J.; Avery, A.C.; Lana, S.E.; Hillers, K.R.; Bachand, A.M.; Avery, P.R. Canine lymphoproliferative disease characterized by lymphocytosis: Immunophenotypic markers of prognosis. J. Vet. Int. Med. 2008, 22, 596–601.
48. Gorczyca, W. Immunophenotypic Markers: Differential Diagnosis. In Atlas of Differential Diagnosis in Neoplastic Hematopathology; CRC Press: Boca Raton, FL, USA, 2014; pp. 119–155.
49. Bhargava, P.; Kallakury, B.V.; Ross, J.S.; Azumi, N.; Bagg, A. CD79a is heterogeneously expressed in neoplastic and normal myeloid precursors and megakaryocytes in an antibody clone-dependent manner. Am. J. Clin Pathol. 2007, 128, 306–313.
50. Modiano, J.F. Distinct B-cell and T-cell lymphoproliferative disease prevalence among dog breeds indicates heritable risk. Cancer Res. 2005, 65, 5654–5661.
51. Piccaluga, P.P.; Tabanelli, V.; Pileri, S.A. Molecular genetics of peripheral T-cell lymphomas. Int. J. Hematol. 2014, 99, 219–226.
52. Piccaluga, P.P. Gene expression analysis of peripheral T cell lymphoma, unspecified, reveals distinct profiles and new potential therapeutic targets. J. Clin. Invest. 2007, 117, 823–834.
53. Huang, Y. Peripheral T-cell lymphomas with a follicular growth pattern are derived from follicular helper T cells (TFH) and may show overlapping features with angioimmunoblastic T-cell lymphomas. Am. J. Surg. Pathol. 2009, 33, 682–690.
54. Iqbal, J. Molecular signatures to improve diagnosis in peripheral T-cell lymphoma and prognostication in angioimmunoblastic T-cell lymphoma. Blood 2010, 115, 1026–1036.
55. Gallamini, A. Peripheral T-cell lymphoma unspecified (PTCL-U): A new prognostic model from a retrospective multicentric clinical study. Blood 2004, 103, 2474–2479.
56. Savage, K.J.; Ferreri, A.J.; Zinzani, P.L.; Pileri, S.A. Peripheral T-cell lymphoma--Not otherwise specified. Crit Rev. Oncol. Hematol. 2011, 79, 321–329.

57. Vose, J.; Armitage, J.; Weisenburger, D.; International, T.C.L.P. International peripheral T-cell and natural killer/T-cell lymphoma study: Pathology findings and clinical outcomes. J. Clin. Oncol. 2008, 26, 4124–4130.
58. De Leval, L.; Savilo, E.; Longtine, J.; Ferry, J.A.; Harris, N.L. Peripheral T-cell lymphoma with follicular involvement and a CD4+/bcl-6+ phenotype. Am. J. Surg. Pathol. 2001, 25, 395–400.
59. De Leval, L.; Ralfkiaer, E.; Jaffe, E.S. Peripheral T-cell Lymphoma, Not Otherwise Specified. In Hematopathology; Jaffe, E.S., Harris, N.L., Vardiman, J.W., Campo, E., Arber, D.A., Eds.; Elsevier Saunders: St. Louis, MO, USA, 2011; pp. 541–553.
60. Piccaluga, P.P.; Agostinelli, C.; Tripodo, C.; Gazzola, A.; Bacci, F.; Sabattini, E.; Pileri, S.A.; European T-cell Lymphoma Study Group. Peripheral T-cell lymphoma classification: The matter of cellular derivation. Expert Rev. Hematol. 2011, 4, 415–425.
61. Went, P. Marker expression in peripheral T-cell lymphoma: A proposed clinical-pathologic prognostic score. J. Clin. Oncol. 2006, 24, 2472–2479.
62. Onaindia, A. CD30 Expression by B and T Cells: A Frequent Finding in Angioimmunoblastic T-Cell Lymphoma and Peripheral T-Cell Lymphoma-Not Otherwise Specified. Am. J. Surg. Pathol. 2016, 40, 378–385.
63. Geissinger, E.; Bonzheim, I.; Roth, S.; Rosenwald, A.; Muller-Hermelink, H.K.; Rudiger, T. CD52 expression in peripheral T-cell lymphomas determined by combined immunophenotyping using tumor cell specific T-cell receptor antibodies. Leuk Lymphoma 2009, 50, 1010–1016.
64. Jiang, L.; Yuan, C.M.; Hubacheck, J.; Janik, J.E.; Wilson, W.; Morris, J.C.; Jasper, G.A.; Stetler-Stevenson, M. Variable CD52 expression in mature T cell and NK cell malignancies: Implications for alemtuzumab therapy. Br. J. Haematol2009, 145, 173–179.
65. Weisenburger, D.D. Peripheral T-cell lymphoma, not otherwise specified: A report of 340 cases from the International Peripheral T-cell Lymphoma Project. Blood 2011, 117, 3402–3408.
66. Seelig, D.M.; Avery, P.; Webb, T.; Yoshimoto, J.; Bromberek, J.; Ehrhart, E.J.; Avery, A.C. Canine T-Zone Lymphoma: Unique Immunophenotypic Features, Outcome, and Population Characteristics. J. Vet. Intern. Med. 2014, 28, 878–886.
67. Godde-Salz, E.; Schwarze, E.W.; Stein, H.; Lennert, K.; Grote, W. Cytogenetic findings in T-zone lymphoma. J. Cancer Res. Clin. Oncol. 1981, 101, 81–89.
68. Valli, V.E.; Vernau, W.; de Lorimier, L.P.; Graham, P.S.; Moore, P.F. Canine indolent nodular lymphoma. Vet. Pathol.2006, 43, 241–256.
69. Martini, V.; Marconato, L.; Poggi, A.; Riondato, F.; Aresu, L.; Cozzi, M.; Comazzi, S. Canine small clear cell/T-zone lymphoma: Clinical presentation and outcome in a retrospective case series. Vet. Comp. Oncol. 2015.
70. Teske, E. Cytology of the lymphoid tissues. In Manual of Diagnostic Cytology of the Dog and Cat; Dunn, J., Ed.; Wiley Blackwell: West Sussex, UK, 2014; pp. 330–356.
71. Helbron, D.; Brittinger, G.; Lennert, K. T-zone lymphoma--clinical symptoms, therapy, and prognosis (author's transl). Blut 1979, 39, 117–131.
72. Lennert, K.; Feller, A.C. Histopathology of Non-Hodgkin's Lymphoma (Based on the Updated Kiel Classification); Springer: Berlin, Germany, 1992.
73. Martini, V.; Poggi, A.; Riondato, F.; Gelain, M.E.; Aresu, L.; Comazzi, S. Flow-cytometric detection of phenotypic aberrancies in canine small clear cell lymphoma. Vet. Comp. Oncol. 2015, 13, 281–287.
74. Nogai, H.; Dorken, B.; Lenz, G. Pathogenesis of non-Hodgkin's lymphoma. J. Clin. Oncol. 2011, 29, 1803–1811.
75. Richards, K.L.; Motsinger-Reif, A.A.; Chen, H.W.; Fedoriw, Y.; Fan, C.; Nielsen, D.M.; Small1, G.W.; Thomas, R.; Smith, C.; Dave, S.S. Gene profiling of canine B-cell lymphoma reveals germinal center and postgerminal center subtypes with different survival times, modeling human DLBCL. Cancer Res. 2013, 73, 5029–5039.
76. Mudaliar, M.A.; Haggart, R.D.; Miele, G.; Sellar, G.; Tan, K.A.L.; Goodlad, J.R.; Milne, E.; Vail, D.M.; Kurzman, I.; Crowther, D.; et al. Comparative gene expression profiling

identifies common molecular signatures of NF-kappaB activation in canine and human diffuse large B cell lymphoma (DLBCL). PLoS ONE 2013, 8, e72591.
77. Frantz, A.M.; Sarver, A.L.; Ito, D.; Phang, T.L.; Karimpour-Fard, A.; Scott, M.C.; Valli, V.E.O.; Lindblad-Toh, K.; Burgess, K.E.; Husbands, B.D.; et al. Molecular profiling reveals prognostically significant subtypes of canine lymphoma. Vet. Pathol. 2013, 50, 693–703.
78. Martelli, M.; Ferreri, A.J.; Agostinelli, C.; Di Rocco, A.; Pfreundschuh, M.; Pileri, S.A. Diffuse large B-cell lymphoma. Crit Rev. Oncol. Hematol. 2013, 87, 146–171.
79. Comazzi, S.; Aresu, L.; Marconato, L. Transformation of Canine Lymphoma/Leukemia to More Aggressive Diseases: Anecdotes or Reality? Front. Vet. Sci. 2015, 2, 42.
80. Mey, U.; Hitz, F.; Lohri, A.; Pederiva, S.; Taverna, C.; Tzankov, A.; Meier, O.; Yeow, K.; Renner, C. Diagnosis and treatment of diffuse large B-cell lymphoma. Swiss. Med. Wkly. 2012, 142, w13511.
81. Aresu, L.; Aricò, A.; Ferraresso, F.; Martini, V.; Comazzi, C.S.; Riondato, F.; Giantin, G.; Dacasto, M.; Guadagnin, E.; Frayssinet, P.; et al. Minimal residual disease detection by flow cytometry and PARR in lymph node, peripheral blood and bone marrow, following treatment of dogs with diffuse large B-cell lymphoma. Vet. J. 2014, 200, 318–324.
82. Sehn, L.H.; Scott, D.W.; Chhanabhai, M.; Berry, B.; Ruskova, A.; Berkahn, L.; Connors, J.M.; Gascoyne, R.D. Impact of concordant and discordant bone marrow involvement on outcome in diffuse large B-cell lymphoma treated with R-CHOP. J. Clin. Oncol. 2011, 29, 1452–1457.
83. Raskin, R.E.; Krehbiel, J.D. Prevalence of leukemic blood and bone marrow in dogs with multicentric lymphoma. J. Am. Vet. Med. Assoc. 1989, 194, 1427–1429.
84. Chan, A.C.L.; Chan, J.K.C. Diffuse Large B-cell Lymphoma. In Hematopathology; Jaffe, E.S., Harris, N.L., Vardiman, J.W., Campo, E., Arber, D.A., Eds.; Elsevier Saunders: St. Louis, MO, USA, 2011; pp. 349–381.
85. Engelhard, M.; Brittinger, G.; Huhn, D.; Gerhartz, H.G.; Meusers, P.; Siegert, W.; Thiel, T.; Wilmanns, W.; Aydemir, U.; Bierwolf, S.; et al. Subclassification of diffuse large B-cell lymphomas according to the Kiel classification: distinction of centroblastic and immunoblastic lymphomas is a significant prognostic risk factor. Blood 1997, 89, 2291–2297.
86. Baars, J.W.; de Jong, D.; Willemse, E.M.; Gras, L.; Dalesio, O.; v Heerde, P.; Huygens, P.C.; vd Lelie, H.; Kr vd Borne, A.E. Diffuse large B-cell non-Hodgkin lymphomas: the clinical relevance of histological subclassification. Br. J. Cancer1999, 79, 1770–1776.
87. Alizadeh, A.A.; Eisen, M.B.; Davis, R.E.; Ma, C.; Lossos, I.S.; Rosenwald, A.; Boldrick, J.C.; Sabet, H.; Tran, T.; Yu, X.; et al. Distinct types of diffuse large B-cell lymphoma identified by gene expression profiling. Nature 2000, 403, 503–511.
88. Rosenwald, A.; Wright, G.; Chan, W.C.; Connors, J.M.; Campo, E.; Fisher, R.I.; Gascoyne, R.D.; Muller-Hermelink, H.K.; Smeland, E.B.; Giltnane, J.M.; et al. The use of molecular profiling to predict survival after chemotherapy for diffuse large-B-cell lymphoma. N. Engl. J. Med. 2002, 346, 1937–1947.
89. Wilson, W.H.; Dunleavy, K.; Pittaluga, S.; Hegde, U.; Grant, N.; Steinberg, S.M.; Raffeld, M.; Gutierrez, M.; Chabner, B.A.; Staudt, L.; et al. Phase II study of dose-adjusted EPOCH and rituximab in untreated diffuse large B-cell lymphoma with analysis of germinal center and post-germinal center biomarkers. J. Clin. Oncol. 2008, 26, 2717–2724.
90. Hans, C.P.; Weisenburger, D.D.; Greiner, T.C.; Gascoyne, R.D.; Delabie, J.; Ott, G.; Müller-Hermelink, H.K.; Campo, E.; Braziel, R.M.; Jaffe, E.S.; et al. Confirmation of the molecular classification of diffuse large B-cell lymphoma by immunohistochemistry using a tissue microarray. Blood 2004, 103, 275–282.
91. Angelopoulou, M.K.; Kalpadakis, C.; Pangalis, G.A.; Kyrtsonis, M.C.; Vassilakopoulos, T.P. Nodal marginal zone lymphoma. Leuk Lymphoma 2014, 55, 1240–1250.
92. Oscier, D.; Owen, R.; Johnson, S. Splenic marginal zone lymphoma. Blood Rev. 2005, 19, 39–51.

93. Korsmeyer, S.J. Immunoglobulin gene rearrangement and cell surface antigen expression in acute lymphocytic leukemias of T cell and B cell precursor origins. J. Clin. Invest. 1983, 71, 301–313.
94. Badine, E.; Richa, E.; El-Sayah, D.; Terjanian, T. A case report of splenic rupture due to marginal zone lymphoma. Ann. Hematol. 2007, 86, 845–846.
95. Stefanello, D.; Valenti, P.; Zini, E.; Comazzi, S.; Gelain, M.E.; Roccabianca, P.; Avallone, G.; Caniatti, M.; Marconato, L. Splenic marginal zone lymphoma in 5 dogs (2001-2008). J. Vet. Intern. Med. 2011, 25, 90–93.
96. O'Brien, D.; Moore, P.F.; Vernau, W.; Peauroi, J.R.; Rebhun, R.B.; Rodriguez, C.O., Jr.; Skorupski, K.A. Clinical characteristics and outcome in dogs with splenic marginal zone lymphoma. J. Vet. Intern. Med. 2013, 27, 949–954.
97. Van den Brand, M.; van der Velden, W.J.; Diets, I.J.; Ector, G.I.; de Haan, A.F.; Stevens, W.B.; Hebeda, K.M.; Groenen, P.J.; van Krieken, H.J. Clinical features of patients with nodal marginal zone lymphoma compared to follicular lymphoma: similar presentation, but differences in prognostic factors and rate of transformation. Leuk Lymphoma2015, 1–8.
98. Jaffe, E.S. Nodal Marginal Zone Lymphoma. In Hematopathology; Jaffe, E.S., Harris, N.L., Vardiman, J.W., Campo, E., Arber, D.A., Eds.; Elsevier Saunders: St. Louis, MO, USA, 2011; pp. 319–332.
99. Nathwani, B.N. Marginal zone B-cell lymphoma: A clinical comparison of nodal and mucosa-associated lymphoid tissue types. Non-Hodgkin's Lymphoma Classification Project. J. Clin. Oncol. 1999, 17, 2486–2492.
100. Molina, T.J.; Lin, P.; Swerdlow, S.H.; Cook, J.R. Marginal zone lymphomas with plasmacytic differentiation and related disorders. Am. J. Clin. Pathol. 2011, 136, 211–225.
101. Berger, F.; Felman, P.; Thieblemont, C.; Pradier, T.; Baseggio, L.; Bryon, P.A.; Salles, G.; Callet-Bauchu, E.; Coiffier, B. Non-MALT marginal zone B-cell lymphomas: A description of clinical presentation and outcome in 124 patients. Blood 2000, 95, 1950–1956.
102. Kojima, M.; Inagaki, H.; Motoori, T.; Itoh, H.; Shimizu, K.; Tamaki, Y.; Murase, T.; Nakamura, S. Clinical implications of nodal marginal zone B-cell lymphoma among Japanese: Study of 65 cases. Cancer Sci. 2007, 98, 44–49.
103. Mollejo, M.; Camacho, F.I.; Algara, P.; Ruiz-Ballesteros, E.; Garcia, J.F.; Piris, M.A. Nodal and splenic marginal zone B cell lymphomas. Hematol Oncol. 2005, 23, 108–118.
104. Traverse-Glehen, A.; Felman, P.; Callet-Bauchu, E.; Gazzo, S.; Baseggio, L.; Bryon, P.A.; Thieblemont, C.; Coiffier, B.; Salles, G.; Berger, F. A clinicopathological study of nodal marginal zone B-cell lymphoma. A report on 21 cases. Histopathology 2006, 48, 162–173.
105. Andersen, N.S.; Jensen, M.K.; de Nully Brown, P.; Geisler, C.H. A Danish population-based analysis of 105 mantle cell lymphoma patients: Incidences, clinical features, response, survival and prognostic factors. Eur. J. Cancer 2002, 38, 401–408.
106. Velders, G.A.; Kluin-Nelemans, J.C.; de Boer, C.J.; Hermans, J.; Noordijk, E.M.; Schuuring, E.; Kramer, M.H.; van Deijk, W.A.; Rahder, J.B.; Kluin, P.M.; et al. Mantle-cell lymphoma: A population-based clinical study. J. Clin. Oncol.1996, 14, 1269–1274.
107. Hitz, F.; Bargetzi, M.; Cogliatti, S.; Lohri, A.; Taverna, C.; Renner, C.; Mey, U. Diagnosis and treatment of mantle cell lymphoma. Swiss. Med. Wkly. 2013, 143, w13868.
108. Campo, E.; Jares, P.; Jaffe, E.S. Mantle Cell Lymphoma. In Hematopathology; Jaffe, E.S., Harris, N.L., Vardiman, J.W., Campo, E., Arber, D.A., Eds.; Elsevier Saunders: St. Louis, MO, USA, 2011; pp. 331–348.
109. Ferrer, A.; Salaverria, L.; Bosch, F.; Villamor, N.; Rozman, M.; Beà, S.; Giné, E.; López-Guillerm, A.; Campo, E.; Montserrat, E. Leukemic involvement is a common feature in mantle cell lymphoma. Cancer 2007, 109, 2473–2480.
110. Vose, J.M. Mantle cell lymphoma: 2015 Update on diagnosis, risk-stratification, and clinical management. Am. J. Hematol. 2015, 90, 739–745.

111. Wasman, J.; Rosenthal, N.S.; Farhi, D.C. Mantle cell lymphoma. Morphologic findings in bone marrow involvement. Am. J. Clin. Pathol. 1996, 106, 196–200.
112. Vasef, M.A.; Medeiros, L.J.; Koo, C.; McCourty, A.; Brynes, R.K. Cyclin D1 immunohistochemical staining is useful in distinguishing mantle cell lymphoma from other low-grade B-cell neoplasms in bone marrow. Am. J. Clin. Pathol.1997, 108, 302–307.
113. Van Stee, L.L.; Boston, S.E.; Singh, A.; Romanelli, G.; Rubio-Guzman, A.; Scase, T.J. Outcome and Prognostic Factors for Canine Splenic Lymphoma Treated by Splenectomy (1995-2011). Vet. Surg. 2015, 44, 976–982.
114. Bertoni, F.; Ponzoni, M. The cellular origin of mantle cell lymphoma. Int J. Biochem. Cell. Biol. 2007, 39, 1747–1753.
115. National Cancer Institute sponsored study of classifications of non-Hodgkin's lymphomas: summary and description of a working formulation for clinical usage. The Non-Hodgkin's Lymphoma Pathologic Classification Project. Cancer1982, 49, 2112–2135.
116. Harris, N.; de Leval, F.; Ferry, J.A. Follicular Lymphoma. In Hematopathology; Jaffe, E.S., Harris, N.L., Vardiman, J.W., Campo, E., Arber, D.A., Eds.; Elsevier Saunders: St. Louis, MO, USA, 2011; pp. 267–290.
117. Hitz, F.; Ketterer, N.; Lohri, A.; Mey, U.; Pederiva, S.; Renner, C.; Taverna, C.; Hartmann, A.; Yeow, K.; Bodis, S.; et al.Diagnosis and treatment of follicular lymphoma. Swiss. Med. Wkly. 2011, 141, w13247.
118. Klapper, W. Pathobiology and diagnosis of follicular lymphoma. Semin. Diagn. Pathol. 2011, 28, 146–160.
119. Mann, R.B.; Berard, C.W. Criteria for the cytologic subclassification of follicular lymphomas: A proposed alternative method. Hematol. Oncol. 1983, 1, 187–192.
120. Raible, M.D.; Hsi, E.D.; Alkan, S. Bcl-6 protein expression by follicle center lymphomas. A marker for differentiating follicle center lymphomas from other low-grade lymphoproliferative disorders. Am. J. Clin. Pathol. 1999, 112, 101–107.

CHAPTER 5

Paratuberculosis Infection in Camel (Camelus dromidarius): Current and Prospective Overview

Ahmed M. Alluwaimi

Department of Microbiology and Parasitology, College of Veterinary Medicine, King Faisal University, Al-Ahsaa, Saudi Arabia

ABSTRACT

Camel (Camelus dromidarius) is an important source of meat and milk and an iconic animal of the Saudi Arabian heritage. The accumulative evidence indicated the spread of paratuberculosis infection in the camel herds. Despite the explicit studies on the details of the disease in camels and methods of its diagnosis, paratuberculosis infection in camels suffers from wide gap of knowledge of the disease pathogenesis, camel immune responses to the infection and factors that enhance camel's resistance to the infection. The review discusses the current available information of the disease pathobiology and the approaches employed in the diagnosis of paratuberculosis infection in camel. Effective control of the disease in camel prompts for urgent innovation of the current approaches in the diagnosis. Efficient policies and inspection tools are becoming vital to tackle the possible threats of Crohn's disease to the public health due to the meat and milk consumption.

Keywords: Camel, Johne's Disease, Paratuberculosis, Saudi Arabia, MAP

1. INTRODUCTION

Dromedary camel (Camelus dromedarius) can survive and produce considerable amount of milk during recurrent and prolonged hot and dry environment [1] . Thus, camel milk is considered one of the most valuable food sources due to its nutritional value and medicinal properties [2] . Camel milk and meat are considered an important source of proteins for wide range of population [3] . It was estimated that world camel milk's market was worth 10 billion dollars [3] .

In Saudi Arabia, the camel population in 2010 was estimated 850,000 heads of different breeds [4] . The daily milk production by the indigenous breeds ranges from 6 - 8 liter/head. The total annual production was estimated 2500 - 4900 liter [5] .

The MAP infection was reported in camel in Saudi Arabia [6] -[12] . In the last years, MAP infection in camel has attracted a considerable attention [6] -[13] .

Mycobacterium avium subspecies paratuberculosis (MAP) causes Johne's disease in domestic and wild ruminant like, cattle, sheep, goats, deer, antelope and bison worldwide [14] . In Saudi Arabia, Johne's disease was reported in sheep, goat, dairy cattle, and camel [6] [10] [15] [16] . Long incubation period is the main characteristic feature of MAP infection. Ingestion of fecal material, milk, or colostrum is the main route of infection [14] . The major symptoms of infection are chronic diarrhea, emaciation, decrease milk production, and infertility [17] [18] . The annual economical losses that John's disease caused to the American dairy industry were estimated over $200 million [19] .

It is becoming clearly evident that MAP has important role in the pathogenesis of Crohn's disease [20] . The accumulated evidence definitely exerts public health concern about consumption of the dairy and meat products.

The host adaptive immune response to MAP infection is somewhat paradoxical [21] . Despite the overwhelming research on the pathogenesis of MAP infection, the detailed mechanism by which MAP maintains its persistence and mediates the immunosuppressive status of the host is still daunting. It was noticed that the multitude of the immune responses to the MAP infection was of steady state progression. The MAP maintains the host immune responses through the regulation of overwhelming numbers of genes in wide range of pathways to secure its survival without the full impairment of the immune system [22] [23] .

Control of Johne's disease is hampered by the incompetent diagnostic tests. Different versions of ELISA and molecular based techniques were introduced in the last decades to overcome the impediments of detecting the subclinical MAP infection [24] [25] .

In this overview, we attempt to review the current clinical and pathological understandings of the Johne's disease in camel and the possible application of the newly introduced techniques in the early diagnosis of the disease. The most important prospective fields in the study of the Johne's disease in camel that participate in designing effective approaches in control of the disease will also be addressed.

2. THE ETIOLOGY OF CAMEL PARATUBERCULOSIS (PTB)

MAP is the causative agent of the paratuberculosis (Ptb) or Johne's disease. MAP is an intracellular slow growing of 0.5 - 1.5 μm, short, red staining rods, arranged in clumps in smear stained with Ziehl Neelsen method [26] . MAP strains have been classified into three groups (cattle, sheep and intermediate types) based on restriction fragment length polymorphism (RFLP) analysis coupled with hybridization to the insertion sequence IS900 (IS900-RFLP) and culture characteristics, mainly production of yellow-orange pigment in ovine isolates (type I) and non production of same pigment in bovine isolates (type II), growing rate and shape of colonies on solid media, and by differences in their cell wall antigen lipoarabinomannan (LAM) [27] [28] . The cattle type, the most common in Europe, has been isolated primarily from cattle, and other domestic and wild ruminants, non ruminant species, and also humans [29] [30] .

The MAP strain implicated in camel Ptb was isolated and fully sequenced [31] . The detail of the gene sequencing has revealed that the isolated MAP is the strain that is circulating in sheep herds, which could suggest the transmission of the infection from the infected sheep to camel herds [31] .

3. THE PATHOLOGY AND PATHOBIOLOGY OF PARATUBERCULOSIS IN CATTLE

The pathology of Ptb in cattle has been extensively studied. To understand the peculiar aspects of the pathological features of Ptb in camel, an overview on the pathological changes in the bovine Ptb will be reviewed.

3.1. The Clinicopathology of Paratuberculosis in Cattle

The major pathological changes of Ptb infection are located in the distal ileum and could extend to the ileocaecal valve. The major gross lesion are chronic enteritis, chronic intestinal lyphangitis and mesenteric lymphadenopathy [18] . In addition, the most prominent lesion of Ptb infection is the corrugation of the intestinal mucosa. It was proposed that there are four different pictures for the bovine Ptb based on the stage of the disease [17] . They are:

Stage I. It is the silent picture of the disease, which is associated with no evident pathological signs. The major group of this stage is the young animals at the early stage of the infection. Despite the absence of any clinical or pathological signs these young animals are considered potential source of infectious organisms in the farm [17] .

Stage II. The condition of the infected animals at this stage is in continuing to the previous stage. Diagnosis with the conventional techniques like fecal culture could skip a large numbers of the low and intermittent shedding animals. In addition, the seroconversion is not fully detectable [17] .

Stage III. This stage pursues the long incubation period. The infected animals start to manifest intermittent diarrhea. The blood biochemical picture of the infected animals indicates decrease in the total protein (TPR), albumin (ALB), triglycerides (TRIG) and cholesterol (CHOL) concentration. The organism can be easily detected by the fecal culture as well as the detection of the antibodies by ELISA. Some of the animals at this stage may revert to the second stage and continue to act as carrier [17] .

Stage IV. The animals at this stage develop typical clinical signs of Johne's disease. The major signs are, pipe stream diarrhea, emaciation, cachexia and intermandibular edema (bottle jaw). The organism can be recovered from organs widely distant from the gastrointestinal system like uterus, fetus, udder, male accessory sex glands and kidneys. The enlargement of the mesenteric lymph nodes and mucosa thickening and corrugation of the ileum dominate the major gross signs of the disease at this stage. Some cattle could manifest evident arteriosclerosis [17] .

3.2. The Histopathology of Johne's Disease in Cattle

Although Ptb induces distinct histopathological changes, their relation to the clinical stages of the disease can not be distinctly established [18] . The histological changes were categorized in three different forms according to the severity of the lesions [18] . The mild lesions consist of scattered numbers of giant cells in the villous lamina properia or paracortical areas of lymph nodes. The mild lesions could also be associated with sporadic epithelioid cells with eosinophilic cytoplasm and oval nuclei. The moderate lesions however, are characterized by few macrophages, giant cells in the villous lamina properia, intestinal submucosa, lymph nodes and liver. Marked lesions consist of large groups of macrophages and giant cells scattered in the submucosa, mucosa, muscle tunic and serosa. The evident lesions are the extensive accumulation of the mucoid materials and neutrophils in the villi and crypt glands. The prominent enlargement and inflammation of the lymphatics and lymph nodes are the significant signs of the marked lesions [18] .

3.3. The Biochemical Changes of Ptb in Cattle

Ptb infection in cattle is associated with progressive decrease in TPR, ALB, TRIG and CHOL concentration. Whereas muscle wasting will result in the elevation of creatine kinase (CK) and diphosphate aldolase (ALD) [17] .

The biochemical changes were monitored in calves experimentally infected with MAP [32] . The TRP, ALB, TRIG and CHOL indicated significant decrease by day 400 post infection (PI). However, the damage of muscles and liver was indicated by the elevation of CK, ALD, lactate dehydrogenase (LDH), a-hydroxyburate dehydrogenase (a-DBDH), aspartate aminotransferase (AST) and alanine aminotransferase (ALT). The biochemical changes were monitored at the intervals of 160, 190, 220, 250, 280, 310, 340, 370 and 400 days PI. The progress of the disease has indicated the gradual decrease in the TPR, ALB, CHOL TRIG and alkaline phosphatase (ALP), which they reached their optimum between 370 and 400 days PI [32] . However, the early changes were seen with the AST and ALT at 160 days PI. On the other hand, LDH and a-DBDH activities indicated gradual increase in which by 2 months of the infection they reached the highest level. ALD and CK however were fluctuated in their levels, whereas sorbitol dehydrogenase (SDH) activity did not show any change.

The overall biochemical changes indicated a distinct change in the group of biochemicals that were related to the early stages of the infection, while the biochemicals that are related to the muscles and liver functions were mostly increased at the late stages of the infection [32] .

4. THE PATHOBIOLOGY OF PARATUBERCULOSIS IN CAMEL

4.1. The Gross and Histology of the Ptb in Camel

The pathological changes in camel were extensively studied and found similar to that of the Ptb in cattle at the clinical stage [8] -[12] [33] . The major histological and gross lesions of camel in the clinical stage were confined to the ileum but in certain cases rectum and colon might be involved. Typically ileum is thickened and the mucosa is folded in similar to the cerebral convolutions. The ileocaecal and mesenteric lymph nodes were enlarged due to the congestion and edema. The lesion in liver was characterized with variable whitish granuloma [8] [9] [11] [12] [33] .

In a peculiar approach, the pathological Ptb changes in live camels were examined using ultrasonagraph [34] . The sonography was highly successful in imaging the pathological changes of the organs affected by the infection. Ultrasonagraphy could be useful in accelerating the tentative diagnosis of the clinical Ptb infection, however it is not useful in any way in the early detection of the infection.

The major histopathological lesions were extensive infiltration of basophilic macrophages, epithelioid cells and lymphocytes into mucosa and submucosa of ileum. Similar to the lesions in cattle mesenteric and ileocaecal lymph nodes sinuses were filled with macrophages, epithelioid cells and scattered acid fact bacilli. The hepatic tissues were also infiltrated with macrophages aggregates, epithelioid cells and lymphocytes that resulted in distinct lepromatous granuloma formation [9] [33] .

4.2. The Clinical Feature of Camel Ptb

The major clinical symptoms of the camel Ptb are greatly similar to that of the Ptb in cattle. The main manifestations of the camels at the clinical stage are intermittent diarrhea, reduce milk production, dehydration, emaciation and intermandibular edema [9] [34] [35] .

There are immense lacks of evidence that elaborate the development of the Ptb infection in camel at different stages. This shortage in the evidence had great negative impact on understanding the pathogenesis of the Ptb in camel. Majority of the studies that have described the clinical and pathological changes were carried out on camels at the early or advanced clinical stage of the disease [8] -[12] [33] . Hence, the present gap of Ptb infection in camel requires intensive research particularly studies on the experimental infection.

4.3. The Biochemical and the Hematological Features of Ptb Infection in Camel

The Ptb infection at the clinical stage resulted in significant reduction of the total erythrocytes (RBCs) count and hemoglobin (HB) concentration, whereas the packed cell volume (PCV) and neutrophils percentage increased significantly [8] .

The activities of the plasma proteins were similar to that of the bovine Ptb infection. The TRP and ALB were decreased while the AST, ALT and ALP were significantly increased. The proinflammatory cytokines, interleukin (IL)-1a, IL-1b, IL-6, IL-10, interferon-g (IFN-g) and tumor necrosis factor-a (TNF-a) as well as acute phase proteins indicated significant increase in their levels [8] . Furthermore, the oxidative stress mediators, superoxide dismutase, glutathione concentration and catalase, malondialdyhyde, were significantly increased.

The biochemical studies were carried out in camels at the advanced stage of the disease in which most of the examined animals have expressed the most prominent signs of the disease. Studies on the biochemical changes at the early stages of the infection are very rare or scarce. For instance the sole report on the early biochemical changes was by Salem et al. (2012) which they have studied the biochemical changes at the subclinical stage of the disease in camels of 1 - 3 years old. The hematological picture indicated decrease in the levels of mean corpuscular hemoglobin concentration (MCHC), RBCS, HB and PCV, whereas the leukocytes count increased significantly. Similar to the advanced stage of the infection, the serum proteins showed moderate changes [36] .

This report was dependent on the PCR in defining the infected animals. It is greatly skeptical to consider the PCR as the sole approach in defining the subclinical infection in the young camels. A study on the relation of shedding to seroconversion in camel has indicated that infected young camels were a potential MAP shedder earlier than the development of the seroconversion [37] . Hence, the high shedding of MAP in young camels in compare to the low percentage of the seroconversion most probably could indicate that young camels were not necessarily succumb to the infection [37] . The changes in the biochemical changes, on the other hand could have been considered genuine if the study had continued the monitoring of the biochemical changes for longer period.

5. THE DIAGNOSIS OF THE PTB INFECTION IN CAMEL

Early diagnosis of MAP infection represents one of the major obstacles in successful control of the disease [38] . Immunodiagnostic techniques were applied to assure the early diagnosis of MAP infection. Nevertheless, the newly advanced techniques were not sensitive enough unless they were used in suitable combination to achieve clear-cut diagnosis [25] [39] . The sensitivity of the diagnostic test is greatly influenced by various factors like infective dose and age of the animals. Based on these factors, three groups of animals in regard to MAP infection can be classified, affected, infectious and infected [24] . The affected animals are those that manifest clinical signs such as diarrhea, reduction of milk production and chronic weight loss. Animal infectious of MAP on the other hand, comprise those animals that shed MAP and considered a source of transmission to the susceptible animals. The third group is the infected animals that lie between the affected and the infectious groups. Therefore, the sensitivity of ELISA is greatly influenced by the type of the animals group. Therefore, animals shedding less than 10 colonies per tube are more likely to be seronegative, while shedding more than 70 colonies per tube are considered strong seropositive. Hence, ELISA sensitivity in cattle naturally infected with MAP could be as low as 15% in low shedding young animals, whereas in the moderately shedding animals the sensitivity could reach to 47% - 48%. While the heavily shedding animals the sensitivity is 88% [40] .

The ELISA and PCR were applied in the diagnosis of the Ptb infection in camel [6] [7] [13] . The commercial ELISA kits for the detection the bovine anti-MAP antibodies was seen prudent for detecting the seroconverted camel. However, ELISA was expressed lack of sensitivity in detecting the anti-MAP antibodies in young camels [6] [7] . In view of the variations in the shedding patterns, PCR was not seen efficient tool in detecting the infected camels [7] , however it was seen a prudent method in detecting the infectious ones [24] [41] .

The low efficiency of the ELISA in detecting the disease in young camels has compelled the researchers to search for alternatives with greater sensitivity. Lately the growth of the mycobacteriophage in the viable MAP cells was developed to detect and enumerate the viable MAP in milk and blood of infected cattle [42] -[44] . Moreover, the current available commercial ELISA kits were evaluated in different modification approaches to enhance their sensitivity to detect the Ptb infection in camel [45] . Four commercial ELISA kits were compared to in-house ELISA kit to detect the anti-MAP antibodies in camel [45] . The commercial kits were 1) check kit from IDEXX, 2) ID Vet screen paratuberculosis indirect from ID vet,

Innovative diagnostics, France 3) Parachek Johne's absorbed EIA from Prionic, Australia and 4) Paratub serum-S and Serum-B from Institute Pourquier, France. Whereas the in-house kit was prepared by coating the plate with the paratuberculosis protoplasmatic antigen 3 (PPA-3) from US strain 18 MAP [45] . The kits were evaluated in three different approaches. The first assessment was the evaluation of the kits according to the manufacturer's directions. Although the kits reagents performed properly, their discriminative potency in detecting the anti-MAP was below the expectation [45] . However replacement of the conjugates with either protein-A horse reddish peroxidase (HRP) or goat (Fab) 2 anti-camel IgG HRP and use of 3,3',5,5'tetramethyl benzidine (TMB) substrate has boosted their sensitivity especially the ID screen and in-house ELISA. The second approach was to examine the possible nonspecific binding of sample antibodies with the solid phase. All the commercial kits recorded poor specificity except the in-house ELISA, which indicated reasonable discriminative specificity. The third assessment was to evaluate the capability of the commercial kits to monitor the rise in antibody level in the vaccinated camels. In general, all of the kits, except the in-house ELISA, showed low competent discrimination of the antibodies level between the pre and post vaccination.

The overall results indicated that modification of the commercial kits with introduction of protein-A HRP conjugate and suitable substrate has greatly improved their performance. Nevertheless, the in-house ELISA kit with protein-A HRP conjugate was superior in its efficiency over all of the commercial kits [45] .

6. THE DISPARITY IN KNOWLEDGE OF THE PTB INFECTION IN CAMEL

Although the research interest in the Ptb infection of camel has shown considerable increase in the last years, there are still major lack of the knowledge related to the Ptb infection and the information related to the nature of the host responses to the infection [46] . Addressing the following questions was seen of great importance in broadening the comprehension of the Ptb infection in camel.

1) What is the nature of the camel immune responses to the MAP infection?
2) How MAP maintains its survival and persistence in the harsh camel environment?
3) Is there breed related susceptibility to the MAP infection?

4) Does consumption of meat and milk of MAP-infected camel state public health risks?

For details of addressing the posted questions see Alluwaimi, 2014 [46] .

7. CONCLUSION

In conclusion, the overwhelming deficiencies in the knowledge of the Ptb infection will not be overcome unless embarking on better organization and long lasting road map to tackle the outnumbered shortages in tools and strategies of this national problem.

CITE THIS PAPER

Ahmed M. Alluwaimi (2015) Paratuberculosis Infection in Camel (*Camelus dromidarius*): Current and Prospective Overview. *Open Journal of Veterinary Medicine*,**05**,153-160.

REFERENCES

1. Bekele, T., Lundeheim, N. and Dahlborn, K. (2011) Milk Production and Feeding Behavior in the Camel (Camelus dromedarius) during 4 Watering Regimens. Journal of Dairy Science, 94, 1310-1317.
2. Lorenzen, P.C., Wernery, R., Johnson, B., Jose, S. and Wernery, U. (2011) Evaluation of Indigenous Enzyme Activities in Raw and Pasteurised Camel Milk. Small Ruminant Research, 97, 79-82.
3. Breulmann, M., Böer, B., Wernery, U., Wernery, R., El Shaer, H., Alhadrami, G., Gallacher, D., Peacock, J., Chaudhary, A. and Brown, G. (2007) The Camel from Tradition to Modern Times. The United Nations Educational, Scientific and Cultural Organization Office, Doha.
4. Aljumaah, R.S., Samara, E.M. and Ayadi, M. (2012) Influence of Introducing Machine Milking on Biothermal Parameters of Lactating Camels (Camelus dromedarius). Italian Journal of Animal Science, 11, e73.
5. Aljumaah, R.S., Almutairi, F.F., Ismail, E., Alshaikh, M.A., Sami, A. and Ayadi, M. (2012) Effects of Production System, Breed, Parity and Stage of Lactation on Milk Composition of Dromedary Camels in Saudi Arabia. Journal of Animal and Veterinary Advances, 11, 141-147.
6. Alluwaimi, A.M. (2008) The Efficiency of Bovine ELISA in Detection of the Mycobacterium avium Subspecies Paratuberculosis (MAP) Infection in Camel (Camelus dromedaries) at Different Ages. Journal of Camel Practice and Research, 15, 163-165. [Citation Time(s):5]
7. Alhebabi, A.M. and Alluwaimi, A.M. (2010) Paratuberculosis in Camel (Camelus dromedarius): The Diagnostic Efficiency of ELISA and PCR. Open Veterinary Science Journal, 4, 41-44.

8. El-Deeb, W., Fouda, T. and El-Bahr, S. (2014) Clinico-Biochemical Investigation of Paratuberculosis of Dromedary Camels in Saudi Arabia: Proinflammatory Cytokines, Acute Phase Proteins and Oxidative Stress Biomarkers. Pakistan Veterinary Journal, 34, 484-488.
9. Hereba, A., Hamouda, M. and Al-hizab, F. (2014) Johne's Disease in Dromedary Camel: Gross Findings, Histpathology and PCR. Journal of Camel Practice and Research, 21, 83-88.
10. Gameel, A., Ali, A., Razig, S. and Brown, J. (1994) A Clinico-Pathological Study on Spontaneous Paratuberculosis in Camels (Camelus dromedarius) in Saudi Arabia. Pakistan Veterinary Journal, 14, 15.
11. Tharwat, M., Al-Sobayil, F. and El-Magawry, S. (2013) Clinicobiochemical and Postmortem Investigations in 60 Camels (Camelus dromedarius) with Johne's Disease. Journal of Camel Practice and Research, 20, 145-149.
12. Alharbi, K.B., Al-Swailem, A., Al-Dubaib, M.A., Al-Yamani, E., Al-Naeem, A., Shehata, M., Hashad, M.E., Albusadah, K.A. and Mahmoud, O.M. (2012) Pathology and Molecular Diagnosis of Paratuberculosis of Camels. Tropical Animal Health and Production, 44, 173-177.
13. Hussain, M.H., Saqib, M., Al-Maawali, M.G., Al-Makhladi, S., Al-Zadjali, M.S., Al-Sidairi, T., Asubaihi, S., Al- Rawahi, A. and Mansoor, M.K. (2014) Seroprevalence of Mycobacterium avium Subspecies Paratuberculosis (MAP) and Evaluation of Risk Factors in Camels of the Sultanate of Oman. Tropical Animal Health and Production, 47, 383- 389.
14. Stabel, J.R. (1997) Johne's Disease: A Hidden Threat. Journal of Dairy Science, 81, 283-288.
15. Alluwaimi, A., Hatem, M. and Almousa, J. (1999) The Efficacy of Gel Immunodiffusion and Fecal Smear Tests for Diagnosis of Ovine Paratuberculosis in Sheep in Saudi Arabia. The Egyptian Journal of Immunology, 7, 29-32.
16. Ahmed, S. and Towfik, A. (1995) Johne's Disease among Sheep and Goat of El-Qassiem Area. 16th Annual Meeting of Saudi Biological Society.
17. Whitlock, R.H. and Buergelt, C. (1996) Preclinical and Clinical Manifestations of Paratuberculosis (Including Pathology). Veterinary Clinics of North America: Food Animal Practice, 12, 345-356.
18. Clarke, C. (1997) The Pathology and Pathogenesis of Paratuberculosis in Ruminants and Other Species. Journal of Comparative Pathology, 116, 217-261.
19. Ott, S.L., Wells, S.J. and Wagner, B.A. (1999) Herd-Level Economic Losses Associated with Johne's Disease on US Dairy Operations. Preventive Veterinary Medicine, 40, 179-192.
20. Alluwaimi, A.M. (2007) The Etiology of Mycobacterium avium Subspecies Paratuberculosis in Crohn's Disease. Saudi Medical Journal, 28, 1479-1484.
21. Sweeney, R., Jones, D., Habecker, P. and Scott, P. (1998) Interferon-Gamma and Interleukin 4 Gene Expression in Cows Infected with Mycobacterium paratuberculosis. American Journal of Veterinary Research, 59, 842-847.
22. MacHugh, D.E., Taraktsoglou, M., Killick, K.E., Nalpas, N.C., Browne, J.A., Park, S., Hokamp, K., Gormley, E. and Magee, D. (2012) Pan-Genomic Analysis of Bovine Monocyte-Derived Macrophage Gene Expression in Response to in Vitro Infection with Mycobacterium avium Subspecies Paratuberculosis. Veterinary Research, 43, 25.
23. Shin, M.K., Park, H.T., Shin, S.W., Jung, M., Young, B.I., Park, H.E., Cho, Y.I. and Yoo, H.S. (2015) Whole-Blood Gene-Expression Profiles of Cows Infected with Mycobacterium avium Subsp. Paratuberculosis Reveal Changes in Immune Response and Lipid Metabolism. Journal of Microbiology and Biotechnology, 25, 255-267.
24. Nielsen, S.S. and Toft, N. (2008) Ante Mortem Diagnosis of Paratuberculosis: A Review of Accuracies of ELISA, Interferon-γ Assay and Faecal Culture Techniques. Veterinary Microbiology, 129, 217-235.

25. Collins, M.T., Wells, S.J., Petrini, K.R., Collins, J.E., Schultz, R.D. and Whitlock, R.H. (2005) Evaluation of Five Antibody Detection Tests for Diagnosis of Bovine Paratuberculosis. Clinical and Diagnostic Laboratory Immunology, 12, 685-692.
26. Cocito, C., Gilot, P., Coene, M., de Kesel, M., Poupart, P. and Vannuffel, P. (1994) Paratuberculosis. Clinical Microbiology Reviews, 7, 328-345.
27. Stevenson, K., Hughes, V.M., de Juan, L., Inglis, N.F., Wright, F. and Sharp, J.M. (2002) Molecular Characterization of Pigmented and Nonpigmented Isolates of Mycobacterium avium Subsp. Paratuberculosis. Journal of Clinical Microbiology, 40, 1798-1804.
28. Valentin-Weigand, P. and Goethe, R. (1999) Pathogenesis of Mycobacterium avium Subspecies Paratuberculosis Infections in Ruminants: Still More Questions than Answers. Microbes and Infection, 1, 1121-1127.
29. Pavlik, I., Bejcková, L., Pavlas, M., Rozsypalova, Z. and Koskova, S. (1995) Characterization by Restriction Endonuclease Analysis and DNA Hybridization Using IS900 of Bovine, Ovine, Caprine and Human Dependent Strains of Mycobacterium paratuberculosis Isolated in Various Localities. Veterinary Microbiology, 45, 311-318.
30. Pavlik, I., Horvathova, A., Dvorska, L., Bartl, J., Svastova, P., du Maine, R. and Rychlik, I. (1999) Standardisation of Restriction Fragment Length Polymorphism Analysis for Mycobacterium avium Subspecies Paratuberculosis. Journal of Microbiological Methods, 38, 155-167.
31. Ghosh, P., Hsu, C., Alyamani, E.J., Shehata, M.M., Al-Dubaib, M.A., Al-Naeem, A., Hashad, M., Mahmoud, O.M., Alharbi, K.B. and Al-Busadah, K. (2012) Genome-Wide Analysis of the Emerging Infection with Mycobacterium avium Subspecies Paratuberculosis in the Arabian Camels (Camelus dromedarius). PloS ONE, 7, e31947.
32. Körmendy, B., Szilágyi, M., Tuboly, S. and Nagy, G. (1990) Some Diagnostic Features of the Pathogenesis of Bovine Paratuberculosis (Johne's Disease) and Serum Biochemical Changes after Oral Reinfection. Journal of Veterinary Medicine, Series B, 37, 229-235.
33. Almujalli, A. and Al Ghamdi, G. (2012) Clinicopathological Findings of Partuberclosis in Camels Possible Steps for Control Strategy. Research Journal of Biological Sciences, 7, 128-131.
34. Tharwat, M., Al-Sobayil, F., Ali, A., Hashad, M. and Buczinski, S. (2012) Clinical, Ultrasonographic, and Pathologic findings in 70 Camels (Camelus dromedarius) with Johne's Disease. The Canadian Veterinary Journal, 53, 543-548.
35. Zaghawa, A., Housawi, F., Al-Naeem, A. and Hammoda, M. (2012) Clinical Investigations on Paratuberculosis in Camels in Al-Hasa, KSA. 11th International Colloquium on Paratuberculosis, International Association for Paratuberculosis, The University of Sydney, Sydney, 5-10 February 2012.
36. Salem, M., El-Sayed, A, Fayed, A, Abo El-Hassan, D.G. (2012) Subclinical Infection of Paratuberculosis among Camels in Egypt. Journal of American Science, 8, 1141-1147.
37. Alhebabi, A.M., Badi, F.A. and Alluwaimi, A.M. (2012) The Relation of Mycobacterium avium Subspecies Paratuberculosis Shedding to the Seroconversion in the Camels. Journal of Camel Practice and Research, 19, 255-258.
38. Valentin-Weigand, P. (2002) Johne's Disease: Pathogenesis and Problems Related to Diagnosis. In: Recent Developments and Perspectives in Bovine Medicine, XXII World Buiatrics Congress, Hanover, 18-23.
39. Paolicchi, F.A., Zumarraga, M.J., Gioffre, A., Zamorano, P., Morsella, C., Verna, A., Cataldi, A., Alito, A. and Romano, M. (2003) Application of Different Methods for the Diagnosis of Paratuberculosis in a Dairy Cattle Herd in Argentina. Journal of Veterinary Medicine, Series B, 50, 20-26.
40. Whitlock, R., Wells, S., Sweeney, R.W. and Van Tiem, J. (2000) ELISA and Fecal Culture for Paratuberculosis (Johne's Disease): Sensitivity and Specificity of Each Method. Veterinary Microbiology, 77, 387-398.

41. Nielsen, S.S. (2008) Transitions in Diagnostic Tests Used for Detection of Mycobacterium avium Subsp. Paratuberculosis Infections in Cattle. Veterinary Microbiology, 132, 274-282.
42. Swift, B. and Rees, C.E. (2013) Detecting Mycobacteria in Cattle Blood. The Veterinary Record, 173, 522-523.
43. Stanley, E.C., Mole, R.J., Smith, R.J., Glenn, S.M., Barer, M.R., McGowan, M. and Rees, C.E. (2007) Development of a New, Combined Rapid Method Using Phage and PCR for Detection and Identification of Viable Mycobacterium paratuberculosis Bacteria within 48 Hours. Applied and Environmental Microbiology, 73, 1851-1857.
44. Swift, B.M., Denton, E.J., Mahendran, S.A., Huxley, J.N. and Rees, C.E. (2013) Development of a Rapid Phage-Based Method for the Detection of Viable Mycobacterium avium Subsp. Paratuberculosis in Blood within 48 h. Journal of Microbiological Methods, 94, 175-179.
45. Wernery, U., Abraham, A., Joseph, S., Thomas, R., Syriac, G., Raghavan, R. and Baker, T. (2011) Evaluation of 5 Indirect Elisa for the Detection of Antibodies to Paratuberculosis in Dromedaries. Journal of Camel Practice and Research, 18, 47-52. [Citation Time(s):5]
46. Alluwaimi, A.M. (2014) The Mycobacterium avium Subspecies Paratuberculosis (MAP) Infection in Camel (Camelus dromidarius) in Saudi Arabia: Questions Wait to be Addressed. Paratuberculosis Newsletter, September 2014, 37-42.

CHAPTER 6

Gliomatosis Cerebri in the Brain of a Cat

Stephanie Shrader [1,*], Serene Lai [2], Kelsey Cline [2] and Rachel Moon [2]

[1]*Department of Pathobiology, College of Veterinary Medicine, Auburn University, Auburn, AL 36849, USA*

[2]*Department of Clinical Sciences, College of Veterinary Medicine, Auburn University, Auburn, AL 36849, USA*

ABSTRACT

An eight-year-old, neutered, female, long-haired cat was presented with a three-week history of progressive lethargy, unlocalized pain in the cervical and lumbar spine, and unwillingness to move. An MRI (magnetic resonance imaging) of the brain revealed poorly circumscribed regions of non-contrast-enhancing heterogeneous T2 hyperintensity within the ventral forebrain and midbrain. A mass effect and evidence of increased intracranial pressure, including transtentorial herniation of the midbrain and herniation of the cerebellar vermis through the foramen magnum, were also observed. Due to progressive clinical decline and MRI results, the cat was humanely euthanized. Gross examination of the brain confirmed caudal transtentorial and foramen magnum herniation. The ventral aspect of the forebrain, midbrain, and brainstem were soft and had loss of detail, but lacked a grossly discernible mass. Histopathological examination found a poorly delineated neoplastic mass composed of hyperchromatic cells with indistinct cytoplasm, ovoid to elongate or curved nuclei, and indistinct nucleoli. The cells lacked immunoreactivity for Olig2, GFAP, Iba1, CD3, and Pax5. Based on the cellular morphology, immunolabeling characteristics, and anatomical location, a diagnosis of gliomatosis cerebri was made. Although uncommon, gliomatosis cerebri should be considered as a differential diagnosis in cats with central nervous system disease.

Keywords: cat; gliomatosis cerebri; brain; T2 hyperintensity

1. INTRODUCTION

Gliomatosis cerebri (GC) is an uncommon glial neoplasm, first reported by Nevin [1]. Although it is characterized by neoplastic glial cells that infiltrate the neuroparenchyma, a grossly discernible mass is often lacking [2]. The 2007 World Health Organization (WHO) classification defines human GC as an astrocytic tumor [3]; however, scholars debate whether GC represents a separate entity or a subset of diffuse glioma [4]. Clonal analysis of gliomas has found that GC begins as an oligoclonal process or may result from collision gliomas. This differs from the monoclonality typically observed in low grade and malignant gliomas [5]. In veterinary medicine, GC has been classified as either a miscellaneous type of tumor [6] or as a neuroepithelial neoplasm of unknown origin [7].

2. CASE HISTORY

An eight-year-old, neutered, female, long-haired cat was presented to the Wilford & Kate Bailey Small Animal Teaching Hospital at Auburn University's College of Veterinary Medicine (AUCVM) with a three-week history of progressive lethargy, unlocalized pain in the cervical and lumbar spine, and unwillingness to move. Two weeks prior to presentation at AUCVM, the cat was presented to a local emergency veterinary clinic and reportedly diagnosed with diskospondylitis (based on thoracic radiographs). She was treated with steroids (prednisone, dosage unknown) and butorphanol (0.09 mg/kg PO as needed) resulting in improvement of clinical signs. Over the next couple weeks, the cat experienced a recurrence of cervical pain and a progressive decline in mentation. She presented to the referring veterinarian and was treated with prednisone (increased from 0.9 mg/kg every 48 to 24 h; last dose three days prior to AUCVM presentation), Depo-Medrol (dosage unknown; last dose one day prior to AUCVM presentation), and intermittent doses of butorphanol (0.09 mg/kg PO as needed; last dose two days prior to AUCVM presentation). The cat's mentation declined despite treatment and on the morning of presentation to AUCVM, she experienced a tonic seizure that lasted for 20 s. The cat was taken to the referring veterinarian who administered midazolam (dosage unknown) and referred the cat to AUCVM.

On presentation to AUCVM, the cat had a body weight of 5.7 kg, a body condition score of 4/5, and a rectal temperature of 102.3°F. A neurologic examination found that the cat was dull but responsive, with mild proprioceptive ataxia in all four limbs. Cranial nerve examination and segmental reflexes were normal. Postural reactions were considered to be normal, with the exception of possible delayed hopping in the right thoracic limb. There was hyperesthesia associated with palpation of the cranial cervical vertebrae and auditory bullae; however, it was difficult to determine whether this response was behavioral or pain-related. Bloodwork and radiographs, performed by the referring veterinarian, were reviewed. Hematological and serum biochemical parameters were within reference intervals; spinal radiographs showed spondylosis deformans at L3-L4, but no evidence of diskospondylitis.

Based on clinical findings, a forebrain lesion was considered most likely and magnetic resonance imaging (MRI) of the brain was performed using a 1.5 T scanner and a human knee coil (Infinion: Philips Medical Systems, Andover, MA, USA). Sagittal images were obtained with T2-weighting (T2) (TR 4000 TE 118.3, 3 mm). Transverse images included T2 (TR 4000, TE 145.6, 3 mm), proton density (TR 4000, TE 15, 3 mm), fluid attenuated inversion recovery (FLAIR) (TR 6000, TE 111.6, 3 mm), gradient recalled echo (GRE) (TR 1188, TE 23.5, 3 mm), and T1-weighting (T1) (TR 400, TE 11.5, 3 mm). Post-intravenous contrast T1 images were obtained in all three standard planes approximately 5 min after manual intravenous injection of gadolinium-based contrast (transverse, sagittal and dorsal: TR 400, TE 11.5, 3 mm) (0.11 mmol/kg body weight, gadoteridol (Prohance; Bracco Diagnostic, Inc., Township, NJ, USA)).

On MRI scanning of the brain, an intra-axial, diffuse, non-contrast-enhancing, T2-weighted, hyperintense mass effect extending through the ventral forebrain (including the piriform and temporal lobes, thalamus, crus cerebri, and colliculi) and midbrain (Figure 1) was found. The caudal aspects of the cerebellum were compressed against the occipital bone. There was also caudal transtentorial herniation of the midbrain and herniation of the cerebellar vermis through the foramen magnum (Figure 2). A linear, T2 hyperintense area was present in the dorsal aspect of the C2-C5 spinal cord; transverse images of this area showed dilation of the central canal, T2 hyperintensity, and non-contrast-enhancing T1 hypointensity in the dorsal spinal cord, most consistent with syrinx formation. Based on clinical findings and MRI results, differential diagnoses included multicentric neoplasia or an infectious etiology such as feline infectious peritonitis (FIP), feline immunodeficiency virus (FIV), feline leukemia virus (FeLV), or cryptococcosis. Due to poor prognosis and rapid clinical decline, the cat was humanely euthanized.

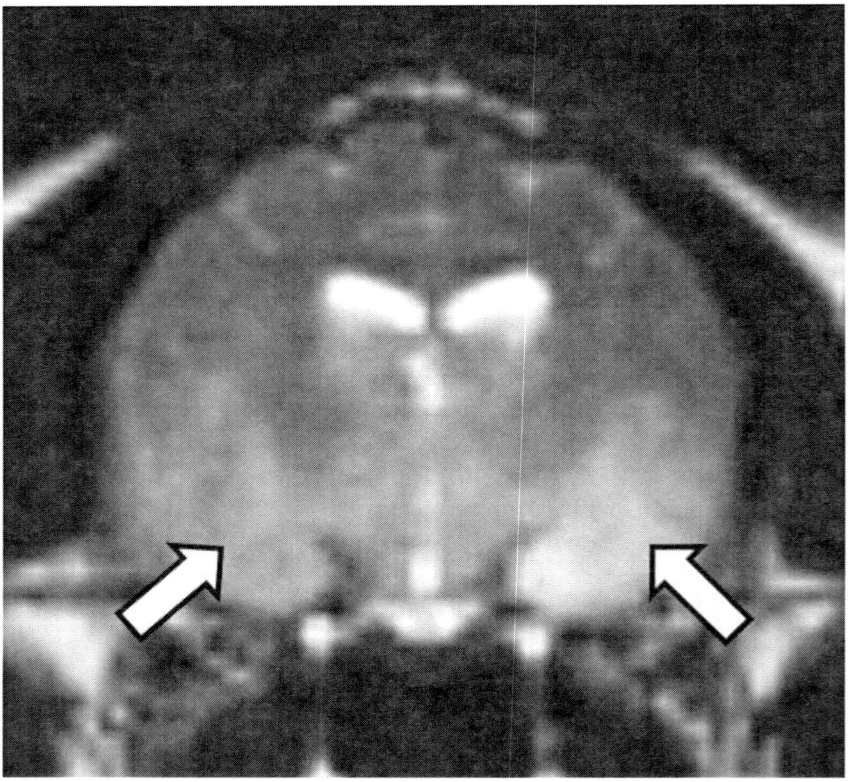

Figure 1. T2 weighted transverse image at the level of the piriform lobe showing the ill-defined, ventrally distributed T2-weighted hyperintensity affecting the piriform lobes, temporal lobes, hypothalamus, and portions of the caudal colliculus and internal capsule (arrows). MRI scan.

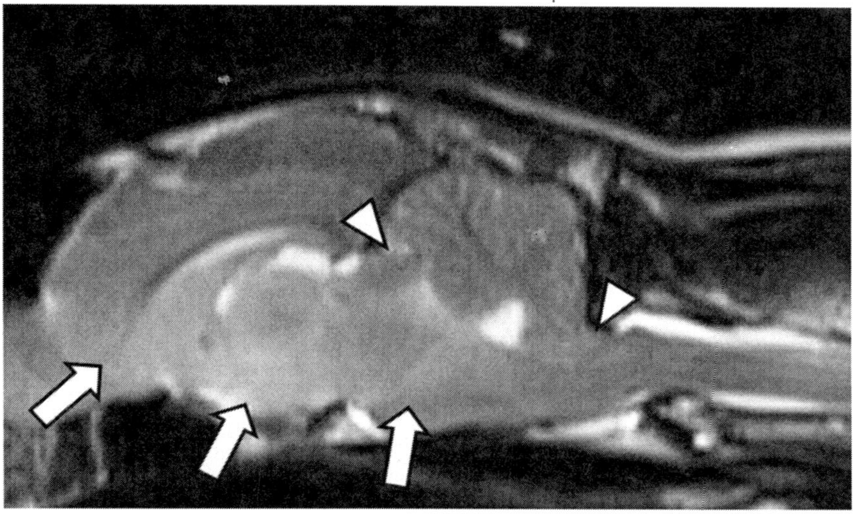

Figure 2. T2-weighted midline sagittal image showing caudal transtentorial herniation of the midbrain through the osseous tentorium and caudal herniation of the cerebellum through the foramen magnum (arrowheads). Additionally, there is an ill-defined, mildly hyperintense mass effect in the ventral aspect of forebrain including the caudal nucleus of the thalamus, interthalamic adhesion, colliculi, and midbrain (arrows). MRI scan.

On post-mortem examination, changes were confined to the central nervous system. Gross examination of the brain confirmed compression of the caudal cerebellum, caudal transtentorial herniation of the midbrain, and herniation of the cerebellar vermis. There was mild to moderate flattening of the straight, suprasylvian, ectosylvian, sylvian, dentate, and occipital gyri with narrowing of the rostral rhinal, lateral rhinal, suprasylvian, and ectosylvian sulci. On the cut surface, the corpus striatum, thalamus, and hippocampus were soft and had loss of detail, but lacked a grossly discernible mass (Figure 3). The spinal cord was unremarkable. Mild spondylosis deformans was present at L3-L4; however, there was no gross evidence of diskospondylitis.

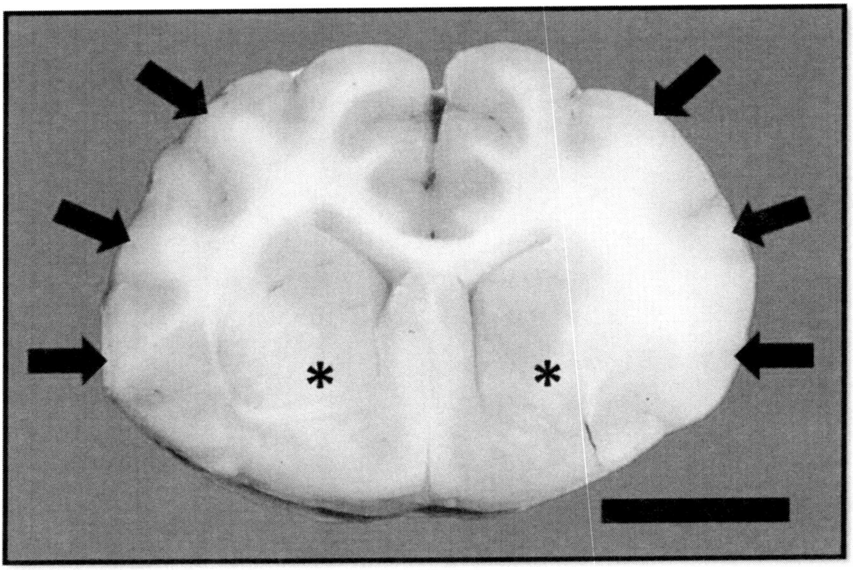

Figure 3. Transverse section of formalin-fixed brain at the level of the septal nuclei. There is mild to moderate flattening of the suprasylvian, ectosylvian, and sylvian gyri (arrows) with narrowing of the associated sulci. The ventral aspects of the corpus striatum (astrices) have bilateral but asymmetrical loss of detail. Bar = 1 cm.

The entire brain, spinal cord, thoracic and abdominal viscera were fixed in 10% neutral buffered formalin, processed routinely, and embedded in paraffin wax. Sections (5 μm) were stained with hematoxylin and eosin (HE). Selected sections were subjected to immunohistochemistry (IHC) for detection of oligodendrocyte transcription factor 2 (Olig2; rabbit monoclonal antibody, 1:100 dilution, Abcam, Cambridge, MA, USA), glial fibrillary acidic protein (GFAP; pre-diluted rabbit polyclonal antibody, Dako, Carpinteria, CA, USA), ionized calcium-binding adapter molecule 1 (Iba-1; rabbit polyclonal antibody, 1:300 dilution, Biocare Medical, Concord, CA, USA), cluster of differentiation 3 (CD3; mouse monoclonal antibody, 1:25 dilution, Dako, Carpinteria, CA, USA), and paired box 5 (Pax5; mouse monoclonal antibody, 1:15 dilution, Dako, Carpinteria, CA, USA) using an automated system (Autostainer Link, Dako, Carpinteria, CA, USA) with a polymer-based horseradish peroxidase-conjugated detection kit with 3, 3′ diaminobenzidine tetrahydrochloride as the chromogen (EnVision™ FLEX high-pH kit, Dako, Carpinteria, CA, USA). Antigen retrieval was performed by heating in citrate buffer and endogenous peroxidase activity was quenched via immersion in a phosphate buffer (EnVision™ FLEX Peroxidase-blocking Reagent, Dako, Carpinteria, CA, USA). Slides were counterstained with hematoxylin. Sections of normal canine brain were used as positive controls for Olig2 and GFAP; sections of normal canine lymph node were used as positive controls for Iba-1, CD3, and Pax5. For negative reagent controls, primary antibodies were replaced with equivalent non-homologous immune sera. Negative internal tissue controls in immunolabeled slides consisted of non-glial, non-lymphoid, and non-histiocytic cells such as vascular endothelium, which did not demonstrate immunoreactivity with any of the antibodies.

On histopathologic examination, ventral grey and white matter of the cerebrum (including the olfactory nuclei, olfactory tracts, corpus striatum, lateral septal nuclei, medial septal nuclei, thalamic nuclei and hippocampus), cerebellum, and brain stem (predominantly affecting the olivary nuclei) were bilaterally but asymmetrically infiltrated and expanded by dense sheets of neoplastic cells that merged gradually into the adjacent parenchyma, resulting in indistinct margins between normal and abnormal tissue (Figure 4a). The neoplastic cells had indistinct cytoplasm, ovoid to elongated or curved hyperchromatic nuclei, and indistinct nucleoli (Figure 4b). Anisokaryosis was mild and there was one mitotic figure per 10 400× fields. Multifocally, the neoplastic cells formed secondary structures of Scherer, including foci of subpial spread (predominantly along the ventral aspect of the brainstem; Figure 5a), subependymal aggregates (Figure 5b) with multifocal extension into and expansion of the leptomeninges, and minimal perineuronal satellitosis (Figure 5c). The neoplastic cell population was negative for Olig2, GFAP, Iba1, CD3, and Pax5. Scattered among these cells were variable numbers of GFAP-positive astrocytes, Olig2-positive oligodendrocytes, and Iba1-positive microglia. All examined sections of spinal cord were histologically unremarkable. Based on the cellular morphology, immunolabeling characteristics, and anatomical location, a diagnosis of gliomatosis cerebri was made.

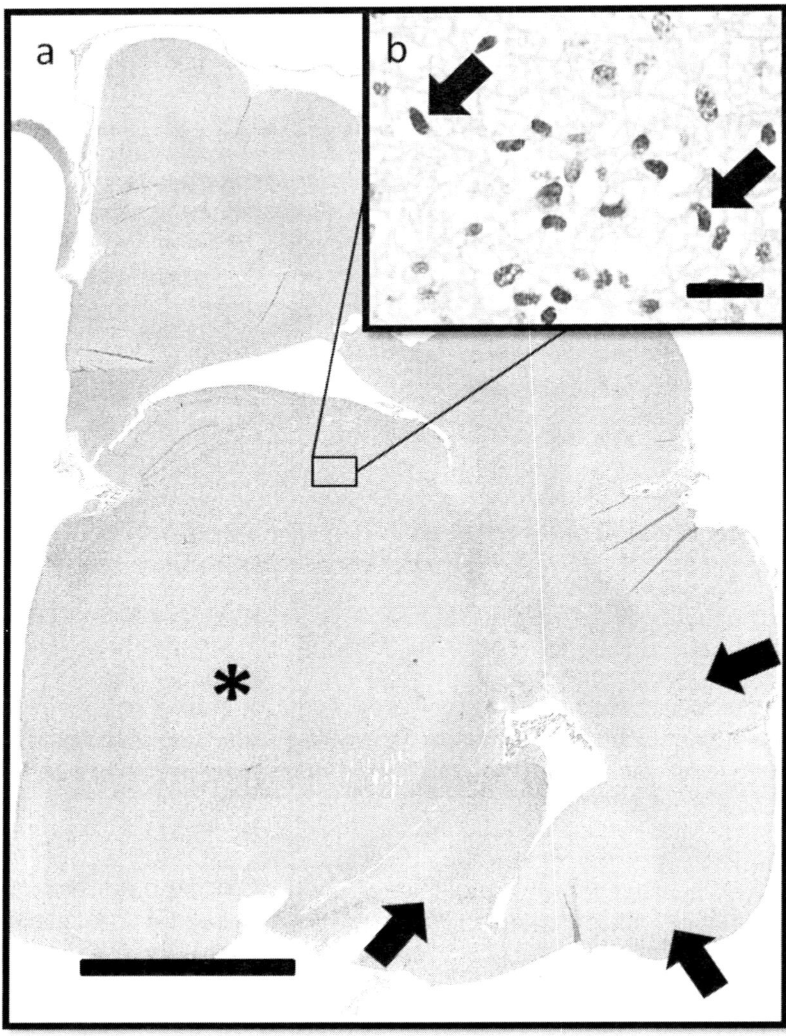

Figure 4. Transverse section of the brain at the level of the interthalamic adhesion. (**a**) Subgross photomicrograph showing loss of general anatomical detail within the thalamic nucleus and internal capsule due to a poorly delineated neoplastic mass (asterisk). Neoplastic cells also obscure the sylvian gyrus, hippocampus, and intervening parenchyma (arrows). HE. Bar = 5 mm. (**b**) Inset of (**a**). The neoplastic cells have indistinct cytoplasm, ovoid to elongated or curved hyperchromatic nuclei, and indistinct nucleoli (arrows). HE. Bar = 20 μm.

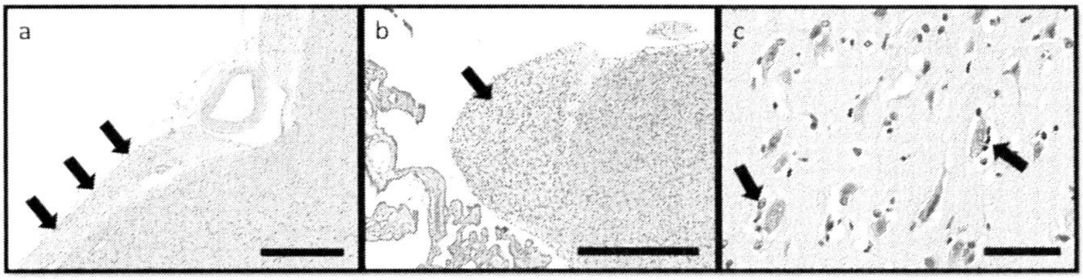

Figure 5. Secondary structures of Scherer, demonstrating migration of glioma cells through the brain, include: (**a**) subpial spread of neoplastic cells (arrows) along the ventral aspect of the brainstem, HE. Bar = 500 µm; (**b**) subependymal aggregation (arrow), HE. Bar = 500 µm; and (**c**) minimal perineuronal satellitosis (arrows). HE. Bar = 50 µm.

3. DISCUSSION

In the present case, gliomatosis cerebri was diagnosed in an eight-year-old, neutered, female, long-haired cat by magnetic resonance imaging, and histological and immunohistochemical examination. Clinical signs in this patient are consistent with the neuroanatomic location of the neoplasm. The mild proprioceptive ataxia was likely due to neoplastic infiltration of the cerebellum and brainstem (predominantly affecting the olivary nucleus). The cerebellum is responsible for coordinating motor function to maintain appropriate posture and movements. To accomplish this, it uses proprioceptive information to modulate the activity of the upper motor neuron (UMN) nuclei and motor cortex, which then influence lower motor neuron (LMN) activity [8]. Cells from the olivary nucleus contribute by extending into the cerebellum to aid in voluntary muscle movement [9]. An additional clinical sign in this patient was cervical spinal hyperesthesia, a common finding in dogs and cats with intracranial neoplasia [10].

MRI of the spinal cord revealed a linear T2 hyperintense, non-enhancing T1 hypointense area in the dorsal aspect of the C2-C5 cord with dilation of the central canal. These imaging characteristics are consistent with a syrinx. Although the entire spinal cord was evaluated grossly and histopathologically, the presence of a syrinx was not able to be confirmed. The reason for the discrepancy is unknown; however, it is possible that the lesion was more subtle than imaging suggested and that collapse occurred during the post-mortem interval.

GC has been reported in the brains of people [1,11], dogs [12,13], and in a single cat [14]. According to the 2007 WHO classification, GC is considered to be a Grade III astrocytoma, a malignant neoplasm characterized by glial cells that invade adjacent normal tissue [3]. Data suggest that men are affected more commonly than women [2] and although it has been reported in all age groups, it is most common in the fifth decade of life [15]. As seen in this case, the T2-weighted MRI of the brain (in both people and dogs) usually shows vague, heterogeneous, non-contrast-enhancing hyperintensities resembling inflammatory disease, vasculitis, leukoencephalopathies, or venous infarcts [2,13,16]. GC MRI findings are generally considered to be characteristic of an astrocytoma, but with wider cerebral dissemination [17]. Human GC is typically characterized by widespread, diffuse, bilateral neoplastic infiltrates in at least three cerebral hemispheric lobes. There is also frequent involvement of the cerebellum, brainstem, and spinal cord [17]. Although imaging was not performed in the previously reported case of GC in the brain of a cat, histopathologic evaluation confirmed a diffuse neoplastic cell infiltrate in the cerebrum with minimal infiltrates in the metencephalon; neoplastic cells were not observed in the cerebellum or spinal cord [14].

GC in the brains of people and dogs, and in the single case report in a cat, is characterized histologically by infiltrates of cells with ovoid to slender nuclei and indistinct cytoplasm admixed with astrocytes, oligodendrocytes, and microglia. Typically, the infiltrating cells fail to form a grossly evident mass [12,14]. In people, the neoplastic cells are often positive for GFAP, although some fusiform cells are GFAP-negative [18]. In dogs, the neoplastic cells are positive for vimentin and nestin [19] and negative for pan-cytokeratin, neuron specific enolase (NSE), neurofilament, CD133, S100, GFAP, CD18, CD3, CD45RA, CD79α, and B lymphocyte antigen 36 (BLA.36) [12,19]. GC has been reported in the feline spinal cord, with a similar hyperintense T2-weighted MRI signal as described in the brains of people and dogs. Feline brain and spinal cord GC has been characterized by neoplastic infiltrates that are GFAP-negative [14,20,21]. In feline spinal cord GC, the neoplastic cells are also reported to be negative for Ki-67, Pax5, and CD45, with strong cytoplasmic immunoreactivity for BLA.36 [21]. Additional microscopic features of GC in people and dogs include perineuronal and perivascular satellitosis, subpial and subependymal neoplastic aggregates, and migration of neoplastic cells within white matter tracts (intrafascicular spread) [9]. These patterns of glioma cell infiltration are referred to as the secondary structures of Scherer; they were not observed in the previously reported case of GC in the feline brain [14,22].

Histopathologic differentials for gliomatosis cerebri include other diffusely infiltrating astrocytomas (diffuse astrocytoma, anaplastic astrocytoma, and glioblastoma), oligodendroglial tumors, mixed gliomas, and lymphoma. Diffuse astrocytomas typically arise supratentorially in the frontal and temporal lobes [17]. Similar to the neoplasm in this patient, diffuse astrocytomas usually have a lack of necrosis, foci of microvascular proliferation, and secondary structures of Scherer [23]. Unlike the neoplasm in this patient, however, diffuse astrocytomas are composed of fairly uniform infiltrating fibrillary astrocytes that have elongated, enlarged hyperchromatic nuclei, scant cytoplasm, and strong GFAP and vimentin cytoplasmic immunoreactivity [17]. Anaplastic astrocytomas typically arise from the same locations as diffuse astrocytomas; however, they have greater overall cellularity and neoplastic cell density. Additionally, unlike diffuse astrocytomas, cells of anaplastic astrocytomas have minimal cytoplasm with nuclear atypia, a greater mitotic index, and higher proliferative activity. The neoplastic cells of both anaplastic and diffuse astrocytomas generally have strong GFAP or vimentin immunoreactivity. Glioblastoma often occurs in the frontal and temporal lobes. Grossly, there are typically well-delineated areas of necrosis and hemorrhage. Histologically, these neoplasms are hypercellular and pleomorphic with anisokaryosis, numerous mitotic figures (some of which are abnormal), intratumoral microvascular proliferation, pseudopalisades of glial cells, and foci of necrosis. In the dog, glioblastomas typically have strong cytoplasmic GFAP immunoreactivity, like other astrocytomas. [17]. Based on cellular morphology, immunolabeling characteristics (lack of GFAP immunoreactivity), and a lack of secondary tumoral characteristics (such as microvascular proliferation), the neoplasm in the present report is not consistent with the above-mentioned astrocytoma subtypes.

As mentioned, oligodendroglial tumors, mixed gliomas, and lymphoma were additional possible histopathologic differentials in this case. Oligodendrogliomas often occur in the frontal, parietal, and temporal lobes; meningeal invasion/metastasis and dissemination within the ventricular system can also occur. In people, dogs, and cats, these tumors are usually well-demarcated and have a grey gelatinous consistency [17,24,25]. Histologically, oligodendrogliomas are characterized by densely packed, fairly uniform cells with round nuclei and well-delineated cytoplasmic borders. Multifocally, there are typically lakes of myxoid material and thin-walled branching capillaries that produce a characteristic “chicken-wire” appearance. Grade III oligodendrogliomas have increased nuclear atypia, foci of necrosis, thrombosis, hemorrhage, and prominent microvascular proliferation. Neoplastic oligodendroglial cells have nuclear Olig2 immunoreactivity and lack GFAP immunoreactivity [17]. Mixed gliomas have astrocytic and oligodendroglial components and are therefore known as oligoastrocytomas. Both cell populations typically have strong cytoplasmic GFAP immunoreactivity, minimal pleomorphism, and few mitoses (unless it is an anaplastic variant). Anaplastic variants, in addition to having anisocytosis, anisokaryosis, and increased numbers of mitotic figures, also have foci of microvascular proliferation [17]. CNS lymphoma can be primary, intravascular, or metastatic. Primary CNS lymphoma often forms a single mass lesion and is composed histologically of sheets of pleomorphic round cells which extend from perivascular aggregates [17,26]. Metastatic lymphomas are typically confined to the leptomeninges and dura, with foci of parenchymal invasion. CD3 and Pax5 immunolabeling can help determine whether these lymphomas are of T-cell or B-cell origin, respectively. [17]. Cellular morphology, lack of immunoreactivity for GFAP, CD3 and Pax5, and lack of secondary tumoral characteristics (such as microvascular proliferation) preclude the diagnosis of an oligodendroglial tumor, mixed glioma, or lymphoma in this patient.

4. CONCLUSIONS

Based on MRI findings, gross examination, anatomic location and behavior, neoplastic cellular morphology, and immunolabeling characteristics, this patient was diagnosed with GC of the brain. Although GC has previously been reported in the brain of a cat, this current case report is unique because it describes MRI findings, details histopathologic findings, and characterizes neoplastic cellular immunoreactivity. Although uncommon, gliomatosis cerebri should be considered as a differential diagnosis for brain neoplasia in the cat.

ACKNOWLEDGMENTS

The authors received no funding for writing or publishing this manuscript.

AUTHOR CONTRIBUTIONS

Serene Lai performed the patient's neurologic examination. Kelsey Cline and Rachel Moon performed the patient's MRI and evaluated the results. Stephanie Shrader performed the post-mortem examination and histopathologic evaluation. Stephanie Shrader, Serene Lai, and Kelsey Cline wrote the paper. All authors contributed to revisions and final editing.

CONFLICTS OF INTEREST

The authors declare no conflict of interest.

REFERENCES

1. Nevin, S. Gliomatosis cerebri. Brain 1938, 61, 170–191.
2. Yerramneni, V.K.; Vinjamuri, S.R.; Purohit, A.K.; Sundaram, C. Gliomatosis cerebri: Case series of six cases with review of literature. J. Neurosci. Rural Pract. 2015, 6, 602–606.
3. Louis, D.N.; Ohgaki, H.; Wiestler, O.D.; Cavanee, W.K.; Burger, P.C.; Jouvet, A.; Scheithauer, B.W.; Kleihues, P. The 2007 WHO Classification of Tumours of the Central Nervous System. Acta Neuropathol. 2007, 114, 97–109.
4. Mawrin, C. Molecular genetic alterations in gliomatosis cerebri: What can we learn about the origin and course of the disease? Acta Neuropathol. 2005, 110, 527–536.
5. Kattar, M.M.; Kupsky, W.J.; Shimoyama, R.K.; Vo, T.D.; Olson, M.W.; Bargar, G.R.; Sarkar, F.H. Clonal analysis of gliomas. Hum. Pathol. 1997, 28, 1166–1179.
6. Summers, B.A.; Cummings, J.F.; Delahunta, A. Tumors of the central nervous system. In Veterinary Neuropathology; Mosby: St. Louis, MO, USA, 1995; pp. 351–401.
7. Koestner, A.; Bilzer, T.; Fatzer, R.; Schulman, R.Y.; Summers, B.A.; Van Winkle, T.J. Histological classification of tumours of the nervous system of domestic animals. In World Health Organization International Histological Classification of Tumours of Domestic Animals; Schulman, F.Y., Ed.; WHO, Armed Forces Institute of Pathology: Washington, DC, USA, 1999; Volume 5, pp. 17–27.
8. Thomson, C.; Hahn, C.; Johnson, C. Veterinary Neuroanatomy: A Clinical Approach; Saunders Elsevier: Edinburgh, UK, 2012; pp. 59–63.
9. Snell, R.S. Clinical Neuroanatomy; Wolters Kluwer Health/Lippincott Williams & Wilkins: Philadelphia, PA, USA, 2010; p. 204.
10. Withrow, S.J.; David, M. Vail. Withrow & MacEwen's Small Animal Clinical Oncology; Saunders Elsevier: St. Louis, MO, USA, 2007; p. 662.
11. Rajz, G.; Nass, D.; Talianski, E.; Pfeffer, R.; Spiegelmann, R.; Cohen, Z. Presentation patterns and outcome of gliomatosis cerebri. Oncol. Lett. 2012, 3, 209–213.
12. Porter, B.; DeLahunta, A.; Summers, B. Gliomatosis cerebri in six dogs. Vet. Pathol. 2003, 40, 97–102.
13. Martin-Vaquero, P.; da Costa, R.C.; Wolk, K.E.; Premanandan, C.; Oglesbee, M.J. MRI features of gliomatosis cerebri in a dog. Vet. Radiol. Ultrasound 2012, 53, 189–192.
14. Stierstorfer, B.; Janowetz, B.; Schmahl, W. Gliomatosis cerebri in a cat. A case report. Tierarztliche Prax. Ausg. Kleintiere Heimtiere 2002, 30, 282–285.
15. Kandler, R.; Smith, C.; Broome, J.; Davies-Jones, G. Gliomatosis cerebri: A clinical, radiological and pathological report of four cases. Br. J. Neurosurg. 1991, 5, 187–193.
16. Ghostine, S.; Raghavan, R.; Michelson, D.; Holshouser, B.; Tong, K. Gliomatosis cerebri mimicking Rasmussen encephalitis. J. Neurosurg. Pediatr. 2007, 107, 143–146.
17. Vandevelde, M.; Higgins, R.J.; Oevermann, A. Veterinary Neuropathology: Essentials of Theory and Practice; Wiley-Blackwell: Ames, IA, USA, 2012; pp. 132–150.
18. Galatioto, S.; Marafioti, T.; Cavallari, V.; Batolo, D. Gliomatosis cerebri; clinical, neuropathological, immunohistochemical and morphometric studies. Zent. Pathol. 1993, 139, 261–267.

19. Plattner, B.; Kent, M.; Summers, B.; Platt, S.; Freeman, A.; Blas-Machado, U.; Clemans, J.; Cheville, N.F.; Garcia-Tapia, D. Gliomatosis cerebri in two dogs. J. Am. Anim. Hosp. Assoc. 2012, 48, 359–65.
20. Jelinek, F. Gliomatosis of the spinal cord in a cat: A case report. Vet. Med. 2013, 6, 331–337.
21. Hammond, J.; de Lahunta, A.; Glass, E.; Kent, M.; Summers, B.; Miller, A. Feline spinal cord gliomas: Clinicopathologic and diagnostic features of seven cases. J. Vet. Diagn. Investig. 2014, 26, 513–520.
22. Holland, E.C. Glioblastoma multiforme: The terminator. Proc. Natl. Acad. Sci. USA 2000, 97, 6242–6244.
23. Tove, L.; Hansson, H.A.; Stein, S.; Sverre, H.T. Prognostic value of histological feature sin diffuse astrocytomas WHO grade II. Int. J. Clin. Exp. Pathol. 2012, 5, 152–158.
24. Dickinson, P.J.; Keel, M.K.; Higgins, R.J.; Koblik, P.D.; LeCouteur, R.A.; Naydan, D.K.; Bollen, A.W.; Vernau, W. Clinical and pathologic features of oligodendrogliomas in two cats. Vet. Pathol. 2000, 37, 160–167.
25. Engelhard, H.H.; Stelea, A.; Mundt, A. Oligodendroglioma and anaplastic oligodendroglioma: Clinical features, treatment, and prognosis. Surg. Neurol. 2003, 60, 443–456.
26. Fondevila, D.; Vilafranca, M.; Pumarola, M. Primary central nervous system T-cell lymphoma in a cat. Vet. Pathol.1998, 35, 550–553.

CHAPTER 7

A Review of Enrofloxacin for Veterinary Use

Tessa Trouchon, Sébastien Lefebvre

USC 1233 INRA-Vetagro Sup, Veterinary School of Lyon, Marcy l'Etoile, France

ABSTRACT

This review outlines the current knowledge on the use of enrofloxacin in veterinary medicine from biochemical mechanisms to the use in the field conditions and even resistance and ecotoxicity. The basics of biochemistry, the mechanisms of action and resistance and pharmacokinetics are presented. Then an overview of available veterinary products, their efficacy and their toxicity against target species, human and environment is provided.

Keywords: Enrofloxacin, Antibiotic Resistances, Veterinary

1. INTRODUCTION

Enrofloxacin (Figure 1(c)), or 1-Cyclopropyl-6-fluoro-7-(4-ethyl-1-piperazinyl)-1,4-dihydro-4-oxo-3-quinoli- necarboxylic acid, belongs to fluoroquinolone family which is a subfamily of quinolone. The first quinolone is the Nalidixic acid (Figure 1(a)) used in animal at the beginning of 1980s, enrofloxacin is the first fluoroquinolone patented in 1984 [1] . The huge evolution in the quinolone family is the addition of a fluor atom on the 6th position which improves quinolones' antibacterial spectrum [2] and creates the fluoroquinolone subfamily. Quinolones have an action on bacterial topoisomerase. The marketing authorization reports a large antimicrobial spectrum for enrofloxacin, which is efficient on most gram-negative and gram-positive bacteria but not efficient on anaerobic bacteria [3] . But with 3.6 tons sold per year in France for animal use [4] , fluroroquinolones are an important family in veterinary medicine that increases the probability of selecting resisting bacteria.

2. ACTION MECHANISM OF ENROFLOXACIN

2.1. Important Physicochemical Properties of Enrofloxacin

Enrofloxacin is a zwitterionic molecule with a pKa_1 = [5.88 - 6.06] and a pKa_2 = [7.70 - 7.74] [5] . The lowest pKa is due to the carboxyl acid group and the second to the basic tertiary amine. So enrofloxacin doesn't bear charge between this two pH. Moreover it is a lipophilic molecule with a logP of 4.70 at a pH of 7 [6] , but 1.88 ± 1.43 with ACDLabs.

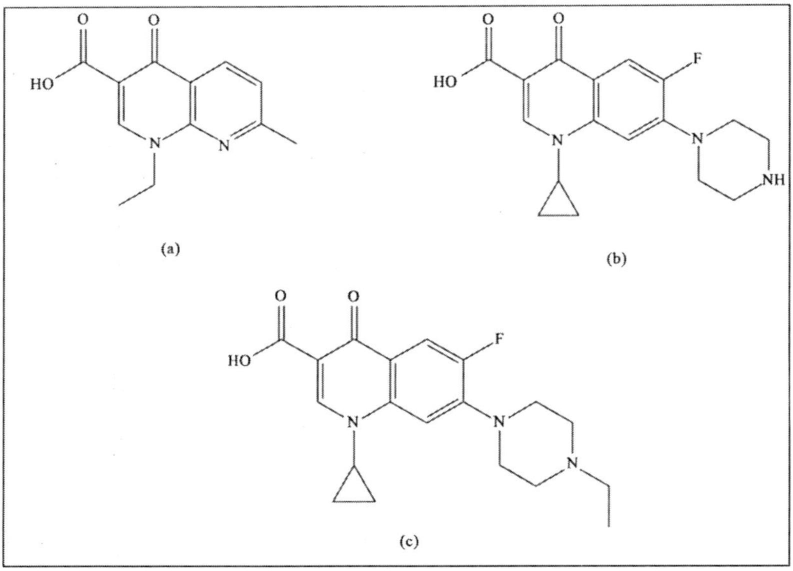

Figure 1. Structure of (a) Nalidixic Acid; (b) Ciprofloxacin and (c) Enrofloxacin.

The Active metabolite of enrofloxacin, ciprofloxacin (Figure 1(b)) or 1-Cyclopropyl-6-fluoro-1, 4-dihydro-7- (1-piperazinyl)-4-oxo-3-quinoline carboxylic acid, which is available as a drug product on veterinary and human market, is a multiple acid with $pKa_1 = 5.15$ and $pKa_2 = 8.25$ [7] . Ciprofloxacin is less lipophilic than enrofloxacin with a logP of −1.11 at a pH of 7.4 [8] .

2.2. Targets' Mechanism of Action

Enrofloxacin, like the other quinolones, has two mains targets of the topoisomerase family. Although these proteins exist in eukaryotes cells, quinolones have less affinity for eukaryotes' topoisomerases than for the DNA Topoisomerase II (Gyrase) and the DNA Topoisomerase IV (Topo IV) two major bacterial topoisomerase [9] . The Gyrase and the Topo IV are two tetramers (Figure 2(a)) formed respectively of two GyrA and two GyrB and of two ParC and two ParE. Moreover GyrA and GyrB are homologous respectively with ParC and ParE [10] .

The Gyrase has an important role in bacteria's life by modifying the topology of the spiral DNA. Indeed, the positive supercoiling stabilizes the DNA and the strands' separation becomes more difficult [11] . Moreover transcription generates a positive supercoiling accumulation that can stop transcription. This positive supercoiling can be released by the Gyrase, which enhances transcription [12] . To do this, the Gyrase binds and wraps around itself a strand of DNA (Figure 2(b)) with the help of the C-terminal domain [13] and cleaves it (Figure 2(c)) with mediating of the catalytic tyrosine Tyr 122 [14] of each GyrA. Tyrosines form covalent phosphotyrosyls with each 5' phosphoryl terminus of both strands, which remain binding during the reaction [15] . This reaction forms a gap in sequence (called G-DNA). The GyrB part catches the other DNA sequence (called transported DNA or T-DNA). The T-DNA is passed through the opened G-DNA (Figure 2(d)) [16] . The G-DNA is closed in an ATP-dependent reaction (Figure 2(e)) [17] . Of this reaction results an adding of negative coiling.

The Gyrase is able to do an intermolecular strand passage at the end of the replication like the Topo IV but without wrapping the DNA as the Topo IV, this is the major difference between reactions of both enzymes [18] . Topo IV catalyzes the segregation of the two daughter DNA molecules after the replication [19] more efficiently than Gyrase [20] .

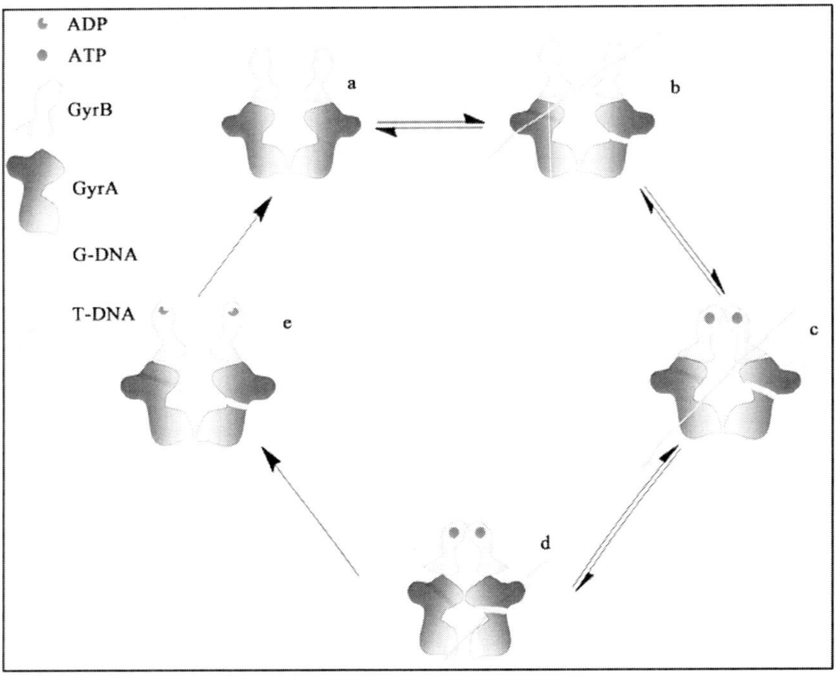

Figure 2. Mechanism of action of Gyrase, with intervention of Gyrase part A (GyrA) and B (GyrB), gap DNA sequence (G-DNA), transported DNA sequence (T-DNA), ATP and GDP. (a) Gyrase with GyrA dimer and GyrB dimer; (b) Interaction of Gyrase with two DNA sequences; (c) Fixation of an ATP molecule and cleavage of the G-DNA strand; (d) Passage of the T-DNA strand through G-DNA; (e) Closing of the G-DNA strand with ATP degradation.

2.3. Interactions between Fluoroquinolones, DNA and Its Target

The penetration in the bacteria differs between Gram-positives and Gram-negatives bacteria. For Gram-negative bacteria, fluoroquinolones pass the outer membrane mainly through porins [21] [22] . The trimeric OmpF porin is mainly used by fluoroquinolone, this pathway can be modulated by Mg^{2+} [23] . But it seems that some quinolones can promote a pathway by interacting with the Lipopolysaccharide (LPS) and create a lipophilic passage [24] . For Gram-positive bacteria the diffusion process is the main uptake pathway [25] .

In the case of mycobacteria the high lipid level of the membrane allows fluoroquinolones to diffuse through membrane [26] .

After the entrance in the bacteria, quinolones have two effects on the bacteria: bacteriostatic at low concentration level or bactericide at high concentration level. Rapidly after the Gyrase has formed a complex with the DNA strand, two quinolones bind with this complex, before the DNA's cleavage [27] . This binding is reversible and induces a conformation modification in Gyrase [28] . This modification induces a cleavage of DNA in a particular location and forms a cleaved complex [29] [30] . At this step the action is still reversible, but reduces rapidly the activity of replication and has a bacteriostatic effect [31] .

The next two pathways are possible one for bacteriostatic concentration and the other for bactericide concentration. For bacteriostatic concentrations, quinolones induce the SOS regulon controlled by the repressor lexA [32] [33] . Activation of the regulon induces the sfiA gene which codes for an inhibitor of cell division and bacteria form a long filamentous structure. This effect is still reversible [34] . This filamentous structure contributes to the death of bacteria [35] .

For bactericide concentration quinolones induce a chromosome fragmentation by creation of a suicide factor or by destabilization of a GyrA dimer [35] . The fragmentation can induce a rapid death or, by reassociation of the GyrA dimer, genomic mutation in bacteria [36] .

2.4. Resistance Mechanism

The resistance to fluoroquinolones in general, is today a main subject for the use of antibiotic substances in veterinary medicine and was the subject of an EMA report [4] . Many studies show an augmentation in the development of resistances to enrofloxacin [37] [38] . Moreover, another study shows the increase of resistance of Escherichia coli to quinolones and fluoroquinolones after using of enrofloxacin in calves [39] .

The mechanism of resistance to fluoroquinolones is based on many pathways: the change of targets, the protection of target, the lower expression of target, the improvement of the efflux pumps, and inactivation of fluoroquinolones. These mechanisms can be chromosomal or plasmide-mediated (Table 1) [40] .

The most common mechanism is the mutation of Gyrase. In the subpart GyrA the mutation is mainly between Ala67 and Gln 106 for E.coli and other bacteria [41] [42] . The mutation is near the GyrA active site which is called Quinolone Resistance-Determining Region (QRDR). For GyrB two mutation are reported at Asp426 and Lys447 [43] . Only one Gyrase mutation is able to increase the Minimum Inhibitory Concentration (MIC) up to 64 times [41] , but the GyrB mutation seems to be less effective in resistance against nalidixic acid than the GyrA mutation [44] .

The mutation of Topo IV is also described but, until today, always with at least one Gyrase mutation. Therefore, these mutations seem to be developed in a second time to increase the resistance more than only Gyrase's mutations. A QRDR can be also characterized in the parC of topoisomerase IV between Tyr 57 and Glu84 [45] .

The protection of target can be attributed to a qnr gene often present on a Plasmide-Mediated Quinolone Resistance (PMQR) or in the chromosomal DNA. The Qnr protein is a dimere which binds with Gyrase and decreases the binding of quinolones with DNA-Gyrase-Qnr complex. But it seems that the Qnr protein reduces the interaction between DNA and Gyrase [46] . Another theory is that the Qnr protein destabilizes the DNA-Gyrase-quinolone complex and promotes the DNA repair after cleavage [47] .

An hypothesis of resistance is the slow growth of cell, bacteria growing slowly seem to resist better, but this mechanism is not known today [40] [48] .

Although resistances are often attributed to the Gyrase or Topo IV mutations, decrease of influx and increase of efflux mechanisms are important pathways of resistance common to a large number of unrelated antibiotics [49] . This defines a Multiple Antimicrobial Resistance (MAR) phenotype. E. coli bearing a MAR operon has higher resistance for the quick death by fluoroquinolones [50] . This operon can induce a decrease of OmpF pumps [51] and an increase of efflux with AcrAB protein [52] . The decrease of porins can also be attributed to PMQR [53] .

2.5. Recommendations to Prevent Resistance

Given the rapidity of development of resistance, we have to consider with the MIC, the Mutant Prevention Concentration (MPC). MPC is a recent indicator which is not well standardized. Commonly MPC50 is the concentration where 50% of the colonies doesn't contain resistant mutant after 72 hours of incubation. To count the mutants, a Polymerase Chain Reaction (PCR) is used but only known resistant mutation is tested [54] . Between MIC and MPC there is a window where enrofloxacin is effective but also where mutants are selected.

The data recorded in Table 2 show that the window between the MIC and the MPC can be very important with a MPC/MIC between 1.6 and 64. This high difference can explain the explosion of resistance against enrofloxacin and might be a base to define new guidelines and dosage for enrofloxacin.

Table 1. Possible origins of resistances to fluoroquinolones.

Chromosomal resistance	Plasmide-mediated resistance
Mutation of target	Lower expression of target
Lower expression of target	Protection of target Improvement of efflux pumps
Protection of target Improvement of efflux pumps	Inactivation of fluoroquinolones

Table 2. MIC and MPC of enrofloxacin for some bacteria.

Bacteria	MIC (µg/mL)	MPC (µg/mL)	(MPC/MIC)	References
Escherichia coli	[0.022 - 0.03]	[0.17 - 0.5]	[7.8 - 16]	[67] [107]
Staphylococcus pseudintermedius	0.13	0.27	2	[108]
Rhodococcus equi	[0.5 - 1]	[8 - 64]	[16 - 64]	[109]
Mannheimia haemolytica	0.16	0.25	1.6	[110]
Salmonella Typhimurium	[0.06 - 0.125]	[1 - 2]	16	[67]
Pseudomonas aeruginosa	2	[16 - 32]	[8 - 16]	[67]

3. PHARMACOKINETIC OF ENROFLOXACIN

3.1. Absorption

There is high variation of bioavailability after oral administration of enrofloxacin between polygastric and monogastric animals from 10% to 80% [55] [56] . This has a direct consequences on the galenic, oral presentations are reserved for pigs, poultry, calves, and carnivores and only injectable solutions are available for cattle. Moreover, bioavailability depends on whether the animals are fed or fasted [55] [57] , and of the presence of ion [58] . These interferences can be explained by the formation of a complex between cation and fluoroquinolone which cannot permeate through the digestive barrier [59] . Another consequence of this complex is the influence of the hardness of water notably in poultry farming where the dilution in water is used [60] . But lipophilic compounds in food can enhance oral bioavailability of fluoroquinolone [61] .

In addition to passive diffusion allowed by enrofloxacin lipophilicity, active transporters have also an important role in intestinal absorption [62] . But these transporters are also important to eliminate fluoroquinolones [62] .

The intramuscular bioavailability is 96% with a maximum concentration 3 hours after administration [3] . Intramuscular bioavailability can be enhanced by solid lipophilic nanoparticles, this technic enhances the duration of enrofloxacin in plasma [3] .

3.2. Metabolism

After administration a high part of enrofloxacin is metabolized into ciprofloxacin in most of species (Table 3). Enrofloxacin has an active metabolite, ciprofloxacin [63] , obtained by deethylation of the ethyl on the piperazin ring. Other metabolites are obtained but they don't have antimicrobial effects [64] .

Only poultry does not have a huge part of enrofloxacin metabolized in ciprofloxacin. The first hepatic pass doesn't have a lot of effect, only 7% of enrofloxacin is metabolized [65] . Moreover ciprofloxacin seems a more potent drug than enrofloxacin [66] [67] . These elements are in favor to grant a more important place to ciprofloxacin in the effects of enrofloxacin-based drugs.

3.3. Distribution

Distribution of enrofloxacin and ciprofloxacin to tissues depends on drug's free concentration, which depends itself on the concentration of protein and the strength of this biding.

The difference between ciprofloxacin and enrofloxacin protein binding (Table 4) are in favor of the theory of enrofloxacin's role of prodrug. Indeed, in some species ciprofloxacin is less bound with proteins, so it is able to be more available to be effective. Moreover, enrofloxacin can interact with other protein-binding drugs for instance it increases the clearance of flunixin meglubine [68] .

As described in Table 5, the volumes of distribution of enrofloxacin and ciprofloxacin in different species are all higher than 1. So enrofloxacin and its metabolite diffuse strongly in tissues, and molecules can be present in cells, so being inactive. Contrary to what we might expect considering its protein binding, ciprofloxacin doesn't diffuse in tissues more than enrofloxacin. Data of Table 6suggest a high affinity of drugs for lung and kidney, even if data confirm enrofloxacin's and ciprofloxacin's good diffusion.

Table 3. Percent of ciprofloxacin after enrofloxacin given.

Species	Ciprofloxacin percent of enrofloxacin plasma concentration	References
Dogs	40	[65]
Dairy cows	59	[75]
Steers	64	[75]
Chickens	<10	[111]
Pigs	51	[112]
Goats	34	[71]

Table 4. Protein binding of enrofloxacin and ciprofloxacin.

Species	Percent of bond enrofloxacin (%)	Percent of bond ciprofloxacin (%)	References
Dogs	34	18	[113]
Dairy cows	59.4	33.7	[75]
Steers	60.8	49.6	[75]
Chickens	23	ND	[114]
Pigs	[31.1 - 37.1]	35	[115]

Table 5. Pharmacokinetics parameters of enrofloxacin and its metabolite ciprofloxacin.

Species	$t_{1/2}$ (h) *		Cl (mL/min/kg) **		V_{ss} (L/kg) ***		References
	Enro	Cipro	Enro	Cipro	Enro	Cipro	
Dogs	2.3	2.8	12.16	7.8	2.45	1.92	[65]
Dairy cows	3.69	2.96	24.16	ND	1.56	ND	[75]
Steers	5.5	7.60	11.6	ND	1.59	ND	[75]
Chickens	6.99	3.11	3.30	15.45	1.98	4.04	[111]
Pigs	26.6	2.60	3.0	17.30	6.40	3.80	[69]
Goats	1.39	[1.82-2.72]	22.18	19.59	1.27	3.33	[71] [116]

*elimination half-live. **clearance. ***volume of distribution at the steady state. Enro: enrofloxacine. Cipro: ciprofloxacine.

Table 6. Maximal concentration (C) and area under the curve (AUC) of enrofloxacin (ENR) and ciprofloxacin (CIP) after an intravenous injection of 5 mg/kg of enrofloxacin [117] .

	Plasma†		Muscle‡		Liver‡		Spleen‡		Lung‡		Kidney‡	
	ENR	CIP	ENR	CIP	ENR	CIP	ENR	CIP	ENR	CIP	ENR	CIP
C	4.44	0.18	3.48	0.19	3.67	2.95	11.04	0.25	3.82	0.86	8.98	2.78
AUC	3.52	0.61	5.61	0.62	5.76	2.02	10.44	2.08	9.44	4.79	9.03	4.91

†Unit of C is µg/mL and unit of AUC is µg.h/mL; ‡Unit of C is µg/g and unit of AUC is µg.h/g.

3.4. Elimination

Elimination parameters show great difference between species (Table 5), especially for pigs where the elimination half-life is high with 26 h [69] . Moreover, for chickens and pigs, clearance of ciprofloxacin is five times higher than enrofloxacin's clearance and their enrofloxacin clearances are lower than other species. This differences might be attributed to a difference of elimination way.

Enrofloxacin's elimination way is mainly renal. This has been proved in rat by nephrecotomizing and comparing with CCl_4 hepatic impairment [70] and in goat with probenecid [71] . Probenecid has been shown to reduce the renal clearance of fluoroquinolones [72] . On the other hand, ciprofloxacin's elimination is both hepatic and renal [73] . For both molecules there is an intestinal recirculation via the bile excretion, moreover, two hour after given, there is no significant difference in the concentration of enrofloxacin in the intestinal content between oral and intramuscular administration of enrofloxacin [74] . Lactation can influence significantly enrofloxacin's and ciprofloxacin's elimination by increasing two folds clearance in dairy cows comparative to steers [75] . It might be explained by an ionic trap.

4. VETERINARY MEDICINAL PRODUCTS

4.1. Available Veterinary Medicines

The first veterinary product based on enrofloxacin was launched by the laboratory Bayer in 1991 (marketing authorisation in 1991) under the trade name Baytril® and it was an oral form for poultry [3] . Now, many veterinary products on the basis of enrofloxacin are available on the market with at least thirty four veterinary medicines under different forms including oral and injectable forms, tablets or bolus [3] . There are many target species including domestic carnivores (dogs and cats), farm animals (cattle, pigs, poultry) and even exotic pets since 2010 [3] .

4.2. Therapeutic Indications and Uses in the Field Conditions

Enrofloxacin is indicated in the treatment of local and systemic diseases caused by a wide range of Gram-negative and Gram-positive bacteria [4] [76] . The most important indication of enrofloxacin in all of the species is the treatment of respiratory infections but it is also indicated in the treatment of digestive, urinary, joint, genital, mammary and dermal infections [3] [77] .

However, enrofloxacin is a third generation fluoroquinolone with a very large spectrum of activity so it has to be reserved to second intention. Indeed, in order to avoid fluoroquinolone resistances, it is important to reserve the use of enrofloxacin to infections resistant to over antibacterial agents and if possible under a susceptibility study [4] .

In the field conditions, enrofloxacin is often off-label used empirically to prevent uterine infections in susceptible embryo-transfer mares. In fact, a conventional dose (5 mg/kg body weight) given pre-breeding followed by two further doses at 36 - 48 h post breeding are supposed to prevent bacterial adherence and provide effective bactericidal concentrations in utero [78] .

Enrofloxacin is often used in aquaculture in Indonesia, Thailand and Vietnam. Indeed, because of the non-hygienic and stressful conditions in aquaculture facilities, the risk of bacterial infections is high and motivates the widely use of antibiotics in fish feed for prophylactic and therapeutic purpose [79] .

5. THERAPEUTIC EFFICACY

Enrofloxacin is a powerful antimicrobial which have shown efficacy against a lot of bacterial diseases [4] [80] . The effectiveness of enrofloxacin against some bacterial infections in cattle, poultry, domestic carnivores (dogs and cats), rodents, lagomorphs and crustaceans has been assessed in many published studies, as well during natural infections as during experimental infections (Table 7). Among these studies, some relate to classical infections contained in the Summary of Product Characteristics (SPC) of veterinary products but some evaluate the efficacy of enrofloxacin in species for which there is no marketing authorisation or against specific bacteria such as Anaplasma marginale in cattle, Ehrlichia canis and Brucella canis in dogs, Bartonella henselae or Bartonella clarridgeiae and Chlamydophila felis in cats, Toxoplasma gondii in Calomys callosus or even Vibrio harveyi in Artemia franciscana (Table 7).

6. ADVERSE EFFECTS

Overall, the fluoroquinolones are well tolerated with fewer adverse effects that are not very serious, especially when compared to their benefits [80]

[81] . The most common side effects of enrofloxacin are digestive disorders including nausea, abdominal discomfort, vomiting and diarrhoea [81] [82] and inflammatory reaction at the site

Table 7. Twenty published studies examining the efficacy of enrofloxacin.

Target specie(s)	Disease(s) and/or bacterial agent(s)	Type of infection	Regimen of enrofloxacin	Purpose of the study	Number of animals	Result(s)	References
CATTLE							
Dairy cows	Acute clinical mastitis (*Escherichia coli*)	N	5 mg/kg IV then SC 24 h later	Comparative efficacy of a ENR + KET treatment and a KET treatment	132 (64 ENR + KET, no controls)	Inefficacy of ENR to treat acute clinical E. coli mastitis	[118]
Sahiwal cattle	Anaplasma marginale	N	5 mg/kg IV SID for 5 days	Comparative efficacy of ENR, OXY and IMI	60 (15 per group + 15 controls)	Inefficacy of ENR to clear persistent infection	[119]
Holstein calves	Anaplasma marginale	E	7.5 mg/kg in a single dose or twice every 3 days	Comparative efficacy of ENR (2 regimens) and LA-OXY	24 (6 per group + 6 controls)	Superiority of ENR	[120]
Calves	Pneumonic pasteurellosis (Pasteurella haemolytica A1)	E	2.5 mg/kg SC SID for 3 days	Efficacy of ENR	36 (12 ENR + 12 positive controls + 12 negative controls)	Inefficacy of ENR to treat experimentally induced pneumonic pasteurellosis	[121]
Feeder calves	Bovine respiratory disease (Mh, Pam, Hs, Mb)	N	12.5 mg/kg SC in a single injection	Comparative efficacy of ENR and TUL in two different states of USA	500 (125 per group per site)	Superiority of TUL	[122]
POULTRY							
Chickens	*Escherichia coli* (Ec) and Pasteurella multocida (Pam)	E	10 mg/kg BID in drinking water for 5 days	Efficacy of ENR (given immediately or 6 hours after infection)	Ec: 30 (10 per group + 10 controls) Pam: 20 (7 per group + 6 controls)	Efficacy of ENR against Ec and Pam	[123]
Broilers	Mycoplasma gallisepticum (Mg)	N	5 mg/kg in drinking water at 1 - 10 and 22 - 32 days of age	Efficacy of ENR in a treatment program against Mg in offspring of Mg-infected chicken-broiler breeders	45,000 (22,500 ENR + 22,500 controls)	Efficacy of ENR	[124]
Chicks (White Leghorn)	Salmonella enterica Serovar Typhimurium DT104	E	10 mg/kg for 5 days (1) or 25 mg/kg for 2 days (2) or 50 mg/kg for 1 day (3) (continuously or pulsed in water or gavage)	Efficacy of ENR (comparative efficacy of high dose short duration treatments and conventional treatment)	481 (151 in (1) + 141 in (2) + 89 in (3) + 100 controls)	Efficacy and best compromise for the 2-day 2.5 dosing treatment	[125]
Broilers (male chicks)	Colibacillosis (*Escherichia coli*)	E	50 ppm in drinking water for 7 days	Comparative efficacy of ENR and bacteriophage (individually and in combination)	320 (40 for each one of the 8 treatments)	Superiority of ENR but synergy between ENR and bacteriophage	[126]
Turkeys (poults)	Respiratory infections due to Ornithobacterium rhinotracheale (associated with avian metapneumovirus)	E	10 mg/kg in drinking water during 20 hours for 5 days or 50 mg/kg during 5, 10 or 20 hours in a single day	Efficacy of ENR (comparative efficacy of 4 regimens)	80 (16 per group + 16 controls)	Superiority of the 10 mg/kg 5-day ENR treatment	[127]

DOMESTIC CARNIVORES							
Dogs	Ehrlichiosis (*Ehrlichia canis*)	E	5 or 10 mg/kg PO BID for 21 days	Comparative efficacy of ENR (2 regimens) and DOX	13	Inefficacy of ENR to clear Ehrlichia canis infection	[128]
Dogs	Recurrent superficial (S) and deep (D) pyoderma	N	5 mg/kg PO SID continued up to 1 (S) or 2 (D) weeks after clinical recovery	Efficacy of ENR	9 (S) + 3 (D)	Efficacy, safety and convenience of ENR	[129]
Dogs	Uncomplicated urinary tract infections	N	18 - 20 mg/kg PO SID for 3 days	Comparative efficacy of ENR (high dose short duration treatment) and AMO-CLA	68 (35 ENR + 33 AMO-CLA)	Non-inferiority of a high dose short duration ENR treatment	[82]
Dogs	Brucellosis (*Brucella canis*)	N	5 mg/kg PO BID for 30 days	Efficacy of ENR for the eradication of Brucella canis in a kennel	12	Incomplete efficacy of ENR but safety use during gestation	[130]
Cats	Chronic bartonellosis (*B. henselae* and *B. clarridgeiae*)	N (25 cats) + E (18 cats)	22.7 mg PO BID for 14 or 28 days	Comparative efficacy of ENR and DOX (each one with 2 treatment durations)	43 (23 ENR + 17 DOX + 3 controls)	Efficacy of high dose long duration ENR treatment (4 to 6 weeks)	[131]
Cats	Conjunctivitis (*Chlamydophila felis*)	N	5 mg/kg SC SID for 3 days then PO SID for 11 days	Comparative efficacy of ENR and DOX	25 (14 ENR + 11 DOX)	Equal improvement of ENR and DOX in clinical signs and infection status	[132]
RODENTS							
Mice	Systemic infections (*Escherichia coli*)	E	5 mg/kg SC or PO	Efficacy of ENR (2 different routes of administration)	NP	Efficacy of ENR and superiority of the injectable form (SC)	[133]
Calomys callosus	Toxoplasmosis (*Toxoplasma gondii*)	E	3 mg/kg SC SID for 3 days	Comparative efficacy of ENR and SUL	15 (5 ENR + 5 SUL + 5 controls)	Efficacy of ENR as a potential alternative drug	[134]
LAGOMORPHS							
Rabbits	Pasteurella multocida (Pam)	N (11 rabbits) + E (12 rabbits)	5 mg/kg SC BID for 10 days	Efficacy of ENR in the elimination of Pam from asymptomatically infected rabbits	11 (N) + 12 (E)	Inefficacy of ENR to eliminate Pasteurella multocida	[135]
CRUSTACEANS							
Artemia franciscana (nauplii)	Vibrio harveyi (strain PN9801)	E	ENR 4h before infection (A) or 24 h after infection (B)	Comparative efficacy of ENR (2 regimens)	600	Efficacy of ENR to stop the course of a bacterial infection in Artemia franciscana	[136]

of injection for injectable forms, particularly in pigs [3] . However, some more serious adverse effects of enrofloxacin could appear targetting the juvenile joints, the reproductive system, the ocular system and the central nervous system.

The best known adverse effects of enrofloxacin concern the joints of young animals and result in arthropathy, articular cartilage degeneration, tendonitis and other forms of tendon injury [80] [81] [83] [84] . In order to try to understand quinolone related-arthropathy, a study was carried out in 1994 on juvenile New Zealand White Rabbits indicating that quinolones stimulate the cellular respiratory burst of immature articular chondrocytes which results in the production of oxygen-derived compounds highly toxic for the cartilage [83] . However the mechanisms underlying fluoroquinolones-induced tendinopathy and cartilage degeneration remained incompletely understood. Thus, further studies were necessary such as the study carried out in 2008 on canine Achilles tendon cells and chondrocytes suggesting that enrofloxacin-induced tendinopathy and cartilage damage could be explained by the inhibition of cell proliferation, induction of apoptosis and DNA fragmentation [84] . In young chickens, a study led in 2009 to determine the chondrotoxic effects of enrofloxacin on avian articular cartilage indicates that only very high dosage of enrofloxacin can induce toxic effects in articular cartilage of growing chickens and that the intensity of chondrotoxicity is dose- and time-dependent. Thus it is suggested that quinolone-induced arthropathy is far less expressed in birds than in mammals [85] .

To assess the effects of enrofloxacin on adult joints, a study was carried out in 2000. The effects of long-term administration of an injectable enrofloxacin solution were evaluated by the monitoring of physical and musculoskeletal variables in adult horses. Adverse effects were only detected with high doses and consist of lameness, cellulitis, tendinitis, sheath effusion and even transient neurologic signs [86] .

The adverse effects of enrofloxacin on the reproductive system were mainly investigated in males in order to assess the impact of this antibiotic on the fertility parameters. A study carried out in 2008 in male chickens suggested that enrofloxacin at therapeutical dose does not affect the sperm motility, the weight of testes, wattles and combs and the testicular concentration of testosterone, ascorbic acid, total protein and cholesterol [87] . On the opposite, a study led in the same year to evaluate toxic effects of enrofloxacin on sperm quality in male mice indicates that a fixed 150 mg/kg dose of enrofloxacin could lead to structural damages in the testicular tissue resulting in disruption of spermatogenesis in the testes with deterioration of motility, content and morphology of sperms [81] . Although the toxicity of enrofloxacin on the female reproductive tract was less investigated, a recent study on the use of intrauterine enrofloxacin infusion in healthy mares reveals acute effects including endometrial ulceration, necrosis and haemorrhage and chronic effects including fibrosis and inflammation [88] .

The ocular toxicity of enrofloxacin has been suggested by the association of enrofloxacin, retinal degeneration and blindness in cats [89] [90] . A retrospective clinical study carried out to assess the possible connection between the administration of parenteral enrofloxacin and the onset of acute retinal degeneration in cats highlights the potential retinotoxicity of parenteral enrofloxacin which can result in acute and diffuse retinal degeneration, particularly with dosages exceeding 5 mg/kg once daily (which is the manufacturer's current dosage recommendation). Mydriasis and blindness were frequently observed but some cats may recover their sight [89] .

At slightly higher doses, central nervous system signs of lethargy, anorexia and hypersalivation were observed in dogs as shown in a recent study carried out to assess a high dose short duration enrofloxacin protocol in dogs with uncomplicated urinary tract infections [82] .

7. RESIDUES AND TOXICITY FOR CONSUMERS

In many animal species, the use of enrofloxacin lead to its de-ethylation to its primary metabolite ciprofloxacin and both enrofloxacin and ciprofloxacin would be found as drug residues in animal muscle and tissue [76] [91] .

The consumption of meat containing these residues represent a significant threat against human health because it may result in disruption of the colonization barrier, development of drug-resistant bacterial strains or even allergies [91] [92] . In this regard, a recent study provides information on in vitro testing to determine if concentrations of veterinary antimicrobial agent residues entering the human colon remain microbiologically active [92] . Thus, the authorities have defined Acceptable Daily Intakes (ADI) of antimicrobial veterinary residues by human. The European Medicines Agency (EMA) have set the overall ADI value for enrofloxacin at 6.2 μg/kg body weight [92] [93] corresponding to the microbiological ADI because it is lower than the toxicological ADI of 12 μg/kg body weight which was calculated by applying a safety factor of 100 to the No Observed Effect Level (NOEL) of 1.2 mg/kg body weight per day [94] .

In order to protect consumer's health, many countries have defined Maximum Residue Limits (MRL) of enrofloxacin and ciprofloxacin in animal-derived products [76] . In the European Union (EU), the MRLs of enrofloxacin and ciprofloxacin in muscle tissues and milk of all species are 100 μg/L [93] but as there is no MRL for enrofloxacin in eggs, enrofloxacin is forbidden in animals form which eggs are produced for human consumption [93] .

To measure the enrofloxacin and ciprofloxacin residues in animal-derived foods including milk, eggs, honey and even mane and tail hair in horses, many analytical methods have been developed. Among these methods, we can mention High Performance Liquid Chromatography (HPLC); HPLC-Diode Array Detection method combined with liquid chromatography-mass spectrometry; HPLC-Ultraviolet Diode Array Detection method; molecularly imprinted solid-phase extraction procedure; liquid chromatography using a metal chelate affinity column; finally, the method of ChemiLuminescence Enzyme ImmunoAssay [76] [77] [91] [95] -[97] .

8. ECOTOXICITY

The release of enrofloxacin and ciprofloxacin in the environment is mainly due to the direct discharge of aquaculture products and the excretion in urine and feces of livestock animals [98] . It results in the contamination of soil, surface water, sediment, ground water and biota [99] . Once released into the environment the behavior of a chemical substance is determined by its tendency to partition from the aqueous phase to the atmosphere which is expressed by the Henry's Law Constant (HLC) and its affinity to adsorb on solid which is expressed by the Octanol-Water Partition Constant (Kow or log Kow) [99] . Enrofloxacin have a very low HLC at ambient temperature (<10 - 15 $atm{\cdot}m^3{\cdot}mol^{-1}$) resulting in a negligible volatile loss and its Kow is low (0.83) while enrofloxacin have a high affinity for sludge, soils and sediments [99] .

In the environment, enrofloxacin and ciprofloxacin can undergo degradations by different processes including photolysis, biodegradation and oxidation by mineral oxides but they are not sensitive to hydrolysis [99] . Despite these degradation mechanisms, environmental half-life time of enrofloxacin and ciprofloxacin are very long (half-life time estimated between 1155 and 3466 days for ciprofloxacin in a mesocosm soil study performed by Walters et al. in 2010, indicating an important persistence in soil matrices [100] .

This long environmental persistence of enrofloxacin and ciprofloxacin can affect the growing or the activity of the soil microbial communities [101] . Indeed, although some edaphic organisms may utilize and decompose enrofloxacin as a nutrient, it is suggested that enrofloxacin or ciprofloxacin concentrations exceeding 0.2 mg/kg result in significant toxicity for the edaphon [101] [102] . Thus, edaphic ammonification and nitrification are affected and the edaphon community structure is modified, which can impact the soil fertility [102] . Furthermore, a quite recent study carried out in 2008 indicates that the exposure of whole earthworms and their different tissues to various concentrations of enrofloxacin can lead to changes in catalase activity and to a lesser extent, in growth rate [103] .

The toxicity of enrofloxacin and ciprofloxacin for aquatic ecosystems has been assessed in many studies, as well on microorganisms as on algae or even aquatic vertebrates and invertebrates [104] [105] . Among these studies, three were carried out recently in zebrafish embryos [106] , in tropical freshwater ecosystems with the monitoring of macro-invertebrates, zooplankton, phytoplankton, periphyton, bacteria, organic matter decomposition and nitrogen cycling [105] and in three tropical aquatic species (the green-algae Chlorella sp., the micro-inver- tebrate Moina macrocopa and the Nile tilapia Oreochromis niloticus) collected in a stream receiving effluents from a Pangasius catfish farm which uses enrofloxacin [104] . The results of these three studies are similar, suggesting that residual concentrations of enrofloxacin and ciprofloxacin in aquatic environment are not likely to result in direct or indirect severe toxic effects on aquatic ecosystems [104] -[106] .

9. CONCLUSION

The fluoroquinolones are one of the most useful classes of antibiotics, as well in human medicine as in veterinary medicine. Thanks to their broad spectrum of activity against a wide range of bacteria and their physicochemical properties, their use is increasing. Enrofloxacin, which is a third generation fluoroquinolone only available in veterinary medicine, is thus used in many species with few adverse effects. However, there are recent concerns about the emergence of quinolone-resistant bacterial strains and the impact on the environment of the overuse of these drugs. Thus, there is now an important need to use fluoroquinolones with caution to preserve their effectiveness for many years. In veterinary medicine, it is essential to reserve these drugs for cases requiring a powerful antibiotic and to prescribe and/or administer them only under a good clinical assessment and with appropriate regimens.

ACKNOWLEDGEMENTS

We thank Prs. Marc Le Borgne, Plamen Kirilov and Caroline Prouillac for helpful suggestions and comments and Virginie Durand for its proofreading.

CITE THIS PAPER

Tessa Trouchon, Sébastien Lefebvre (2016) A Review of Enrofloxacin for Veterinary Use. *Open Journal of Veterinary Medicine*,**06**,40-58.

REFERENCES

1. Roque, A. and Gomez-Gil, B. (2003) Therapeutic Effects of Enrofloxacin in an Experimental Infection with a Luminescent Vibrio harveyi in Artemia franciscana Kellog 1906. Aquaculture, 220, 37-42.
2. Mahler, M., Stunkel, S., Ziegowski, C. and Kunstyr, I. (1995) Inefficacy of Enrofloxacin in the Elimination of Pasteurella multocida in Rabbits. Laboratory Animals, 29, 192-199.
3. Barbosa, B.F., Gomes, A.O., Ferro, E.A.V., Napolitano, D.R., Mineo, J.R. and Silva, N.M. (2012) Enrofloxacin Is Able to Control Toxoplasma gondii Infection in Both in Vitro and in Vivo Experimental Models. Veterinary Parasitology, 187, 44-52.
4. Slate, A.R., Bandyopadhyay, S., Francis, K.P., Papich, M.G., Karolewski, B., Hod, E.A., et al. (2014) Efficacy of Enrofloxacin in a Mouse Model of Sepsis. Journal of the American Association for Laboratory Animal Science, 53, 381-386.
5. Gerhardt, N., Schulz, B.S., Werckenthin, C. and Hartmann, K. (2006) Pharmacokinetics of Enrofloxacin and Its Efficacy in Comparison with Doxycycline in the Treatment of Chlamydophila felis Infection in Cats with Conjunctivitis. Veterinary Record, 159, 591-594.
6. Kordick, D.L., Papich, M.G. and Breitschwerdt, E.B. (1997) Efficacy of Enrofloxacin or Doxycycline for Treatment of Bartonella henselae or Bartonella clarridgeiae Infection in Cats. Antimicrobial Agents and Chemotherapy, 41, 2448-2455.
7. Wanke, M.M., Delpino, M.V. and Baldi, P.C. (2006) Use of Enrofloxacin in the Treatment of Canine Brucellosis in a Dog Kennel (Clinical Trial). Theriogenology, 66, 1573-1578.
8. Reddy, S., Kumari, K.N., Rao, V.V., Rayulu, V.C. and Sivajothi, S. (2014) Efficacy of Enrofloxacin in the Treatment of Recurrent Pyoderma in Dogs. Journal of Advanced Veterinary and Animal Research, 4, 108-112.
9. Neer, T.M., Eddlestone, S.M., Gaunt, S.D. and Corstvet, R.E. (1999) Efficacy of Enrofloxacin for the Treatment of Experimentally Induced Ehrlichia Canis Infection. Journal of Veterinary Internal Medicine, 13, 501-504.
10. Garmyn, A., Martel, A., Froyman, R., Nauwynck, H., Duchateau, L., Haesebrouck, F., et al. (2009) Efficacy of Four Enrofloxacin Treatment Regimens against Experimental Infection in Turkey Poults with Avian Pneumovirus and Ornithobacterium rhinotracheale. Avian Pathology, 38, 287-292.
11. Huff, W.E., Huff, G.R., Rath, N.C., Balog, J.M. and Donoghue, A.M. (2004) Therapeutic Efficacy of Bacteriophage and Baytril (Enrofloxacin) Individually and in Combination to Treat Colibacillosis in Broilers. Poultry Science, 83, 1944-1947.
12. Randall, L.P., Cooles, S.W., Coldham, N.C., Stapleton, K.S., Piddock, L.J.V. and Woodward, M.J. (2006) Modification of Enrofloxacin Treatment Regimens for Poultry Experimentally Infected with Salmonella enterica Serovar Typhimurium DT104 To Minimize Selection of Resistance. Antimicrobial Agents and Chemotherapy, 50, 4030-4037.
13. Barbour, E.K., Hamadeh, S., Talhouk, R., Sakr, W. and Darwish, R. (1998) Evaluation of an Enroflox-acin-Treatment Program against Mycoplasma gallisepticum Infection in Broilers. Preventive Veterinary Medicine, 35, 91-99.

14. Rawiwet, V., Chansiripornchai, P. and Chansiripornchai, N. (2010) Comparison of the Efficacy of Enrofloxacin against Escherichia coli or Pasteurella multocida Infection in Chickens. The Thai Journal of Veterinary Medicine, 40, 297-301.
15. Robb, E.J., Tucker, C.M., Corley, L., Bryson, W.L., Rogers, K.C., Sturgess, K., et al. (2007) Efficacy of Tulathromycin versus Enrofloxacin for Initial Treatment of Naturally Occurring Bovine Respiratory Disease in Feeder Calves. Veterinary Therapeutics, 8, 127-135.
16. Thompson, P.N., Van Amstel, S.R. and Henton, M. (1998) The Clinical Efficacy of Enrofloxacin in the Treatment of Experimental Bovine Pneumonic Pasteurellosis. Onderstepoort Journal of Veterinary Research, 65, 105-112.
17. Facury-Filho, E.J., de Carvalho, A.ú., Ferreira, P.M., Moura, M.F., Apolinário, B.C., Santos, L. de P.H., et al. (2012) Effectiveness of Enrofloxacin for the Treatment of Experimentally-Induced Bovine Anaplasmosis. Revista Brasileira de Parasitologia Veterinária, 21, 32-36.
18. Atif, F.A., Khan, M.S., Khan, M.A., Ashraf, M. and Avais, M. (2012) Chemotherapeutic Efficacy of Oxytetracycline, Enrofloxacin and Imidocarb for the Elimination of Persistent Anaplasma marginale Infection in Naturally Infected Sahiwal Cattle. Pakistan Journal of Zoology, 44, 449-456.
19. Suojala, L., Simojoki, H., Mustonen, K., Kaartinen, L. and Py&oumlr&aumllä, S. (2010) Efficacy of Enrofloxacin in the Treatment of Naturally Occurring Acute Clinical Escherichia coli Mastitis. Journal of Dairy Science, 93, 1960-1969.
20. Tang, J., Yang, X., Zheng, Z., Yu, W., Hu, K. and Yu, H. (2006) Pharmacokinetics and the Active Metabolite of Enrofloxacin in Chinese Mitten-Handed Crab (Eriocheir sinensis). Aquaculture, 260, 69-76.
21. Garcia Ovando, H., Gorla, N., Poloni, G., Trotti, N., Prieto, G. and Errecalde, C. (2000) Intravenous Pharmacokinetics of Ciprofloxacin in Goats. International Journal of Antimicrobial Agents, 15, 77-79.
22. Messenger, K.M., Papich, M.G. and Blikslager, A.T. (2012) Distribution of Enrofloxacin and Its Active Metabolite, Using an in Vivo Ultrafiltration Sampling Technique after the Injection of Enrofloxacin to Pigs. Journal of Veterinary Pharmacology and Therapeutics, 35, 452-459.
23. Bugyei, K., Black, W.D. and McEwen, S. (1999) Pharmacokinetics of Enrofloxacin Given by the Oral, Intravenous and Intramuscular Routes in Broiler Chickens. Canadian Journal of Veterinary Research, 63, 193-200.
24. Bidgood, T.L. and Papich, M.G. (2005) Plasma and Interstitial Fluid Pharmacokinetics of Enrofloxacin, Its Metabolite Ciprofloxacin, and Marbofloxacin after Oral Administration and a Constant Rate Intravenous Infusion in Dogs. Journal of Veterinary Pharmacology and Therapeutics, 28, 329-341.
25. Anadón, A., Martínez-Larra&ntildeaga, M.R., Díaz, M.J., Fernández-Cruz, M.L., Martínez, M.A., Frejo, M.T., et al. (1999) Pharmacokinetic Variables and Tissue Residues of Enrofloxacin and Ciprofloxacin in Healthy Pigs. American Journal of Veterinary Research, 60, 1377-1382.
26. García Ovando, H., Gorla, N., Luders, C., Poloni, G., Errecalde, C., Prieto, G., et al. (1999) Comparative Pharmacokinetics of Enrofloxacin and Ciprofloxacin in Chickens. Journal of Veterinary Pharmacology and Therapeutics, 22, 209-212.
27. Blondeau, J.M., Borsos, S., Blondeau, L.D., Blondeau, B.J.J. and Hesje, C.E. (2012) Comparative Minimum Inhibitory and Mutant Prevention Drug Concentrations of Enrofloxacin, Ceftiofur, Florfenicol, Tilmicosin and Tulathromycin against Bovine Clinical Isolates of Mannheimia haemolytica. Veterinary Microbiology, 160, 85-90.
28. Berghaus, L.J., Giguère, S. and Guldbech, K. (2013) Mutant Prevention Concentration and Mutant Selection Window for 10 Antimi-crobial Agents against Rhodococcus equi. Veterinary Microbiology, 166, 670-675.
29. Gebru Awji, E., Tassew, D.D., Lee, J.-S., Lee, S.-J., Choi, M.-J., Reza, M.A., et al. (2012) Comparative Mutant Prevention Concentration and Mechanism of Resistance to Veterinary Fluoroquinolones in Staphylococcus Pseudintermedius. Ve-terinary Dermatology, 23, 376-e69.

30. Beri, S., Sidhu, P.K., Kaur, G., Chandra, M. and Rampal, S. (2014) Comparative Mutant Prevention Concentration and Antibacterial Activity of Fluoroquinolones against Escherichia coli in Diarrheic Buffalo Calves. Journal of Chemotherapy, 27, 312-316.
31. Carlsson, G., Patring, J., Kreuger, J., Norrgren, L. and Oskarsson, A. (2013) Toxicity of 15 Veterinary Pharmaceuticals in Zebrafish (Danio rerio) Embryos. Aquatic Toxicology, 126, 30-41.
32. Rico, A., Dimitrov, M.R., Van Wijngaarden, R.P.A., Satapornvanit, K., Smidt, H. and Van den Brink, P.J. (2014) Effects of the Antibiotic Enrofloxacin on the Ecology of Tropical Eutrophic Freshwater Microcosms. Aquatic Toxicology, 147, 92-104.
33. Andrieu, M., Rico, A., Phu, T.M., Huong, D.T.T., Phuong, N.T. and Van den Brink, P.J. (2015) Ecological Risk Assessment of the Antibiotic Enrofloxacin Applied to Pangasius Catfish Farms in the Mekong Delta, Vietnam. Chemosphere, 119, 407-414.
34. Gao, Y., Sun, X., Sun, Z., Zhao, N. and Li, Y. (2008) Toxic Effects of Enrofloxacin on Growth Rate and Catalase Activity in Eisenia fetida. Environmental Toxicology and Pharmacology, 26, 177-180.
35. Zhou, X., Chen, C., Yue, L., Sun, Y., Ding, H. and Liu, Y. (2008) Excretion of Enrofloxacin in Pigs and Its Effect on Ecological Environment. Environmental Toxicology and Pharmacology, 26, 272-277.
36. Girardi, C., Greve, J., Lamsh&oumlft, M., Fetzer, I., Miltner, A., Sch&aumlffer, A., et al. (2011) Biodegradation of Ciprofloxacin in Water and Soil and Its Effects on the Microbial Communities. Journal of Hazardous Materials, 198, 22-30.
37. Walters, E., McClellan, K. and Halden, R.U. (2010) Occurrence and Loss Over Three Years of 72 Pharmaceuticals and Personal Care Products from Biosolids-Soil Mixtures in Outdoor Mesocosms. Water Research, 44, 6011-6020.
38. Van Doorslaer, X., Dewulf, J., Van Langenhove, H. and Demeestere, K. (2014) Fluoroquinolone Antibiotics: An Emerging Class of Environmental Micropollutants. Science of the Total Environment, 500-501, 250-269.
39. Boxall, A., Kolpin, D., Holling-Sorensen, B. and Tolls, J. (2003) Are Veterinary Medicines Causing Environmental Risks? Environmental Science & Technology, 15, 286A-294A.
40. Yatsukawa, Y., Ito, H., Matsuda, T., Nakamura, M., Watai, M. and Fujita, K. (2011) Determination of Residual Fluoroquinolones in Honey by Liquid Chromatography Using Metal Chelate Affinity Chromatography. Journal of AOAC International, 94, 1319-1327.
41. Gorla, N., Chiostri, E., Ugnia, L., Weyers, A., Giacomelli, N., Davicino, R., et al. (1997) HPLC Residues of Enrofloxacin and Ciprofloxacin in Eggs of Laying Hens. International Journal of Antimicrobial Agents, 8, 253-256.
42. Dunnett, M., Richardson, D.W. and Lees, P. (2004) Detection of Enrofloxacin and Its Metabolite Ciprofloxacin in Equine Hair. Research in Veterinary Science, 77, 143-151.
43. Committee for Medicinal Products for Veterinary Use (CVMP) (1998) Enrofloxacin, Summary Report (2).
44. Committee for Medicinal Products for Veterinary Use (CVMP) (2002) Enrofloxacin, Summary Report (5).
45. Ahn, Y., Linder, S.W., Veach, B.T., Steve Yan, S., Haydée Fernández, A., Pineiro, S.A., et al. (2012) In Vitro Enrofloxacin Binding in Human Fecal Slurries. Regulatory Toxicology and Pharmacology, 62, 74-84.
46. Yan, H., Tian, M. and Row, K.H. (2008) Determination of Enrofloxacin and Ciprofloxacin in Milk Using Molecularly Imprinted Solid-Phase Extraction. Journal of Separation Science, 31, 3015-3020.
47. Giuliano, E.A. and van der Woerdt, A. (1999) Feline Retinal Degeneration: Clinical Experience and New Findings (1994-1997). Jaaha-American Animal Hospital Association, 35, 511-514.

48. Gelatt, K.N., Van Der Woerdt, A., Ketring, K.L., Andrew, S.E., Brooks, D.E., Biros, D.J., et al. (2001) Enrofloxacin-Associated Retinal Degeneration in Cats. Veterinary Ophthalmology, 4, 99-106.
49. Rodriguez, J.S., Han, S., Nielsen, S., Pearson, L.K., Gay, J.M. and Tibary, A. (2012) Consequences of Intrauterine Enrofloxacin Infusion on Mare Endometrium. Journal of Equine Veterinary Science, 32, 106-111.
50. Al-Nazawi, M.H. (2008) Effects of Enrofloxacin and Marbofloxacin Administration on Some Fertility Parameters of Male Chicken. Journal of Taibah University for Science, 1, 1-4.
51. Bertone, A.L., Tremaine, W.H., Macoris, D.G., Simmons, E.J., Ewert, K.M., Herr, L.G., et al. (2000) Effect of Long-Term Administration of an Injectable Enrofloxacin Solution on Physical and Musculoskeletal Variables in Adult Horses. Journal of the American Veterinary Medical Association, 217, 1514-1521.
52. Maslanka, T., Jaroszewski, J.J., Mikolajczyk, A. and Rotkiewicz, T. (2009) Effect of Increasing Doses of Enrofloxacin on Chicken Articular Cartilage. Polish Journal of Veterinary Sciences, 12, 21-33.
53. Lim, S., Hossain, M.A., Park, J., Choi, S.H. and Kim, G. (2008) The Effects of Enrofloxacin on Canine Tendon Cells and Chondrocytes Proliferation in Vitro. Veterinary Research Communications, 32, 243-253.
54. Hayem, G., Petit, P.X., Levacher, M., Gaudin, C., Kahn, M.F. and Pocidalo, J.J. (1994) Cytofluorometric Analysis of Chondrotoxicity of Fluoroquinolone Antimicrobial Agents. Antimicrobial Agents and Chemotherapy, 38, 243-247.
55. Westropp, J.L., Sykes, J.E., Irom, S., Daniels, J.B., Smith, A., Keil, D., et al. (2012) Evaluation of the Efficacy and Safety of High Dose Short Duration Enrofloxacin Treatment Regimen for Uncomplicated Urinary Tract Infections in Dogs. Journal of Veterinary Internal Medicine, 26, 506-512.
56. Aral, F., Kara&ccedilal, F. and Baba, F. (2008) The Effect of Enrofloxacin on Sperm Quality in Male Mice. Research in Veterinary Science, 84, 95-99.
57. Sarkozy, G. (2001) Quinolones: A Class of Antimicrobial Agents. Veterinární Medicína (Praha), 46, 257-274.
58. Sapkota, A., Sapkota, A.R., Kucharski, M., Burke, J., McKenzie, S., Walker, P., et al. (2008) Aquaculture Practices and Potential Human Health Risks: Current Knowledge and Future Priorities. Environment International, 34, 1215-1226.
59. González, C., Moreno, L., Fumuso, E., García, J., Rivulgo, M., Confalonieri, A., et al. (2009) Enrofloxacin-Based Therapeutic Strategy for the Prevention of Endometritis in Susceptible Mares. Journal of Veterinary Pharmacology and Therapeutics, 33, 287-294.
60. Cinquina, A.L., Roberti, P., Giannetti, L., Longo, F., Draisci, R., Fagiolo, A., et al. (2003) Determination of Enrofloxacin and Its Metabolite Ciprofloxacin in Goat Milk by High-Performance Liquid Chromatography with Diode-Array Detection: Optimization and Validation. Journal of Chromatography A, 987, 221-226.
61. Yu, F., Yu, S., Yu, L., Li, Y., Wu, Y., Zhang, H., et al. (2014) Determination of Residual Enrofloxacin in Food Samples by a Sensitive Method of Chemiluminescence Enzyme Immunoassay. Food Chemistry, 149, 71-75.
62. Idowu, O.R., Peggins, J.O., Cullison, R. and von Bredow, J. (2010) Comparative Pharmacokinetics of Enrofloxacin and Ciprofloxacin in Lactating Dairy Cows and Beef Steers Following Intravenous Administration of Enrofloxacin. Research in Veterinary Science, 89, 230-235.
63. Wiuff, C., Lykkesfeldt, J., Aarestrup, F.M. and Svendsen, O. (2002) Distribution of Enrofloxacin in Intestinal Tissue and Contents of Healthy Pigs after Oral and Intramuscular Administrations. Journal of Veterinary Pharmacology and Therapeutics, 25, 335-342.
64. Martinez, M., McDermott, P. and Walker, R. (2006) Pharmacology of the Fluoroquinolones: A Perspective for the Use in Domestic Animals. The Veterinary Journal, 172, 10-28.

65. Neuman, M. (1986) Comparative Pharmacokinetic Parameters of New Systemic Fluoroquinolones. International Journal of Clinical Pharmacology Research, 7, 173-179.
66. Rao, G.S., Ramesh, S., Ahmad, A.H., Tripathi, H.C., Sharma, L.D. and Malik, J.K. (2002) Pharmacokinetics of Enrofloxacin and Its Metabolite Ciprofloxacin in Goats Given Enrofloxacin Alone and in Combination with Probenecid. The Veterinary Journal, 163, 85-93.
67. Hwang, Y.H., Kim, M.S., Song, I.B., Lim, J.H., Park, B.K. and Yun, H.I. (2009) Altered Pharmacokinetics of Enrofloxacin in Experimental Models of Hepatic and Renal Impairment. Veterinary Research Communications, 33, 481-487.
68. Nouws, J.F.M., Mevius, D.J., Vree, T.B., Baars, A.M. amd Laurensen, J. (1988) Pharmacokinetics, Renal Clearance and Metabolism of Ciprofloxacin Following Intravenous and Oral Administration to Calves and Pigs. Veterinary Quarterly, 10, 156-163.
69. Ogino, T., Mizuno, Y., Ogata, T. and Takahashi, Y. (2005) Pharmacokinetic Interactions of Flunixin Meglumine and Enrofloxacin in Dogs. American Journal of Veterinary Research, 66, 1209-1213.
70. Pasquali, F. and Manfreda, G. (2007) Mutant Prevention Concentration of Ciprofloxacin and Enrofloxacin against Escherichia coli, Salmonella Typhimurium and Pseudomonas Aeruginosa. Veterinary Microbiology, 119, 304-310.
71. Küng, K., Riond, J.-L. and Wanner, M. (1993) Pharmacokinetics of Enrofloxacin and Its Metabolite Ciprofloxacin after Intravenous and Oral Administration of Enrofloxacin in Dogs. Journal of Veterinary Pharmacology and Therapeutics, 16, 462-468.
72. Cester, C.C. and Toutain, P.L. (1997) A Comprehensive Model for Enrofloxacin to Ciprofloxacin Transformation and Disposition in Dog. Journal of Pharmaceutical Sciences, 86, 1148-1155.
73. Riviere, J.E., and Papich, M.G. (2013) Veterinary Pharmacology and Therapeutics. John Wiley & Sons, Hoboken, 993.
74. Tyczkowska, K., Hedeen, K.M., Aucoin, D.P. and Aronson, A.L. (1989) High-Performance Liquid Chromatographic Method for the Simultaneous Determination of Enrofloxacin and Its Primary Metabolite Ciprofloxacin in Canine Serum and Prostatic Tissue. Journal of Chromatography B: Biomedical Sciences and Applications, 493, 337-346.
75. Dautrey, S., Felice, K., Petiet, A., Lacour, B., Carbon, C. and Farinotti, R. (1999) Active Intestinal Elimination of Ciprofloxacin in Rats: Modulation by Different Substrates, British Journal of Pharmacology, 127, 1728-1734.
76. Coulet, M., Morello, C., Cox, P. and Lohuis, J. (2005) Pharmacoki-netics of Ibafloxacin in Healthy Cats. Journal of Veterinary Pharmacology and Therapeutics, 28, 37-44.
77. Sumano, L.H., Gutierrez, O.L., Aguilera, R., Rosiles, M.R., Bernard, B.M.J., Gracia, M.J. (2004) Influence of Hard Water on the Bioavailability of Enrofloxacin in Broilers. Poultry Science, 83, 726-731.
78. Simon, Z., Katja, B., Darko, U., Marjan, V. and Albin, K. (2010) Metal Cation–Fluoroquinolone Complexes Do Not Permeate through the Intestinal Absorption Barrier. Journal of Pharmaceutical and Biomedical Analysis, 53, 655-659.
79. Ziółkowski, H., Jaroszewski, J.J., Maslanka, T., Grabowski, T., Katolik, K., Paweska, J., et al. (2014) Influence of Oral Co-Administration of a Preparation Containing Calcium and Magnesium and Food on Enrofloxacin Pharmacokinetics. Research in Veterinary Science, 97, 99-104.
80. Steinman, A., Britzi, M., Levi, O., Lavy, E., Lichter, A. and Soback, S. (2006) Lack of Effect of Diet on the Pharmacokinetics of Enrofloxacin in Horses. Journal of Veterinary Pharmacology and Therapeutics, 29, 67-70.
81. Vancutsem, P., Babish, J. and Schwark, W. (1990) The Fluoroquinolone Antimicrobials: Structure, Antimicrobial Activity, Pharmacokinetics, Clinical Use in Domestic Animals and Toxicity. The Cornell Veterinarian, 80, 173-186.

82. Nielsen, P. and Gyrd-Hansen, N. (1997) Bioavailability of Enrofloxacin after Oral Administration to Fed and Fasted Pigs. Pharmacology & Pharmaceutical Medicine, 80, 246-250.
83. Blondeau, J.M., Zhao, X., Hansen, G. and Drlica, K. (2001) Mutant Prevention Concentrations of Fluoroquinolones for Clinical Isolates of Streptococcus Pneumoniae. An-timicrobial Agents and Chemotherapy, 45, 433-438.
84. Martínez-Martínez, L., Pascual, A. and Jacoby, G.A. (1998) Quinolone Resistance from a Transferable Plasmid. The Lancet, 351, 797-799.
85. Tavío, M. del M., Vila, J., Ruiz, J., Ruiz, J., Martín-Sánchez, A.M. and de Anta, M.T.J. (1999) Mechanisms Involved in the Development of Resistance to Fluoroquinolones in Escherichia coli Isolates. Journal of Antimicrobial Chemotherapy, 44, 735-742.
86. Cohen, S.P., McMurry, L.M. and Levy, S.B. (1988) MarA Locus Causes Decreased Expression of OmpF Porin in Multiple-Antibiotic-Resistant (Mar) Mutants of Escherichia coli. Journal of Bacteriology, 170, 5416-5422.
87. Goldman, J.D., White, D.G. and Levy, S.B. (1996) Multiple Antibiotic Resistance (Mar) Locus Protects Escherichia coli from Rapid Cell Killing by Fluoroquinolones. Antimicrobial Agents and Chemotherapy, 40, 1266-1269.
88. George, A.M. and Levy, S.B. (1983) Amplifiable Resistance to Tetracycline, Chloramphenicol, and Other Antibiotics in Escherichia coli: Involvement of a Non-Plasmid-Determined Efflux of Tetracycline. Journal of Bacteriology, 155, 531-540.
89. Dalhoff, A., Matutat, S. and Ullmann, U. (1995) Effect of Quinolones against Slowly Growing Bacteria. Chemotherapy, 41, 92-99.
90. Vetting, M.W., Hegde, S.S., Wang, M., Jacoby, G.A., Hooper, D.C. and Blanchard, J.S. (2011) Structure of QnrB1, a Plasmid-Mediated Fluoroquinolone Resistance Factor. The Journal of Biological Chemistry, 286, 25265-25273.
91. Tran, J.H., Jacoby, G.A. and Hooper, D.C. (2005) Interaction of the Plasmid-Encoded Quinolone Resistance Protein Qnr with Escherichia coli DNA Gyrase. Antimicrobial Agents and Chemotherapy, 49, 118-125.
92. Hopkins, K.L., Davies, R.H. and Threlfall, E.J. (2005) Mechanisms of Quinolone Resistance in Escherichia coli and Salmonella: Recent Developments. International Journal of Antimicrobial Agents, 25, 358-373.
93. Nakamura, S., Nakamura, M., Kojima, T. and Yoshida, H. (1989) gyrA and gyrB Mutations in Quinolone-Resistant Strains of Escherichia coli. Antimicrobial Agents and Chemotherapy, 33, 254-255.
94. Yoshida, H., Bogaki, M., Nakamura, M., Yamanaka, L.M. and Nakamura, S. (1991) Quinolone Resistance-Determining region in the DNA Gyrase gyrA Gene of Escherichia coli. Antimicrobial Agents and Chemotherapy, 35, 1647-1650.
95. Yoshida, H., Bogaki, M., Nakamura, M. and Nakamura, S. (1990) Quinolone Resistance-Determining Region in the DNA Gyrase gyrA Gene of Escherichia coli. Antimicrobial Agents and Chemotherapy, 34, 1271-1272.
96. Drlica, K. and Zhao, X. (1997) DNA Gyrase, Topoisomerase IV, and the 4-Quinolones. Microbiology and Molecular Biology Reviews, 61, 377-392.
97. Ruiz, J., Pons, M.J. and Gomes, C. (2012) Transferable Mechanisms of Quinolone Resistance. International Journal of Antimicrobial Agents, 40, 196-203.
98. Pereira, R.V., Siler, J.D., Ng, J.C., Davis, M.A., Grohn, Y.T. and Warnick, L.D. (2014) Effect of On-Farm Use of Antimicrobial Drugs on Resistance in Fecal Escherichia coli of Pre-weaned Dairy Calves. Journal of Dairy Science, 41, 92-99.
99. Cummings, K.J., Aprea, V.A. and Altier, C. (2014) Antimicrobial Resistance Trends among Escherichia coli Isolates Obtained from Dairy Cattle in the Northeastern United States, 2004-2011. Foodborne Pathogens and Disease, 11, 61-67.
100. Cohn, L.A., Gary, A.T., Fales, W.H. and Madsen, R.W. (2003) Trends in Fluoroquinolone Resistance of Bacteria Isolated from Canine Urinary Tracts. Journal of Veterinary Diagnostic Investigation, 15, 338-343.
101. Gocke, E. (1991) Mechanism of Quinolone Mutagenicity in Bacteria. Mutation Research/Fundamental and Molecular Mechanisms of Mutagenesis, 248, 135-143.

102. Drlica, K., Malik, M., Kerns, R.J., Zhao, X. (2008) Quinolone-Mediated Bacterial Death, Antimicrob. Antimicrobial Agents and Chemotherapy. 52, 385-392.
103. Maguin, E., Lutkenhaus, J. and D'Ari, R. (1986) Reversibility of SOS-Associated Division Inhibition in Escherichia coli. Journal of Bacteriology, 166, 733-738.
104. Malik, M., Zhao, X. and Drlica (2006) K., Lethal Fragmentation of Bacterial Chromosomes Mediated by DNA Gyrase and Quinolones. Molecular Microbiology, 61, 810-825.
105. Kelley W.L. (2006) Lex Marks the Spot: The Virulent Side of SOS and a Closer Look at the LexA Regulon. Molecular Microbiology, 62, 1228-1238.
106. Wentzell, L.M. and Maxwell, A. (2000) The Complex of DNA Gyrase and Quinolone Drugs on DNA Forms a Barrier to the T7 DNA Polymerase Replication Complex. Journal of Molecular Biology, 304, 779-791.
107. Lockshon, D. and Morris, D.R. (1985) Sites of Reaction of Escherichia coli DNA Gyrase on pBR322 in Vivo as Revealed by Oxolinic Acid-Induced Plasmid Linearization. Journal of Molecular Biology, 181, 63-74.
108. Leo, E., Gould, K.A., Pan, X.-S., Capranico, G., Sanderson, M.R., Palumbo, M., et al. (2005) Novel Symmetric and Asymmetric DNA Scission Determinants for Streptococcus Pneumoniae Topoisomerase IV and Gyrase Are Clustered at the DNA Breakage Site. The Journal of Biological Chemistry, 280, 14252-14263.
109. Kampranis, S.C. and Maxwell, A. (1998) Conformational Changes in DNA Gyrase Revealed by Limited Proteolysis. The Journal of Biological Chemistry, 273, 22606-22614.
110. Kampranis, S.C. and Maxwell, A. (1998) The DNA Gyrase-Quinolone Complex ATP Hydrolysis and the Mechanism of DNA Cleavage. The Journal of Biological Chemistry, 273, 22615-22626.
111. Lambert, P. A. (2002) Cellular Impermeability and Uptake of Biocides and Antibiotics in Gram-Positive Bacteria and Mycobacteria. Journal of Applied Microbiology, 92, 46S-54S.
112. Bedard, J., Wong, S. and Bryan, L.E. (1987) Accumulation of Enoxacin by Escherichia coli and Bacillus Subtilis. Antimicrobial Agents and Chemotherapy, 31, 1348-1354.
113. Bryan, L.E. and Bedard, J. (1991) Impermeability to Quinolones in Gram-Positive and Gram-Negative Bacteria. European Journal of Clinical Microbiology & Infectious Diseases, 10, 232-239.
114. Brauser, A., Schroeder, I., Gutsmann, T., Cosentino, C. and Moroni, A., Hansen, U.-P., et al. (2012) Modulation of Enrofloxacin Binding in OmpF by Mg2+ as Revealed by the Analysis of Fast Flickering Single-Porin Current. The Journal of General Physiology, 140, 69-82.
115. Yamano, Y., Nishikawa, T. and Komatsu, Y. (1999) Outer Membrane Proteins Responsible for the Penetration of β-Lactams and Quinolones in Pseudomonas Aeruginosa. Journal of Antimicrobial Chemotherapy, 26, 175-184.
116. Danilchanka, O., Pavlenok, M. and Niederweis, M. (2008) Role of Porins for Uptake of Antibiotics by Mycobacterium Smegmatis. Antimicrobial Agents and Chemotherapy, 52, 3127-3134.
117. Zechiedrich, E.L. and Cozzarelli, N.R. (1995) Roles of Topoisomerase IV and DNA Gyrase in DNA Unlinking during Replication in Escherichia coli. Genes & Development, 9, 2859-2869.
118. Peng, H. and Marians, K.J. (1993) Decatenation Activity of Topoisomerase IV during oriC and pBR322 DNA Replication in Vitro. Proceedings of the National Academy of Sciences of the United States of America, 90, 8571-8575.
119. Peng, H. and Marians, K.J. (1995) The Interaction of Escherichia coli Topoisomerase IV with DNA. The Journal of Biological Chemistry, 270, 25286-25290.
120. Gellert, M., Mizuuchi, K., O'Dea, M.H. and Nash, H.A. (1976) DNA Gyrase: An Enzyme that Introduces Superhelical Turns into DNA. Proceedings of the National Academy of Sciences of the United States of America, 73, 3872-3876.

121. Wang, J.C., (1998) Moving One DNA Double Helix through Another by a Type II DNA Topoisomerase: The Story of a Simple Molecular Machine. Quarterly Reviews of Biophysics, 31, 107-144.
122. Tse, Y.C., Kirkegaard, K. and Wang, J.C. (1980) Covalent Bonds between Protein and DNA. Formation of Phosphotyrosine Linkage between Certain DNA Topoisomerases and DNA. The Journal of Biological Chemistry, 255, 5560-5565.
123. Horowitz, D.S. and Wang, J.C. (1987) Mapping the Active Site Tyrosine of Escherichia coli DNA Gyrase. The Journal of Biological Chemistry, 262, 5339-5344.
124. Gubaev, A. and Klostermeier, D. (2014) The Mechanism of Negative DNA Supercoiling: A Cascade of DNA-Induced Conformational Changes Prepares Gyrase for Strand Passage. DNA Repair, 16, 23-34.
125. Chong, S., Chen, C., Ge, H. and Xie, X.S. (2014) Mechanism of Transcriptional Bursting in Bacteria. Cell, 158, 314-326.
126. Anselmi, C., Bocchinfuso, G., De Santis, P. and Fuà, M. (1998) A. Scipioni, M. Savino, Statistical Thermodynamic Approach for Evaluating the Writhe Transformations in Circular DNAs. The Journal of Physical Chemistry B, 102, 5704-5714.
127. Kato, J., Nishimura, Y., Imamura, R., Niki, H., Hiraga, S. and Suzuki, H. (1990) New Topoisomerase Essential for Chromosome Segregation in E. coli. Cell, 63, 393-404.
128. Levine, C., Hiasa, H. and Marians, K.J. (1998) DNA Gyrase and Topoisomerase IV: Biochemical Activities, Physiological Roles during Chromosome Replication, and Drug Sensitivities. Biochimica et Biophysica Acta (BBA)-Gene Structure and Expression, 1400, 29-43.
129. Takács-Novák, K., Józan, M., Hermecz, I. and Szász, G. (1992) Lipophilicity of Antibacterial Fluoroquinolones. International Journal of Pharmaceutics, 79, 89-96.
130. Lin, C.-E., Deng, Y.-J., Liao, W.-S., Sun, S.-W., Lin, W.-Y. and Chen, C.-C. (2004) Electrophoretic Behavior and pKa Determination of Quinolones with a Piperazinyl Substituent by Capillary Zone Electrophoresis. Journal of Chromatography A, 1051, 283-290.
131. Lizondo, M., Pons, M., Gallardo, M. and Estelrich, J. (1997) Physicochemical Properties of Enrofloxacin. Journal of Pharmaceutical and Biomedical Analysis, 15, 1845-1849.
132. Barbosa, J., Barrón, D., Jiménez-Lozano, E. and Sanz-Nebot, V. (2001) Comparison between Capillary Electrophoresis, Liquid Chromatography, Potentiometric and Spectrophotometric Techniques for Evaluation of pKa Values of Zwitterionic Drugs in Acetonitrile-Water Mixtures. Analytica Chimica Acta, 437, 309-321.
133. Committee for Medicinal Products for Veterinary Use (CVMP) (2007) Public Statement on the Use of (Fluoro) Quinolones in Food-Producing Animals in the European Union: Development of Resistance and Impact on Human and Animal Health.
134. Fauchier, N. (2013) Med'Vet, le recueil des spécialités à usage vétérinaire. 2014th Edition, Med'com, Paris.
135. Wright, D.H., Brown, G.H., Peterson, M.L. and Rotschafer, J.C. (2000) Application of Fluoroquinolone Pharmacodynamics. Journal of Antimicrobial Chemotherapy, 46, 669-683.
136. Grohe, K., Zeiler, H.-J., Metzger, K.G. and Grohe, K. (1987) 7-Amino-1-Cyclopropyl-4-Oxo-1, 4-Dihydro-Quinoline- and Naphthyridine-3-Carboxylic Acids and Antibacterial Agents Containing These Compounds. US4670444 (A). http://worldwide.espacenet.com/publicationDetails/biblio?FT=D&date=19870602&DB=EPODOC&locale=en_EP&CC=US&NR=4670444A&KC=A&ND=5

CHAPTER 8

Coronary Artery Anomalies in Animals

Brian A. Scansen

Department of Clinical Sciences, Colorado State University, Fort Collins, CO 80523, USA

ABSTRACT

Coronary artery anomalies represent a disease spectrum from incidental to life-threatening. Anomalies of coronary artery origin and course are well-recognized in human medicine, but have received limited attention in veterinary medicine. Coronary artery anomalies are best described in the dog, hamster, and cow though reports also exist in the horse and pig. The most well-known anomaly in veterinary medicine is anomalous coronary artery origin with a prepulmonary course in dogs, which limits treatment of pulmonary valve stenosis. A categorization scheme for coronary artery anomalies in animals is suggested, dividing these anomalies into those of major or minor clinical significance. A review of coronary artery development, anatomy, and reported anomalies in domesticated species is provided and four novel canine examples of anomalous coronary artery origin are described: an English bulldog with single left coronary ostium and a retroaortic right coronary artery; an English bulldog with single right coronary ostium and transseptal left coronary artery; an English bulldog with single right coronary ostium and absent left coronary artery with a prepulmonary paraconal interventricular branch and an interarterial circumflex branch; and a mixed-breed dog with tetralogy of Fallot and anomalous origin of all coronary branches from the brachiocephalic trunk. Coronary arterial fistulae are also described including a coronary cameral fistula in a llama cria and an English bulldog with coronary artery aneurysm and anomalous shunting vessels from the right coronary artery to the pulmonary trunk. These examples are provided with the intent to raise awareness and improve understanding of such defects.

Keywords: heart; congenital; single coronary; veterinary; dog; cat

1. INTRODUCTION

Coronary arteries perfuse the heart and facilitate nutrient and oxygen delivery to the metabolically active myocardium. Myocardial ischemia related to atherosclerosis and coronary artery obstruction is currently the leading cause of human death and is projected to remain so beyond the year 2030 [1]. Myocardial ischemia and infarction related to coronary artery disease also occur in dogs and cats [2,3], but are more often associated with other underlying diseases and are not believed to commonly impact the morbidity and mortality of domesticated species.

Congenital anomalies of coronary artery anatomy in man are important as a cause of non-arteriosclerotic ischemia and present an increased risk of sudden death in young athletes [4,5]. Coronary artery anomalies (CAA) in dogs are rarely reported to be of clinical significance, unless in the setting of pulmonary valve stenosis (PS) when the coronary anatomy limits intervention [6,7]. The aim of this manuscript is to review the coronary arterial circulation of animals, highlighting recognized anomalies and their clinical or comparative anatomical importance.

2. CORONARY ARTERY EMBRYOLOGY AND ANATOMY

Connection of the coronary arteries to their appropriate aortic valvar sinuses occurs late in development, after closure of the interventricular septum and coincident with development of the aortic root and semilunar valves [8]. Early in development, the coronary arteries develop as layered epithelial cells filled with erythrocytes (called sinusoids or blood islands), which proliferate and conjoin to form a network of vascular channels [9]. As venous connections develop, these primitive vascular channels are drained to the coronary sinus. The coronary arteries are eventually connected to the aortic valvar sinuses, allowing antegrade flow from aorta through the coronary vascular bed

[9]. The process by which the coronary arteries connect to the aorta is an area of uncertainty with recent reports based upon murine data describing endocardial strands growing from the aortic root as the outflow tracts are undergoing septation [8]. Previously, it was thought that the developing epicardial coronary arteries grew toward and then into the developing aortic root [9].

Ostia (openings), vessels (individual arteries), and their directional course around and within the heart are all described when considering the coronary artery anatomy. In the normal mammalian heart, the aortic valve sinuses are adjacent (facing) the right and left aspects of the pulmonary valve while the third sinus is non-adjacent. To maintain appropriate anatomical terminology in a congenitally-malformed heart where the position of the great vessels may be altered, the aortic sinuses are therefore named as if they were viewed from the non-adjacent aortic sinus and facing the pulmonary valve giving a right aortic sinus or cusp, a left aortic sinus or cusp, and a non-adjacent sinus or cusp [8]. In the vast majority of animal hearts, the two coronary ostia are located within the left and right aortic valvar sinuses adjacent to the pulmonary trunk (Figure 1). Commonly, a third ostium is present in the right valvar sinus representing a small conus branch (Figure 2C), known as the right accessory coronary artery [10], that arborizes across the right ventricular outflow tract. Two ostia may also be present in the left valvar sinus representing separate origins of the paraconal interventricular (Pc) and circumflex (Cx) arteries [11] or a separate ostium for the septal branch distinct from the left coronary artery (LCA) [12]. The coronary ostia are typically located in the middle of the valvar sinus and below the sinotubular junction (Figure 2). If they originate above the sinotubular junction, by at least 1 cm in a human, they are termed to have a high take-off.

The normal LCA arises from the left coronary ostium, comprises a short main segment, and then bifurcates into the Pc and Cx branches (Figure 1), or trifurcates into the Pc, Cx, and septal branches. In some dogs, there is no discrete LCA and the left coronary branches arise directly from the left coronary ostium (Figure 2A,B). The Pc branch descends obliquely on the left face of the heart and to the ventricular apex within the paraconal interventricular groove. The Cx branch runs within the coronary groove along the atrioventricular junction and encircles the mitral valve annulus. The septal branch differs in the dog as compared to the human

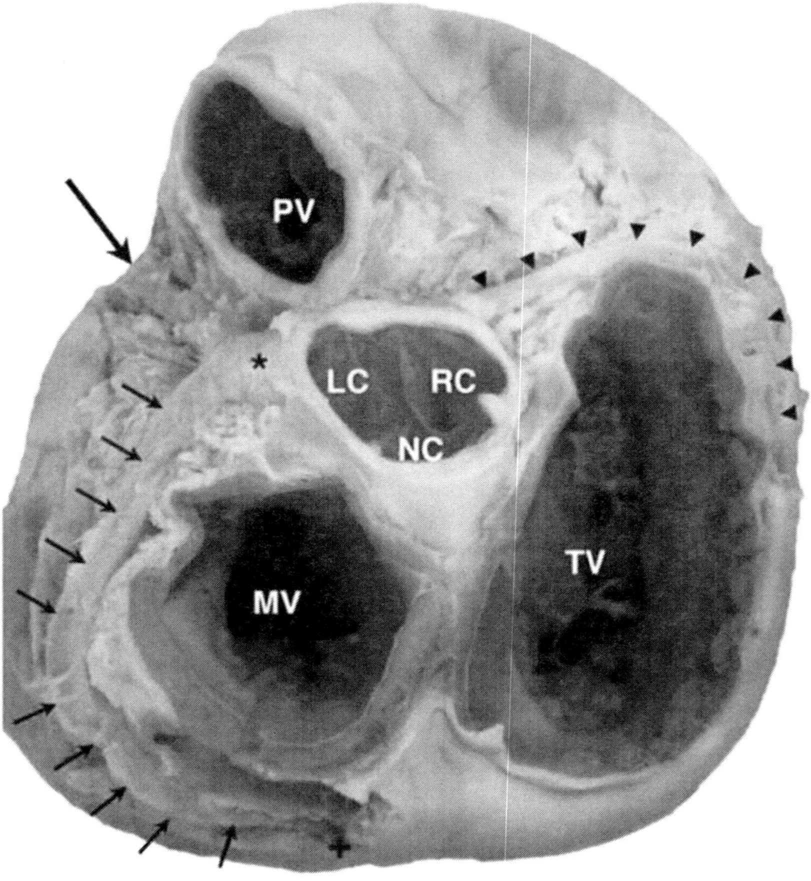

Figure 1. Canine specimen of the heart base, viewed from the dorsal perspective. The left coronary artery (*) can be seen arising from the left-adjacent cusp (LC) of the aortic valve and dividing into the paraconal interventricular branch (large arrow) and circumflex branch (multiple small arrows), which wraps around the mitral valve (MV) annulus to descend as the subsinuosal interventricular branch (+). The right coronary artery (arrowheads) can be seen arising from the right-adjacent cusp (RC) of the aortic valve and passing to the right and caudal of the tricuspid valve (TV) annulus. PV = pulmonary valve; NC = non-adjacent cusp of aortic valve.

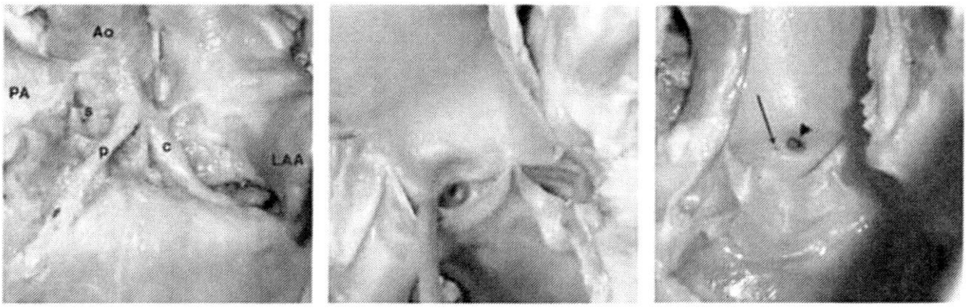

(a) (b) (c)

Figure 2. Variations of normal coronary ostia in the dog. (**a**) Left coronary trifurcation in a dog with a very short (nearly absent) left main coronary artery in whom the left coronary ostium gives rise to a septal branch (s), paraconal interventricular branch (p), and a circumflex branch (c). Ao = aorta; LAA = left auricular appendage; PA = pulmonary artery. (**b**) View of the left aortic valvar sinus from the dog in panel (a) showing the trifurcation of the left coronary system. (**c**) Photograph of the right aortic valvar sinus from a different dog showing a right coronary ostium (arrowhead) with an additional small ostium (arrow) consistent with the right accessory or conus branch.

or equine anatomy because it provides the majority of the blood supply to the interventricular septum and may in rare cases be larger than the Pc branch [12,13]. In the dog, the septal branch arises either off the LCA, directly from the left coronary ostium, or off the Pc branch and descends adjacent to the pulmonary valve annulus lying within the subendocardium along the right ventricular aspect of the septum. The subsinuosal interventricular branch arises variably from the LCA or right coronary artery (RCA), predominately related to species differences discussed below; it descends along the caudal border of the heart within the subsinuosal interventricular groove. The RCA arises from the right coronary ostium and courses to the right and caudally within the right atrioventricular groove to supply the right ventricular mass (Figure 1); as noted above, a small accessory branch often is seen extending cranially from the RCA or directly from the right aortic valvar sinus toward the right ventricular outflow tract. However, in the normal animal heart, the major CAs are not found in a position cranial to the right ventricular outflow tract or pulmonary valve annulus.

The coronary dominance of a heart refers to the coronary artery (LCA or RCA) that perfuses the majority of the myocardial tissue and variable methods to determine this have been reported including which coronary artery supplies the subsinuosal interventricular branch, which coronary artery extends beyond the crux of the heart, the relative lengths and number of branches of the LCA or RCA, and the origination and layout of the arteries at the apex of the heart [14]. The dog normally displays a left dominant coronary circulation [14], the cat is less studied but appears variable with a majority of cases indicating right coronary dominance [15], the rabbit is variable with predominately left coronary dominance [16], and the goat [17], swine [18], horse and donkey [19], cow, camelids [20], and 90% of humans [21] are right dominant. In the mouse [22] and rabbit [23], the subsinuosal interventricular branch is reported to be absent. Notably, a report of 10 chinchillas found that all dissected specimens lacked an RCA suggesting that a single LCA may be normal in that species [24]. In both humans and veterinary species, balanced or co-dominant coronary circulations are observed in which both the LCA and RCA supply the diaphragmatic aspect of the myocardial mass. Species variability in coronary dominance has importance as experimental studies of coronary occlusion in the dog result in different regions of myocardial ischemia and unique collateral pathways than would occur in species with a right dominant coronary circulation.

3. PREVALENCE AND CATEGORIZATION OF CORONARY ARTERY ANOMALIES

Compared to humans, reports of congenital CAA in animals are sparse, presumably because coronary artery disease is less common in veterinary species and therefore diagnostic coronary arteriograms are not routinely performed. The estimated prevalence of CAA in humans is difficult to quantify, due to a lack of agreement on what constitutes an anomaly versus a variant of normal [4]. The prevalence has been cited as high as 5.64% of humans, though many if not most of these defects were not clinically significant [4]. To clarify the uncertainties in prevalence, it has been suggested that only those variants that are present in less than 1% of the general population be considered as true anomalies of the coronary artery circulation [4]. Complications of CAA in humans relate to myocardial ischemia from ostial narrowing, extramural compression, or increased risk of atherosclerotic disease [4]. Coronary artery anomalies are seen with greater incidence in humans with congenital heart disease, particularly in conditions associated with malposition of the aortic root. Up to a third of patients with transposition of the great arteries and a quarter of patients with congenitally-corrected transposition of the great arteries have abnormal coronary artery anatomy; up to 14% of patients with tetralogy of Fallot have abnormal coronary artery branching with

5% of cases having a major coronary artery branch crossing anterior (cranial) to the pulmonary valve annulus; up to 18% of patients with common arterial trunk have a single coronary ostium; and a third or more of cases with double outlet right ventricle display abnormal coronary artery anatomy [25,26]. Studies of coronary artery anatomy from animals with congenital heart disease are not available, but the author's impression, based on evaluation of numerous post mortem specimens, is that a similar proportion of animals, both large and small animal species, with complex congenital heart disease such as double outlet right ventricles and transposition complexes have CAA. Figure 3 shows an example of a cow with double outlet right ventricle and a CAA; in this animal, a single right coronary ostium provided the coronary arterial circulation with the LCA taking a prepulmonary course.

Classification schemes to report CAA are inconsistent in human medicine and even more variable in animal reports. Table 1 attempts to categorize major anomalies that are likely to be associated with clinical signs or cardiac complications as compared to those that are more likely to have minor clinical impact. In people, the major CAAs cause profound myocardial ischemia or high risk of congestive heart failure. Some of the minor anomalies, however, have also been implicated in sudden death, particularly during exertion in young athletes [4,27]. Sudden death appears most often when an anomalous coronary artery origin results in ostial stenosis, an acute angle as it exits the aortic wall, an intramural course (within the aortic wall), or an interarterial course between the great vessels [4,27]. The risk of sudden death with a minor CAA in animals has not been established. Descriptive naming of specific anatomy is now preferred to alphanumeric coding schemes [28].

Veterinary Reports of Coronary Artery Anomalies

Dogs and cattle are the veterinary species most frequently reported with CAA, though reports also exist in the horse and pig. Reports of CAA in dogs include the single right coronary ostium with an anomalous prepulmonary LCA first reported in 1959 in a necropsy study of an asymptomatic mongrel dog [29] and later characterized in association with PS [6,7,30,31,32], single left coronary ostium with an anomalous prepulmonary RCA [33,34], single right coronary ostium with suspected interarterial LCA in an English bulldog [35], single right coronary ostium in a Keeshond with common arterial trunk [36], single right coronary

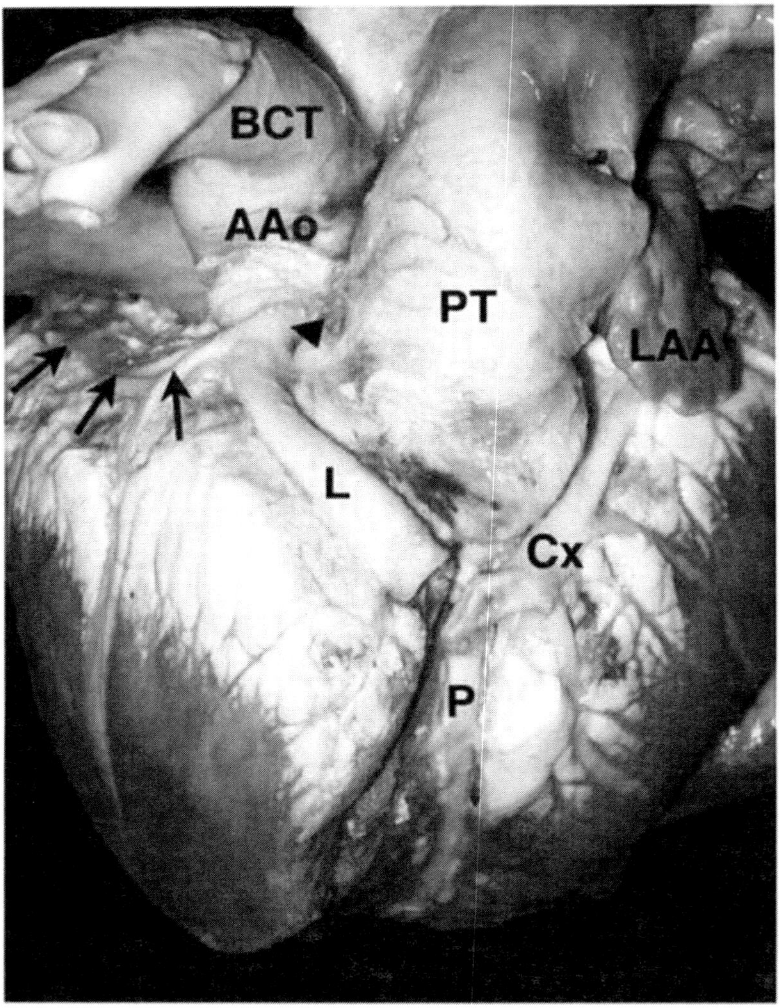

Figure 3. Photograph from a 2-month-old calf with double outlet right ventricle and a coronary artery anomaly. A single right coronary ostium (arrowhead) gives rise to the right coronary artery (arrows) as well as the left main coronary artery (L), which crosses cranial (prepulmonary) to the pulmonary trunk (PT) before dividing into the paraconal interventricular branch (P) and the circumflex branch (Cx). Note that the autopsy cut along the interventricular groove has transected the left coronary artery. AAo = ascending aorta, BCT = brachiocephalic trunk, LAA = left auricular appendage.

Table 1. Categorization of congenital coronary artery anomalies; modified Dodge-Khatami et al. [57].

Major Coronary Artery Anomalies
Anomalous origin from the pulmonary artery
Left coronary artery from PA [a,b]
Right coronary artery from PA
Both coronary arteries from PA
Coronary cameral fistula
Draining to RV [b,c]
Draining to RA
Draining to LV [b]
Draining to LA
Coronary to pulmonary arterial shunt [a,b]
Congenital atresia of the LCA
Minor Coronary Artery Anomalies
Anomalous origin from the aorta
Origin of LCA from right aortic sinus [a,b,d]
Origin of RCA from left aortic sinus [a,e]
Origin of paraconal branch from right aortic sinus [a]
Origin of paraconal branch from RCA
Origin of circumflex branch from right aortic sinus [a]
Origin of circumflex branch from RCA
Single coronary artery
Single left coronary artery
Single right coronary artery
Inverted coronary arteries
Other anomalous aortic origin [a]
Anomalous epicardial course
Prepulmonary course [a,b,d,*]
Interarterial course [a,b,d]
Transseptal course [a]
Retroaortic course [a]
Duplication of a coronary artery
Duplication of paraconal branch
Duplication of circumflex branch
Duplication of RCA
Multiple ostia within left aortic sinus [a]
Multiple ostia within right aortic sinus [a]
High take-off coronary artery [a]
Coronary artery aneurysm [a]
Myocardial bridging of an epicardial coronary artery [a,e]

[a] = canine examples are reported, presented in this paper, or have been observed by the author; [b] = bovine examples are reported, presented in this paper, or have been observed by the author; [c] = camelid examples are reported, presented in this paper, or have been observed by the author; [d] = equine examples are reported; [e] = porcine examples are reported; * this anomaly has major clinical significance in the setting of concurrent pulmonary valve stenosis.

ostium in a Collie with double outlet right ventricle [37], anomalous origin of the LCA from the pulmonary trunk in a miniature poodle [38], LCA aneurysm in a German shepherd dog with subaortic stenosis [39], coronary to pulmonary artery fistula in a German shepherd dog [40], and incidental myocardial bridging in many dogs found on autopsy [41,42]. In the cow, single right coronary ostium with an interarterial LCA [43], single right coronary ostium with a prepulmonary LCA and coronary-to-pulmonary artery fistula [9], LCA to left ventricle fistula [44], dual origin of the LCA [45], left circumflex branch to right ventricle fistula [46], and several cases of anomalous origin of the LCA from the pulmonary trunk [47,48,49] have been described. In the horse, single right coronary ostium with a prepulmonary LCA [50] as well as single right coronary ostium with an interarterial LCA have been reported [51]. In the pig, single left coronary ostium with an anomalous RCA has been reported as an incidental finding [52]. The Syrian hamster has been proposed as an animal model of CAA with numerous forms of anomalous coronary artery origin and course described from inbred research colonies [53,54], as well as LCA from the pulmonary trunk [55] and hypoplastic or rudimentary coronary arteries [56].

4. ANOMALIES OF CORONARY ORIGIN AND COURSE

4.1. Anomalous Origin from the Aorta

Coronary arteries that arise anomalously from the aorta have previously been categorized in human and veterinary literature using the angiographic classification of Lipton [58]. In this system, naming is based upon the vessel's origin from the right (R) or left (L) aortic valvar sinus, whether only one (I), two (II), or three (III) branches of the major coronary arteries are present, and the course of the anomalous vessel, whether anterior (A), between (B), or posterior (P) to the great vessels [59]. This was the naming scheme adapted by Buchanan in his report [6] of the single right coronary ostium with an anomalous prepulmonary left coronary artery (so-called R-II-A anatomy) in four dogs. The description in the title of that report [6] (single coronary artery) differs from current nomenclature as those four dogs did not have single coronary artery anatomy—all coronary arteries were present, but the left was anomalous in origin (arising from the right coronary ostium) and course (prepulmonary). Single coronary artery anatomy (R-I or L-I by this classification) exists when there is only one major artery that terminates in both the right left branches [59] and has not been comprehensively described in an animal, to the author's knowledge. Terminology of anomalous aortic origin, particularly the single coronary ostium that is observed in dogs, has evolved in human medicine with most recent articles referring to these as anomalous aortic origin of the coronary artery (AAOCA) or anomalous origin of a coronary artery from the opposite sinus (ACAOS) with modifiers to describe the artery involved (left or right) and the proximal course [27,60].

While the prepulmonary course of an anomalous coronary artery is considered only a minor anomaly in Table 1 because the anomalous vessel provides normal systemic blood supply to the myocardium, this anatomy in the dog has been associated with greater clinical significance in the setting of PS. The prepulmonary course of an anomalous coronary artery becomes significant if surgical or catheter-based therapies are planned, specifically right ventriculotomy or balloon pulmonary valvuloplasty, as damage to the coronary artery in this location may occur and can be fatal [31,61]. The association between PS and anomalous coronary artery anatomy was first reported by Buchanan and Patterson [62] in an English bulldog and further characterized in subsequent reports [6,30]. Additional reports in English bulldogs and boxer dogs have found alternative variants of the single coronary ostium including a single left coronary ostium with a prepulmonary RCA course [33] and a single right coronary ostium with a presumed interarterial LCA course [35]. As variants of this anatomy appear common in brachycephalic dog breeds, the differing arterial courses associated with a single coronary ostium are shown schematically in Figure 4.

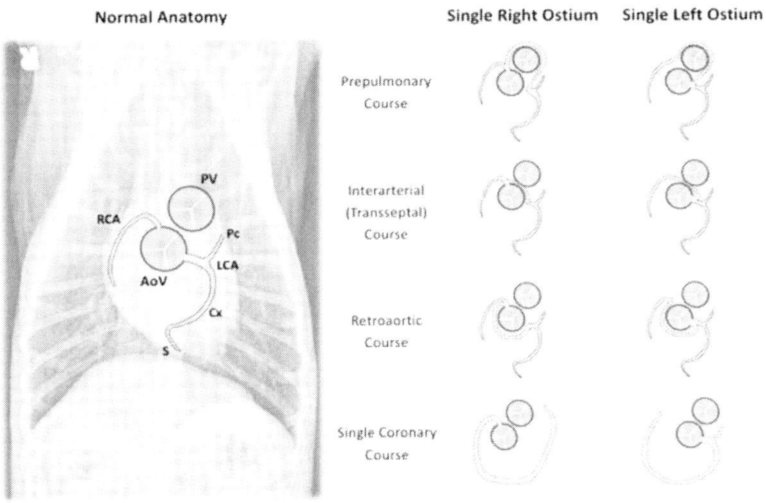

Figure 4. Schematic representation of anomalies of coronary arterial origin and course. The normal anatomy is shown superimposed on a ventrodorsal radiograph and this same orientation is maintained for each drawing. The drawings demonstrate variations in anomalous course (dashed lines) for the major coronary arterial branches in the setting of either a single right or single left coronary ostium. AoV = aortic valve; Cx = circumflex branch; LCA = left coronary artery; Pc = paraconal interventricular branch; PV = pulmonary valve; RCA = right coronary artery; and S = subsinuosal interventricular branch.

The difference between the interarterial and transseptal course reflects the position of the anomalous course—with the interarterial course lying between the pulmonary trunk and aorta and the transseptal traversing lower (more apical) and typically surrounded by ventricular myocardium [28], though some authors dispute this distinction [63,64]. As in humans, it is likely that the anomaly of greatest clinical significance is an interarterial course due to risk of myocardial ischemia [4,43]. If pulmonary stenosis is concurrently diagnosed, the prepulmonary course is significant due to potential for damage during balloon valvuloplasty [61].

Four canine examples of CAAs with anomalous aortic origin not previously described in the veterinary literature are shown in Figure 5. These include an English bulldog with single left coronary ostium and a retroaortic RCA (Figure 5A), an English bulldog with single right coronary ostium and transseptal LCA (Figure 5B), an English bulldog with single right coronary ostium and absent LCA in whom the paraconal interventricular branch takes a prepulmonary course while the circumflex branch takes an interarterial course (Figure 5C), and a mixed-breed dog that had tetralogy of Fallot with anomalous origin of all coronary branches from the brachiocephalic trunk (Figure 5D).

An area of debate surrounds the association between the prepulmonary arterial course and concurrent PS in bracycephalic dogs. Buchanan postulated that the prepulmonary course of the anomalous LCA caused subvalvar PS in these dogs [30]. However, the first reported case of this CAA in a dog was reported from a mixed breed dog with no evidence of pulmonary obstruction [29]. The author has also observed cases of English bulldogs with a single right coronary ostium and a prepulmonary LCA in which the right ventricular outflow tract and pulmonary valve are normal [65]. Humans with a CAA that takes a prepulmonary course do not have concurrent PS [4]. If coronary arteries, even when anomalous in course, are composed of normal cells and undergo normal growth patterns, it seems illogical that their development in a prepulmonary position would be disproportionate to the growth of the right ventricular outflow tract and lead to constriction. As the heart enlarges with growth, so to should the length of the anomalous artery and constriction of the pulmonary annulus or subpulmonary outflow tract would not develop. Given the paucity of coronary arteriograms performed in normal dogs, it may be that the prepulmonary course of a CAA is not causative but is unrelated to PS, being searched for only once PS is diagnosed. The high prevalence of PS in brachycephalic breeds [66], coupled with the apparently high prevalence of CAA in these breeds, may make these co-morbidities likely to be found in combination, even if they are unrelated. The dog whose computed tomography (CT) scan is shown in Figure 5C

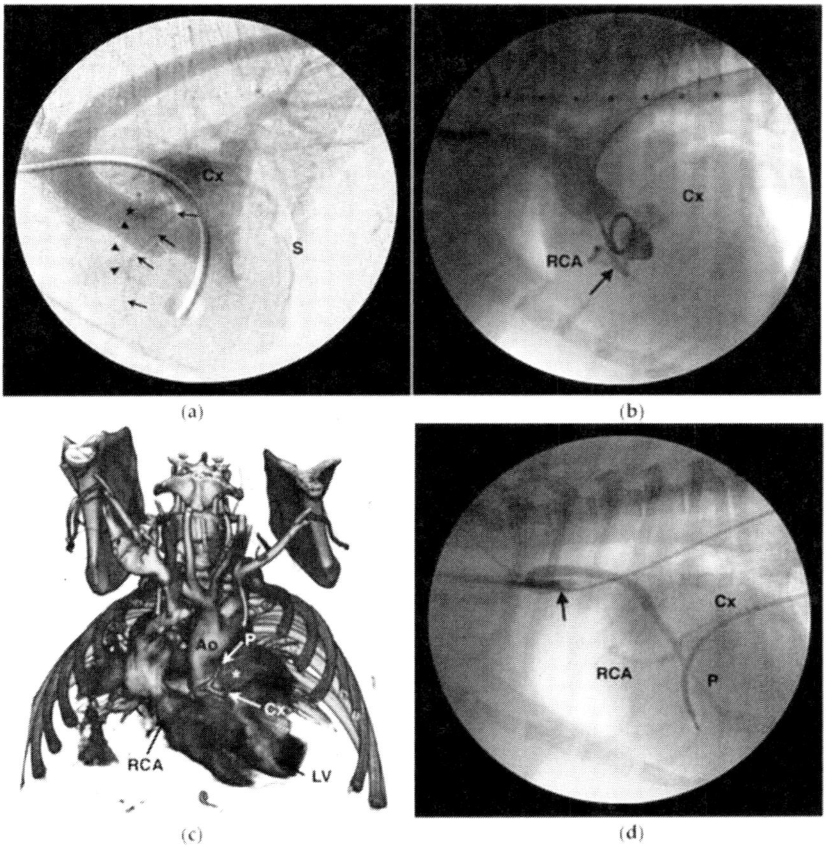

Figure 5. Angiograms of unique coronary artery anomalies in 4 dogs. (**a**) Levophase from a digitally-subtracted right ventriculogram in an English bulldog with pulmonary valve stenosis showing a single left coronary ostium (*) that gives a short left main coronary artery before branching into the paraconal interventricular (arrowheads), circumflex (Cx), and subsinuosal interventricular (S) branches. There is no right coronary ostium and the right coronary artery (arrows) can be seen arising from either the left main or circumflex branch. (**b**) Aortic root angiogram from an English bulldog with a single right coronary ostium that gives the right coronary artery (RCA) as well as a short left main coronary artery (arrow) that passes between the aortic root and right ventricular outflow tract before giving off the circumflex branch (Cx) and small septal or paraconal branches. (**c**) 3-dimensional, volume-rendered reformat of a computed tomography angiogram from an English bulldog with a single right coronary ostium from which arises the right coronary artery (RCA), absence of the left coronary artery, a prepulmonary course for the

paraconal interventricular branch (P), and an interarterial course for the circumflex branch (Cx). A portion of the right ventricular outflow tract has been digitally removed to show the left coronary branches encircling the pulmonary trunk (*). LV = left ventricle, Ao = aorta. (**d**) Selective arteriogram from a mixed breed dog with tetralogy of Fallot whose entire coronary circulation arises off the brachiocephalic trunk (arrow), traverses lateral to the ascending aorta, and then trifurcates caudal to the aortic root into the right coronary artery (RCA) and the paraconal interventricular (P), circumflex (Cx), and subsinuosal interventricular branches.

had no evidence of pulmonary valve stenosis even though a single right coronary ostium with a prepulmonary paraconal interventricular branch was present—indicating that these anomalies can be present distinct from the clinical scenario of PS with which they are more commonly associated. Conversely, the cranial subvalvar filling defect in the right ventricular outflow tract often seen in PS bulldogs that corresponds to the course of the prepulmonary artery does appear to contribute to the severity of obstruction in many of these patients [6,30]. In the last 33 English or French bulldogs with PS seen by the author, nine had a CAA confirmed by CT or angiography, 13 had normal coronary arterial anatomy confirmed by CT or angiography, and 11 did not undergo advanced testing leaving their coronary anatomy uncertain. Comprehensive data on non-brachycephalic breeds is less available, but in the author's experience CAA are exceedingly rare in other breeds with PS. Further understanding of the pathogenesis and relationship of CAA to PS in dogs is needed.

Diagnostic testing for anomalous coronary artery origin or course can include echocardiography (Figure 6), selective angiography (Figure 7), CT (Figure 8), and magnetic resonance imaging (MRI). In human medicine, CAA of minor clinical significance are most often diagnosed incidentally during coronary angiography [4]. When there is a suspicion of CAA, diagnostic testing in humans begins with a detailed echocardiographic evaluation, suspicious findings are confirmed by CT or MRI, and intravascular ultrasound is employed during interventional treatment to characterize the severity of ostial or luminal stenosis [4,60]. Notably, echocardiographic evaluation is not considered definitive, particularly in large persons with suboptimal imaging windows (comparable to the English bulldog) [60]. In the author's experience in animals, transthoracic echocardiography can be suggestive of a single coronary ostium but is not definitive. Transesophageal echocardiography is more capable of determining the origin and course of the coronary vessels than transthoracic imaging [32], but is not as comprehensive as

CT or MRI that allow for volume-rendered 3-dimensional reformatting of the heart (Figure 8) and a thorough evaluation of the major epicardial coronary arteries—both origin and course [34,65]. Selective coronary angiography is often diagnostic in animals, particularly when performed in two or more orthogonal planes, but is more invasive and does not inform the spatial relationship of the CAA to surrounding heart structures as can be obtained with volume-rendered reformats of CT or MRI scans.

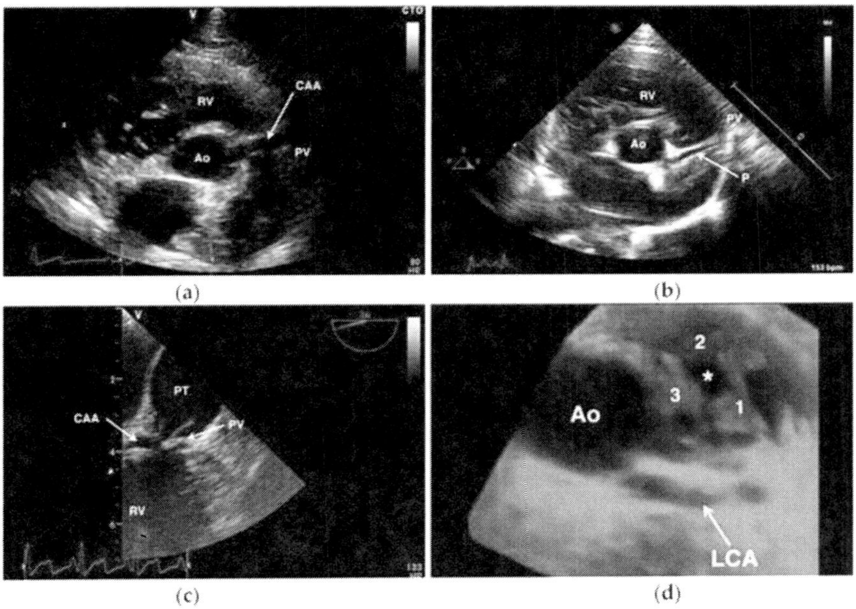

Figure 6. Echocardiographic examples of anomalous coronary artery origin or course. (**a**) Transthoracic short-axis image of the aortic root (Ao) showing a large single right coronary ostium with an anomalous coronary artery (CAA) traversing over the pulmonary valve (PV) annulus. RV = right ventricle. (**b**) Transthoracic short-axis image of the aortic root (Ao) from a dog with normal coronary anatomy demonstrating a normal paraconal interventricular branch (P) that appears to cross cranial to the PV annulus. (**c**) Transesophageal long-axis image of the RV outflow tract of a dog with a CAA seen crossing the PV at the base of the pulmonary trunk (PT). (**d**) Three-dimensional transesophageal image of the Ao from a dog with single right coronary ostium and a prepulmonary left coronary artery (LCA) showing the position of the LCA encircling the three PV leaflets (labeled 1 through 3) and the stenotic PV orifice (*).

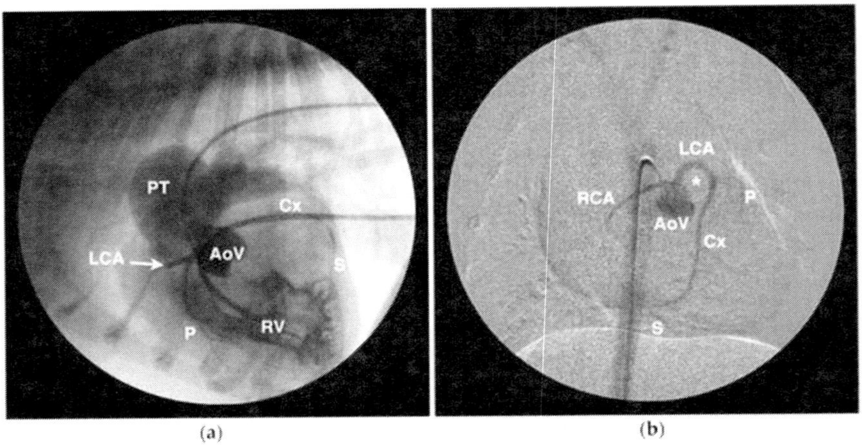

Figure 7. Angiographic examples of anomalous coronary artery origin and course. (**a**) Lateral angiogram from a French bulldog showing simultaneous aortic root (AoV) and right ventricular (RV) injections that demonstrate RV hypertrophy, post-stenotic dilation of the pulmonary trunk (PT), and a single right coronary ostium from which arises the left coronary artery (LCA) that encircles the RV outflow tract before branching into the paraconal interventricular (P), circumflex (Cx), and subsinuosal (S) branches. (**b**) Digitally-subtracted angiogram in a ventrodorsal oblique projection from the same dog as in panel (**a**) showing similar anatomy and highlighting the prepulmonary course of the LCA around the RV outflow tract which occupies the position highlighted by the *.

Coronary artery stent-angioplasty, surgical unroofing of an intramural coronary segment, or osteoplasty to create a new opening in the appropriate valvar sinus are treatment strategies in humans with single coronary ostium in whom myocardial ischemia is present or perceived to be a risk [4].

The appearance of a vessel (parallel double-line echoes) overlying the right ventricular outflow tract on transthoracic echocardiography should be interpreted cautiously. While this finding may suggest a CAA with prepulmonary course (Figure 6A), in the author's opinion it is not definitive. It is possible to create an image where normal coronary vessels appear to overlie or cross the right ventricular outflow tract—both the accessory branch of the RCA as well as the paraconal interventricular branch of the LCA (Figure 6B) can lie in close proximity to the pulmonary valve and right ventricular outflow tract confusing the diagnosis of a prepulmonary coronary arterial course. With

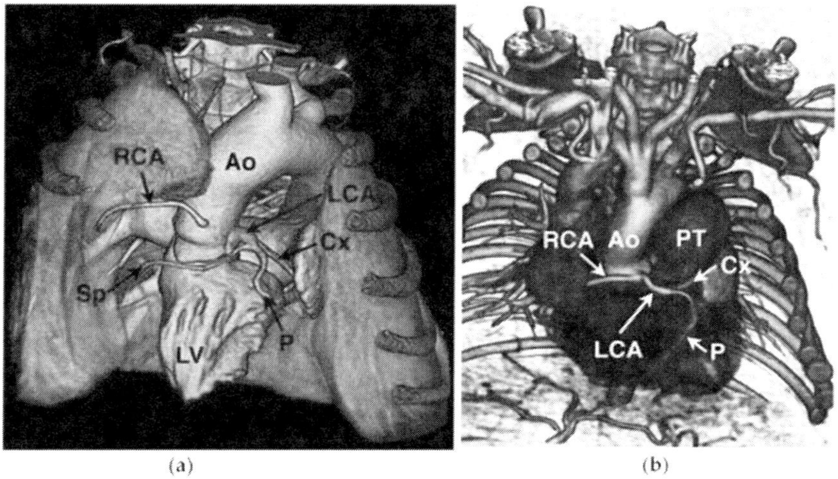

Figure 8. Computed tomography angiography examples of the coronary circulation from two different English bulldogs. (**a**) 3-dimensional volume-rendered reformat showing normal coronary artery anatomy. Ao = aorta, Cx = circumflex branch, LCA = left coronary artery, LV = left ventricle, P = paraconal interventricular branch, RCA = right coronary artery, and Sp = septal branch. (**b**) 3-dimensional volume-rendered reformat showing a single right coronary ostium with prepulmonary left coronary artery encircling the pulmonary valve annulus in a dog with concurrent pulmonary valve stenosis. PT = pulmonary trunk.

transesophageal echocardiography (Figure 6C), it is possible to more thoroughly interrogate the aortic valvar sinuses for presence/absence of coronary ostia and to follow their proximal course. High take-off ostia or suboptimal imaging planes may still limit conclusive imaging with this modality and determining the course of the coronary vessels beyond their proximal origin can be challenging. Considerable experience is necessary to correctly diagnose coronary anatomy by 2-dimensional echocardiography; 3-dimensional imaging in the author's experience provides a better overview and a more complete 'en face' view of the pulmonary annulus and relationship of a prepulmonary arterial course (Figure 6D). In human medicine, 2-dimensional transesophageal echocardiographic diagnosis of the precise CAA present showed poor agreement with surgical findings and marked differences of interpretation when reviewed by a core laboratory [27]; an evaluation of the potential improvement afforded by 3D transesophageal echocardiographic imaging is not available, to the author's knowledge. As the therapeutic options and the prognostic significance of correctly identifying a CAA in the setting of PS differ widely from a mistaken diagnosis, the author advises caution in making a definitive diagnosis by 2-dimensional echocardiography and still recommends angiography or cross-sectional imaging in most cases.

4.2. Anomalous Origin from the Pulmonary Trunk

Anomalous origin of a coronary artery from the pulmonary trunk results in myocardial steal and a left-to-right shunt as the low pulmonary vascular resistance drives blood flow away from the myocardium and to the pulmonary circulation. Reports in animals are rare, being described in a dog [38] and several bovine cases [47,48,49]. Re-implantation to the aorta is the preferred treatment strategy in humans, though ligation of the LCA at the pulmonary valvar sinus may be a feasible treatment strategy in the dog to resolve shunt flow if sufficient collateral flow to the myocardium is provided through the RCA.

5. CORONARY ARTERIAL FISTULAE

Fistulous connections between the coronary arterial system and the cardiac chambers (coronary cameral fistula), the pulmonary circulation, or the vena cavae are occasionally observed in animals. This manuscript specifically refers to congenital CAA, though acquired aortocardiac fistula may present similarly and are most often reported in the horse [67]. Termed coronary cameral fistulae when they drain to a cardiac chamber [68], direct connections of the coronary arteries in humans arise from the RCA in 52%, the paraconal interventricular branch in 30%, and the circumflex branch in 18% of cases [69]. Clinical signs and cardiac remodeling vary by the size of the fistula (amount of flow) and the chambers involved with left heart enlargement expected in most cases and right heart enlargement noted when the fistula drains to the right atrium or right ventricle [68,69]. These lesions may be isolated (Figure 9), but can also be observed in combination with ventricular hypoplasia such as pulmonary atresia with intact ventricular septum when the suprasystemic pressure in the right ventricle forces blood into primitive vascular channels that connect with the coronary circulation [70].

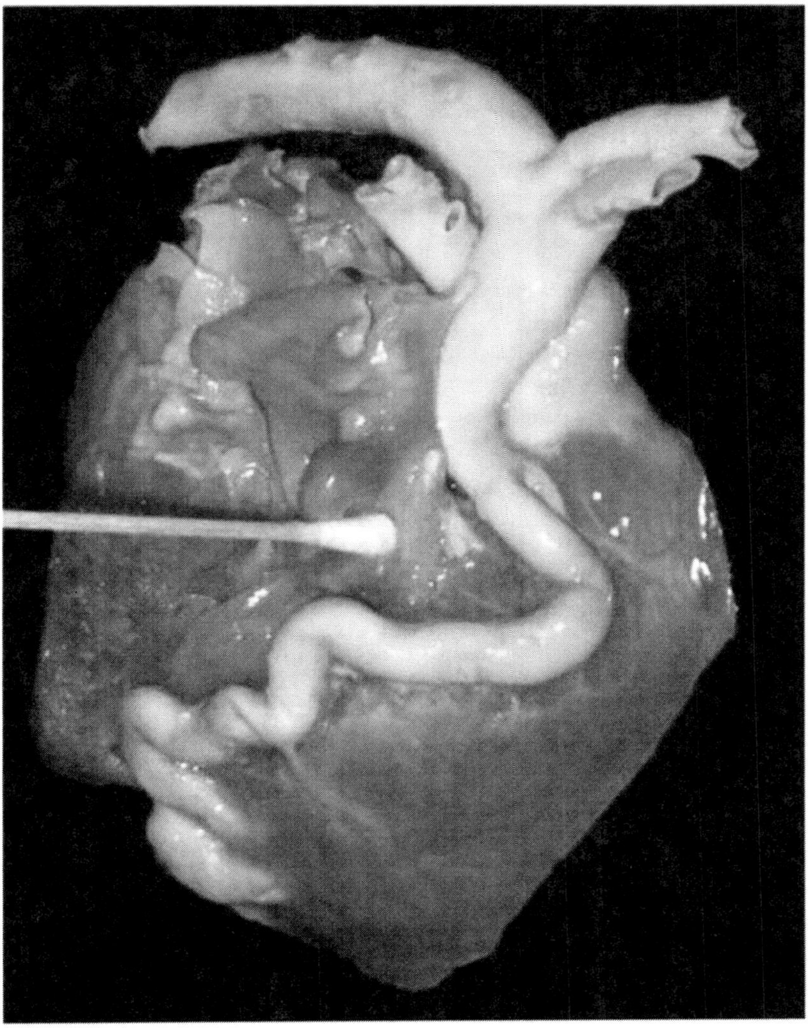

Figure 9. Photograph of the heart from a 6-day-old llama cria showing a severely dilated and tortuous right coronary artery fistula, which drained to the proximal right ventricle.

Coronary to pulmonary artery fistulae to the pulmonary circulation are occasionally observed in animals, most often in the author's experience as part of a larger thoracic arteriovenous malformation [71], though isolated coronary to pulmonary arterial connections also occur (Figure 10).

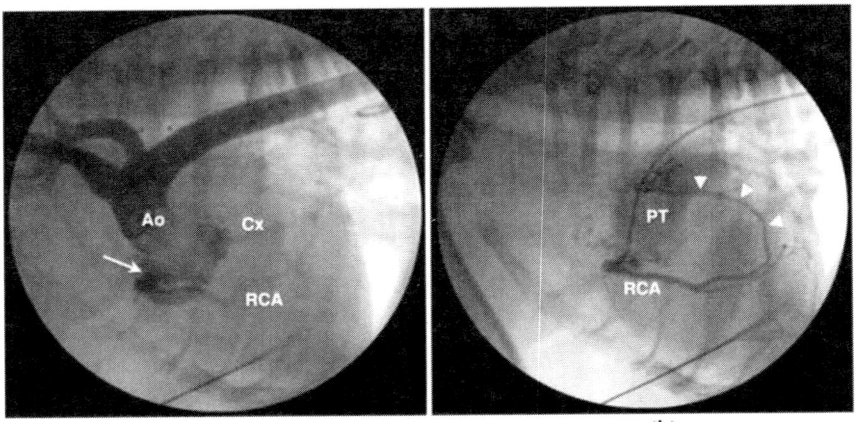

(a) (b)

Figure 10. Angiographic images from an English bulldog with single right coronary ostium, aneurysmal dilation of the prepulmonary left coronary artery and circumflex branch (Cx), and coronary to pulmonary arterial fistulae/malformation. (**a**) The aortic root (Ao) injection shows the single right coronary ostium (arrow), aneurysmal dilation of the Cx, and a more normal diameter right coronary artery (RCA). (**b**) Selective right coronary angiogram from the same dog as in (**a**) showing one of several fistulous connections (arrowheads) from the RCA to the pulmonary trunk (PT).

6. CORONARY ARTERY ANEURYSM

Coronary artery aneuryms, also termed coronary artery ectasia, may be congenital or acquired. Reports in dogs have been in conjunction with subaortic stenosis [39], and fistulous vessels in published reports are often aneurysmal similar to that seen in Figure 9. An example of an aneurysmal RCA is provided in Figure 10 from an English bulldog with a single right coronary ostium, prepulmonary LCA, and fistulous connections to the pulmonary arterial circulation consistent with an arteriovenous malformation. The clinical significance of coronary artery aneurysms is uncertain and therapeutic intervention has not been reported in an animal. In humans, coronary artery aneurysms are typically acquired secondary to atherosclerotic, infectious, or connective tissue disease processes [72]. Therapy in people concentrates on limiting thromboembolic risk with antiplatelet and anticoagulant medications, implantation of covered stents, or surgical excision [72].

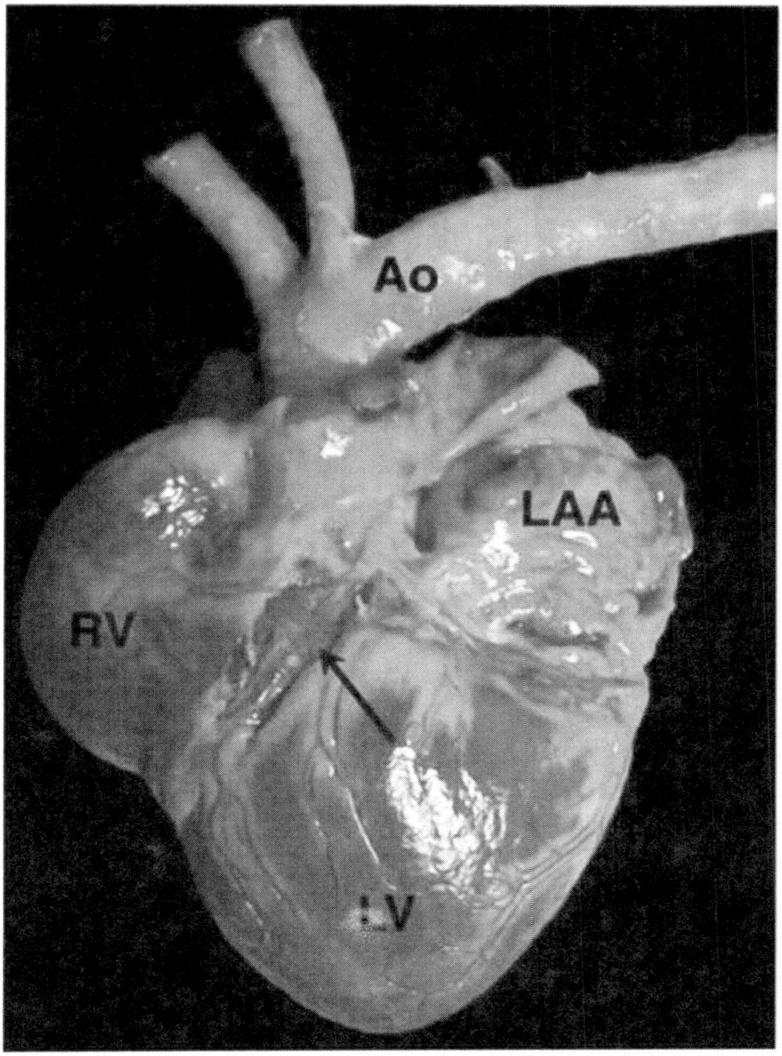

Figure 11. A photograph of a myocardial bridge (arrow) covering a portion of the left paraconal interventricular branch of the left coronary artery in a dog. Ao = aorta, LAA = left auricular appendage, LV = left ventricle, RV = right ventricle.

7. MYOCARDIAL BRIDGING OF EPICARDIAL CORONARY ARTERIES

Myocardial bridging is an anatomical anomaly seen in human [73], canine [42], and porcine [74] hearts where the epicardial coronary arteries tunnel below a section of myocardium. Veterinary studies[42,74] suggest prevalence of up to 30% in dogs and near 50% in swine, which are comparable to human reports [73]. The clinical significance of this anomaly is uncertain as the majority of coronary arterial flow occurs in diastole, while constriction of the coronary artery by a myocardial bridge is predominately a systolic event. However, the size of the tunneled segment, concurrent left ventricular hypertrophy, tachycardia and a shortened diastolic perfusion time, and coronary spasm associated with the myocardial bridge all may play a role in impairing myocardial perfusion and result in clinical signs in humans [73]. Whether similar pathophysiology occurs in animals is presently unknown. An example of a myocardial bridge incidentally detected on post mortem evaluation is shown in Figure 11.

8. CONCLUSIONS

Anomalies of the coronary artery circulation have been observed in nearly all domesticated species both in isolation and in association with concurrent congenital heart disease. Understanding the anatomy, terminology, and clinical implications of these CAA are important both in the treatment of veterinary patients as well as serving as potential models for understanding CAA in humans. As the diagnostic capabilities available to veterinary cardiologists improve, it is likely that additional anomalies of the coronary arterial circulation will be found.

ACKNOWLEDGMENTS

The author gratefully acknowledges the assistance of John Bonagura with the angiograms performed in the dog shown in Figure 10.

CONFLICTS OF INTEREST

The author declares no conflict of interest.

REFERENCES

1. Mathers, C.D.; Loncar, D. Projections of global mortality and burden of disease from 2002 to 2030. PLoS Med. 2006, 3.
2. Driehuys, S.; Van Winkle, T.J.; Sammarco, C.D.; Drobatz, K.J. Myocardial infarction in dogs and cats: 37 cases (1985–1994). J. Am. Vet. Med. Assoc. 1998, 213, 1444–1448.
3. Falk, T.; Jonsson, L. Ischaemic heart disease in the dog: A review of 65 cases. J. Small Anim. Pract. 2000, 41, 97–103.
4. Angelini, P. Coronary artery anomalies: An entity in search of an identity. Circulation 2007, 115, 1296–1305.
5. Basso, C.; Maron, B.J.; Corrado, D.; Thiene, G. Clinical profile of congenital coronary artery anomalies with origin from the wrong aortic sinus leading to sudden death in young competitive athletes. J. Am. Coll. Cardiol. 2000, 35, 1493–1501.
6. Buchanan, J.W. Pulmonic stenosis caused by single coronary artery in dogs: Four cases (1965–1984). J. Am. Vet. Med. Assoc. 1990, 196, 115–120.
7. Fonfara, S.; Martinez Pereira, Y.; Swift, S.; Copeland, H.; Lopez-Alvarez, J.; Summerfield, N.; Cripps, P.; Dukes-McEwan, J. Balloon valvuloplasty for treatment of pulmonic stenosis in english bulldogs with an aberrant coronary artery. J. Vet. Intern. Med. 2010, 24, 354–359.
8. Spicer, D.E.; Henderson, D.J.; Chaudhry, B.; Mohun, T.J.; Anderson, R.H. The anatomy and development of normal and abnormal coronary arteries. Cardiol. Young 2015, 25, 1493–1503.
9. Perloff, J.K.; Marelli, A.J. Congenital anomalies of the coronary circulation. In Perloff's Clinical Recognition of Congenital Heart Disease, 6th ed.; Elsevier Saunders: Philadelphia, PA, USA, 2012; pp. 530–543.
10. Bezuidenhout, A.J. The heart and arteries. In Miller's Anatomy of the Dog, 4th ed.; Evans, H.E., De Lahunta, A., Eds.; Elsevier Saunders: St. Louis, MO, USA, 2013; pp. 428–504.
11. Noestelthaller, A.; Probst, A.; Konig, H.E. Branching patterns of the left main coronary artery in the dog demonstrated by the use of corrosion casting technique. Anat. Histol. Embryol. 2007, 36, 33–37.
12. Donald, D.E.; Essex, H.E. The canine septal coronary artery; an anatomic and electrocardiographic study. Am. J. Physiol. 1954, 176, 143–154.
13. Ozgel, O.; Dursun, N. The arterial vascularization of septum interventriculare in donkeys (Equus asinus L.). Anat. Histol. Embryol. 2005, 34, 80–84.
14. De Oliveira, C.L.S.; David, G.S.; Carvalho, M.; Dornelas, D.; Araujo, S.; da Silva, N.C.; Ruiz, C.R.; Fernandes, J.R.; Wafae, N. Anatomical indicators of dominance between the coronary arteries of dogs. Int. J. Morphol. 2011, 29, 845–849.
15. Corr, P.B.; Pearle, D.L.; Hinton, J.R.; Roberts, W.C.; Gillis, R.A. Site of myocardial infarction. A determinant of the cardiovascular changes induced in the cat by coronary occlusion. Circ. Res. 1976, 39, 840–847.
16. Day, S.B.; Johnson, J.A. The distribution of the coronary arteries of the rabbit. Anat. Rec. 1958, 132, 633–643.
17. Pinto Neto, J.L.; Leão, C.E.S.; Vieira, T.H.M.; Lopes, A.K.M.S.; Vieira, S.R.C.; Silva, N.C.D.; Wafae, G.C.; Ruiz, C.R.; Wafae, N. Anatomical indicators of dominance among the coronary arteries in goats. Braz. J. Vet. Res. Anim. Sci. 2009, 46, 48–53.
18. Vieira, T.H.M.; Moura, P.C., Jr.; Vieira, S.R.C.; Moura, P.R.; Silva, N.C.; Wafae, G.C.; Ruiz, C.R.; Wafae, N. Anatomical indicators of dominance between the coronary arteries in swine. Morphologie 2008, 92, 3–6.
19. Ozgel, O.; Haligur, A.C.; Dursun, N.; Karakurum, E. The macroanatomy of coronary arteries in donkeys (Equus asinus L.). Anat. Histol. Embryol. 2004, 33, 278–283.
20. Taha, A.A.; Abel-Magied, E.M. The coronary arteries of the dromedary camel (Camelus dromedarius). Anat. Histol. Embryol. 1996, 25, 295–299.

21. Gebhard, C.; Fuchs, T.A.; Stehli, J.; Gransar, H.; Berman, D.S.; Budoff, M.J.; Achenbach, S.; Al-Mallah, M.; Andreini, D.; Cademartiri, F.; et al. Coronary dominance and prognosis in patients undergoing coronary computed tomographic angiography: Results from the confirm (coronary ct angiography evaluation for clinical outcomes: An international multicenter) registry. Eur. Heart J. Cardiovasc. Imaging 2015, 16, 853–862.
22. Yoldas, A.; Ozmen, E.; Ozdemir, V. Macroscopic description of the coronary arteries in swiss albino mice (Mus musculus). J. S. Afr. Vet. Assoc. 2010, 81, 247–252.
23. Bahar, S.; Ozdemir, V.; Eken, E.; Tipirdamaz, S. The distribution of the coronary arteries in the angora rabbit. Anat. Histol. Embryol. 2007, 36, 321–327.
24. Ozdemir, V.; Cevik-Demirkan, A.; Turkmenoglu, I. The right coronary artery is absent in the chinchilla (Chinchilla lanigera). Anat. Histol. Embryol. 2008, 37, 114–117.
25. Baraona, F.; Valente, A.M.; Porayette, P.; Pluchinotta, F.R.; Sanders, S.P. Coronary arteries in childhood heart disease: Implications for management of young adults. J. Clin. Exp. Cardiol. 2012.
26. Freire, G.; Miller, M.S. Echocardiographic evaluation of coronary arteries in congenital heart disease. Cardiol. Young2015, 25, 1504–1511.
27. Lorber, R.; Srivastava, S.; Wilder, T.J.; McIntyre, S.; DeCampli, W.M.; Williams, W.G.; Frommelt, P.C.; Parness, I.A.; Blackstone, E.H.; Jacobs, M.L.; et al. Anomalous aortic origin of coronary arteries in the young: Echocardiographic evaluation with surgical correlation. J. Am. Coll. Cardiol. Cardiovasc. Imaging 2015, 8, 1239–1249.
28. Villa, A.D.; Sammut, E.; Nair, A.; Rajani, R.; Bonamini, R.; Chiribiri, A. Coronary artery anomalies overview: The normal and the abnormal. World J. Radiol. 2016, 8, 537–555.
29. Day, S.B. A left coronary artery originating from a single coronary stem in a dog. Anat. Rec. 1959, 134, 55–59.
30. Buchanan, J.W. Pathogenesis of single right coronary artery and pulmonic stenosis in english bulldogs. J. Vet. Intern. Med. 2001, 15, 101–104.
31. Minami, T.; Wakao, Y.; Buchanan, J.; Muto, M.; Watanabe, T.; Suzuki, T.; Takahashi, M. A case of pulmonic stenosis with single coronary artery in a dog. Jpn. J. Vet. Sci. 1989, 51, 453–456.
32. Navalon, I.; Pradelli, D.; Bussadori, C.M. Transesophageal echocardiography to diagnose anomalous right coronary artery type r2a in dogs. J. Vet. Cardiol. 2015, 17, 262–270.
33. Visser, L.C.; Scansen, B.A.; Schober, K.E. Single left coronary ostium and an anomalous prepulmonic right coronary artery in 2 dogs with congenital pulmonary valve stenosis. J. Vet. Cardiol. 2013, 15, 161–169.
34. Laborda-Vidal, P.; Pedro, B.; Baker, M.; Gelzer, A.R.; Dukes-McEwan, J.; Maddox, T.W. Use of ECG-gated computed tomography, echocardiography and selective angiography in five dogs with pulmonic stenosis and one dog with pulmonic stenosis and aberrant coronary arteries. J. Vet. Cardiol. 2016, 18, 418–426.
35. Waterman, M.I.; Abbott, J.A. Novel coronary artery anomaly in an English bulldog with pulmonic stenosis. J. Vet. Intern. Med. 2013, 27, 1256–1259.
36. Van Mierop, L.H.; Patterson, D.F.; Schnarr, W.R. Pathogenesis of persistent truncus arteriosus in light of observations made in a dog embryo with the anomaly. Am. J. Cardiol. 1978, 41, 755–762.
37. Koo, S.T.; LeBlanc, N.L.; Scollan, K.F.; Sisson, D.D. Complete transposition of the great arteries with double outlet right ventricle in a dog. J. Vet. Cardiol. 2016, 18, 179–186.
38. Turk, M.A.; Turk, J.R.; Hopkins, M.G.; Wagner, J.A. Unexpected death in an adult dog with anomalous origin of the left coronary artery from the pulmonary trunk. Cornell Vet. 1984, 74, 344–348.
39. Hernandez, J.L.; Bélanger, M.-C.; Benoit-Biancamano, M.-O.; Girard, C.; Pibarot, P. Left coronary aneurysmal dilation and subaortic stenosis in a dog. J. Vet. Cardiol. 2008, 10, 75–79.

40. Pelosi, A.; Côté, E.; Eyster, G.E. Congenital coronary-pulmonary arterial shunt in a German shepherd dog: Diagnosis and surgical correction. J. Vet. Cardiol. 2011, 13, 153–158.
41. Yamaguchi, M.; Tangkawattana, P.; Muto, M.; Nakade, T.; Taniyama, H.; Miyata, Y.; Nakayama, T.; Hamlin, R.L. Myocardial bridge muscle on left anterior descending coronary artery differs from subepicardial myocardium of the left ventricle in dogs. Acta Anat. 1996, 157, 238–247.
42. Tangkawattana, P.; Muto, M.; Nakayama, T.; Karkoura, A.; Yamano, S.; Yamaguchi, M. Prevalence, vasculature, and innervation of myocardial bridges in dogs. Am. J. Vet. Res. 1997, 58, 1209–1215.
43. Bartoli, C.R.; Wead, W.B.; Giridharan, G.A.; Prabhu, S.D.; Koenig, S.C.; Dowling, R.D. Mechanism of myocardial ischemia with an anomalous left coronary artery from the right sinus of Valsalva. J. Thorac. Cardiovasc. Surg. 2012, 144, 402–408.
44. Dennis, S.M.; Gardiner, M.R. Congenital coronary artery defect in a cow. Vet. Rec. 1972, 90, 281.
45. Vitums, A. Dual origin of the left coronary artery in a newborn calf. Anat. Anz. 1973, 133, 73–75.
46. Bildfell, R.J.; Pringle, J.K.; Miller, L.M. Coronary arterioventricular anomaly in a calf. J. Vet. Diagn. Investig. 1996, 8, 500–502.
47. Sandusky, G.E.; Smith, C.W. Anomalous left coronary artery in a calf. J. Am. Vet. Med. Assoc. 1978, 173, 475–477.
48. Shank, A.M.; Bryant, U.K.; Jackson, C.B.; Williams, N.M.; Janes, J.G. Anomalous origin of the left coronary artery from the pulmonary artery (ALCAPA) in four calves. Vet. Pathol. 2008, 45, 634–639.
49. Tanaka, A.; Yasuda, M.; Murakami, T. Aberrant origin of coronary arteries from pulmonary trunk in cattle. Adv. Anim. Cardiol. 2009, 42, 1–7.
50. Rooney, J.R.; Franks, W.C. Congenital cardiac anomalies in horses. Pathol. Vet. 1964, 1, 454–464.
51. Karlstam, E.; Ho, S.Y.; Shokrai, A.; Agren, E.; Michaelsson, M. Anomalous aortic origin of the left coronary artery in a horse. Equine Vet. J. 1999, 31, 350–352.
52. Matthews, K.A.; Gogas, B.D.; Sumida, A.; Nagai, H.; King Iii, S.B.; Chronos, N.; Hou, D. Anomalous right coronary artery originating from the left sinus of Valsalva in a Yucatan minipig. Comp. Med. 2012, 62, 127–130.
53. Fernandez, M.C.; Duran, A.C.; Real, R.; Lopez, D.; Fernandez, B.; de Andres, A.V.; Arque, J.M.; Gallego, A.; Sans-Coma, V. Coronary artery anomalies and aortic valve morphology in the syrian hamster. Lab. Anim. 2000, 34, 145–154.
54. Duran, A.C.; Fernandez-Gallego, T.; Fernandez, B.; Fernandez, M.C.; Arque, J.M.; Sans-Coma, V. Solitary coronary ostium in the aorta in Syrian hamsters. A morphological study of 130 cases. Cardiovasc. Pathol. 2005, 14, 303–311.
55. Sans-Coma, V.; Arque, J.M.; Duran, A.C.; Cardo, M. Origin of the left main coronary artery from the pulmonary trunk in the syrian hamster. Am. J. Cardiol. 1988, 62, 159–161.
56. Duran, A.C.; Arque, J.M.; Fernandez, B.; Fernandez, M.C.; Fernandez-Gallego, T.; Rodriguez, C.; Sans-Coma, V. Rudimentary coronary artery in Syrian hamsters (Mesocricetus auratus). Anat. Histol. Embryol. 2009, 38, 270–274.
57. Dodge-Khatami, A.; Mavroudis, C.; Backer, C.L. Congenital heart surgery nomenclature and database project: Anomalies of the coronary arteries. Ann. Thorac. Surg. 2000, 69, S270–297.
58. Lipton, M.J.; Barry, W.H.; Obrez, I.; Silverman, J.F.; Wexler, L. Isolated single coronary artery: Diagnosis, angiographic classification, and clinical significance. Radiology 1979, 130, 39–47.
59. Mandal, S.; Tadros, S.S.; Soni, S.; Madan, S. Single coronary artery anomaly: Classification and evaluation using multidetector computed tomography and magnetic resonance angiography. Pediatr. Cardiol. 2014, 35, 441–449.
60. Angelini, P. Is echocardiography adequate to identify the severity of anomalous coronary arteries? J. Am. Coll. Cardiol. Cardiovasc. Imaging 2016, 9, 898–899.

61. Kittleson, M.; Thomas, W.; Loyer, C.; Kienle, R. Single coronary artery (type R2A). J. Vet. Intern. Med. 1992, 6, 250–251.
62. Buchanan, J.W.; Patterson, D.F. Selective angiography and angiocardiography in dogs with congenital cardiovascular disease. J. Am. Vet. Radiol. Soc. 1965, 6, 21–39.
63. Loukas, M.; Groat, C.; Khangura, R.; Owens, D.G.; Anderson, R.H. The normal and abnormal anatomy of the coronary arteries. Clin. Anat. 2009, 22, 114–128.
64. Hlavacek, A.; Loukas, M.; Spicer, D.; Anderson, R.H. Anomalous origin and course of the coronary arteries. Cardiol. Young 2010, 20 (Suppl. 3), 20–25.
65. Scansen, B.A.; Chiappone, G.A.; Layman, R.; White, R.D. Computed tomography angiography of the coronary arterial circulation of bulldogs under conscious sedation is feasible and defines anomalous anatomy. (abstract). J. Vet. Intern. Med. 2014, 28, 1353.
66. Buchanan, J.W. Causes and prevalence of cardiovascular diseases. In Current Veterinary Therapy XI: Small Animal Practice; Kirk, R.W., Bonagura, J.D., Eds.; WB Saunders Co.: Philadelphia, PA, USA, 1992; pp. 647–654.
67. Marr, C.M.; Reef, V.B.; Brazil, T.J.; Thomas, W.P.; Knottenbelt, D.C.; Kelly, D.F.; Baker, J.R.; Reimer, J.M.; Maxson, A.D.; Crowhurst, J.S. Aorto-cardiac fistulas in seven horses. Vet. Radiol. Ultrasound 1998, 39, 22–31.
68. Mangukia, C.V. Coronary artery fistula. Ann. Thorac. Surg. 2012, 93, 2084–2092.
69. Qureshi, S.A. Coronary arterial fistulas. Orphanet J. Rare Dis. 2006, 1.
70. Perloff, J.K.; Marelli, A.J. Pulmonary atresia with intact ventricular septum. In Perloff's Clinical Recognition of Congenital Heart Disease, 6th ed.; Elsevier Saunders: Philadelphia, PA, USA, 2012; pp. 429–438.
71. Culshaw, G.J.; Wagner, T.; Luis Fuentes, V.; Schwarz, T.; Yool, D.A.; French, A.T.; Brockman, D.J. Identification and surgical ligation of aortopulmonic vascular malformation causing left heart volume overload in 4 dogs. J. Vet. Intern. Med. 2013, 27, 583–587.
72. Pahlavan, P.S.; Niroomand, F. Coronary artery aneurysm: A review. Clin. Cardiol. 2006, 29, 439–443.
73. Tarantini, G.; Migliore, F.; Cademartiri, F.; Fraccaro, C.; Iliceto, S. Left anterior descending artery myocardial bridging: A clinical approach. J. Am. Coll. Cardiol. 2016, 68, 2887–2899.
74. Kosinski, A.; Grzybiak, M.; Kozlowski, D. Distribution of myocardial bridges in domestic pig. Pol. J. Vet. Sci. 2010, 13, 689–693.

CHAPTER 9

Inaccurate Assessment of Canine Body Condition Score, Bodyweight, and Pet Food Labels: A Potential Cause of Inaccurate Feeding

Philippa S. Yam *, Gregory Naughton, Christina F. Butowski and Amanda L. Root

School of Veterinary Medicine, College of Medical, Veterinary and Life Sciences, University of Glasgow, 464 Bearsden Road, Glasgow G61 1QH, UK

ABSTRACT

The objectives were to investigate owners' ability to assign the correct bodyweight (BW) and body condition score (BCS) to their dog and to interpret wet and dry pet food labels by estimating how much to feed daily. One hundred and seventy-four questionnaires were completed. Owner estimated BW was compared to actual BW, correct being defined within ±10% of actual BW. Correct interpretation of the total amount of food required was determined by the number of cans (±25% of cans) required for wet food and grams (±20% of grams) for dry food, based on the dog's actual BW, the feeding guidelines on the label, and a comparison with the owner's estimate. Eleven percent of owners overestimated BCS and 19% overestimated BW. Only 48% of owners could correctly estimate their dog's BW. Only 23% and 43% of owners could correctly estimate how much wet and dry food to feed, respectively. Chi-square analysis demonstrated a significant positive association for owners correctly estimating their dog's BW and interpreting the wet pet food label. Many owners are not aware of their pet's BCS and BW and cannot accurately interpret pet food labels. Further owner education to improve these skills is needed if dogs are to be fed correctly.

Keywords: canine; pet food labels; pet food; obesity; body condition score; body weight

1. INTRODUCTION

In the UK, 88% of owners choose a commercially prepared pet food with wet and dry pet food diets accounting for 41% and 59% of sales of commercial food, respectively [1,2]. Similar figures are published in the USA and Australia [3]. All commercial pet foods sold in the UK must state various pieces of information on the label (in accordance with FEDIAF code of practice to ensure that they comply with EU legislation) including feeding guidelines—'The instructions for proper use should indicate how to provide a daily ration in terms of amounts required to be fed relative to the life stage, the life-style and the size of the pet' [4]. Nevertheless, a recent survey of 8000 households claims that only 10% of dog owners read the labels on pet food [2].

It has been estimated that over half of dogs in the UK, as well as in other parts of the world, are overweight or obese [5,6,7]. This is of significance as evidence suggests that overweight and obese animals are predisposed to a number of health conditions, lifespan may be reduced, and health-related quality of life compromised [8,9,10,11]. Understanding why so many dogs are overweight and obese is crucial if it is to be tackled. In this regard, numerous papers have been published focusing on both modifiable and non-modifiable predisposing factors [12] and inaccurate feeding can be considered one such factor.

It is conceivable that failure to feed appropriate amounts of pet food may occur due to either lack of ability of owners to estimate their pet's body condition score (BCS), bodyweight (BW), or difficulty in interpretation of the feeding guidelines. The aim of the current study was to investigate this hypothesis, and assess whether owners can estimate their pets' BCS,

BW, and the amount of food appropriate for their pet by interpreting a dry or wet pet food label.

2. MATERIALS AND METHODS

2.1. Sample Population

Dogs were recruited from The People's Dispensary for Sick Animals (PDSA) PetCheck and microchipping vehicles in four locations—Glasgow, Paisley, Dundee, and Liverpool—and from parks in Glasgow during the summers of 2014 (by GN) and 2015 (by CFB). The PDSA PetCheck vehicles provide a free service open to all dog owning members of the general public, and the organization aims to provide general council on the health and welfare of pet animals. Owners were approached and asked to fill in a paper-based questionnaire and were recruited solely on their willingness to participate with no exclusions applied. All participants gave consent and approval for this study was granted by the University of Glasgow, School of Veterinary Medicine, Ethics and Welfare Committee.

2.2. Data Collection

BCS was assessed using a 5-point scale [13]. 1: emaciated, 2: lean, 3: ideal, 4: overweight, 5: obese. The dogs were weighed on a set of scales (Marsden Group). All body condition scoring and weighing of the dogs was completed in 2014 by GN and in 2015 by CB, who were trained by PY who has extensive experience in this procedure. The interviewer's assessment of BCS and BW of each dog was taken as the gold standard.

2.3. Questionnaire and Assessment of Pet Food Labels

The questionnaire was paper-based and designed at the University of Glasgow (Supplementary Material). It is comprised of four sections: Section 1 asks general questions about the dog including BCS (in both word and picture format) and BW as estimated by the owner. Section 2 asks about the general health and wellbeing of the dog. Section 3 asks the owners to read and interpret a wet pet food label (wet pet food label used was from a 400 g tin of Pedigree complete adult original dog food formula) (2014 and 2015) and a dry pet food label (dry pet food label used was from a 5 kg box of Purina Bakers complete adult dog food formula) (2015) to determine how much of that brand of food they would feed their dog in total in a day if fed exclusively (the question asked was, "If you were feeding just this food to your dog (i.e., no mixer or table scraps), how many grams or tins would you give in total each day?"). Section 4 was used to record the actual BCS and BW of the dog (as assessed by GN and CB).

2.4. Statistical Methods

Data from both collection periods were amalgamated and stored in an Excel spreadsheet (Microsoft Office 2011). Only questionnaires that were completed in full were analysed. Owners were asked to allocate a BCS of their dog in picture and word format. Each of these answers was then compared to the actual BCS and assigned a score of 1 (correct) or 0 (incorrect). Owner estimated BW was compared to actual BW, correct being defined as within ±10% of actual BW. All the dogs were given a breed category; however, to account for the dogs that were not purebred, a category of crossbreed was included. For each specific breed, a single category was made and listed as such.

Microsoft Excel (Analysis-Tool Kit for Excel 2016) and Minitab v.17 (Minitab Inc., State College, PA, USA) were used to generate frequencies, prevalence, descriptive statistics, distribution, and chi-square tests. Descriptive statistics were done on the dogs' age, weight, actual BCS, desired BCS, population BCS, and estimated BCS (word and picture). Statistical significance of the variables was tested using chi-square analysis, which was performed on the quantitative data, BCS and BW, for all breed groups (crossbreeds, pedigrees, and all dogs). The prevalence of misperception was calculated as the proportion of owners who could correctly identify their dog's BCS and BW. The information from the open and closed-ended questions were analysed quantitatively, and themes were identified.

Correct interpretation of the total amount of wet food required to feed their dog in one day was assessed by determining the number of cans required, based on the dog's actual BW and the feeding guidelines on the label, and then comparing it to the owners estimate (correct to within ±25% of cans recommended for actual BW). Correct interpretation of the total amount of dry food required to feed their dog in one day was assessed by determining the number of grams required, based on the dog's actual BW and feeding guidelines on the label, and then comparing it to the owner's estimate (correct to within ±20% of grams recommended on the label, increments being calculated for each kg of BW). Each of these parameters was then compared using the chi-square test, correct interpretation of a wet or dry pet food label compared to actual BW. Statistical significance was defined as $p < 0.05$.

3. RESULTS

Over the sampling period, a total of 189 questionnaires were completed; 89 in 2014, and 100 in 2015. Of these, 15 had missing information providing 174 fully completed questionnaires for further analysis.

The signalment of the dogs included in the analysis are listed in Table 1. The actual BCS were BCS 2, 17% (n = 29); BCS 3, 45% (n = 79); BCS 4, 31% (n = 54), and BCS 5, 7% (n = 12). West Highland White Terriers ($p < 0.01$; $p < 0.02$) and Cocker Spaniels ($p < 0.03$; $p < 0.04$) were the only breeds to be overweight or obese compared to any other pedigree or crossbreed dog, respectively.

Table 1. Signalment of dogs included in study.

	Frequency
SEX	Number (%)
Male Neutered	44 (26%)
Male Entire	57 (33%)
Female Neutered	42 (24%)
Female Entire	30 (17%)
AGE	
0-3 years	94 (54%)
4-7 years	44 (25%)
≥ 8 years	36 (21%)
BREED*	
Bedlington Terrier	3
Bichon Frise	7
Border Collie	10
Border Terrier	5
Boxer	2
Cavalier King Charles Spaniel	3
Cocker Spaniel	11
Crossbreed	54
Golden Retriever	2
Husky	5
Jack Russell	6
Labrador	12
Lakeland Terrier	3
Lhasa Apso	2
Lurcher	2
Maltese terrier	2
Pug	5
Schnauzer	2
Shih Tzu	4
Springer Spaniel	5
Staffordshire bull terrier	3
Tibetan Terrier	3
West Highland White Terrier	8

* There was one each of the following breeds: Beagle, Cairn Terrier, Dachshund, English Bulldog, Eurasier, German Spitz, Greyhound, Hungarian Vizsla, Irish Terrier, Irish Water Spaniel, Italian Spinone, Rottweiler, Welsh Terrier, Whippet, and Yorkshire Terrier.

Just over half of the owners were able to correctly identify their dog's BCS, either by word association (55%) or by picture (51%) (Table 2). Nineteen owners (11%) overestimated their dog's BCS and 66 owners (38%) underestimated their dogs BCS—50 owners by 1 BCS increment, 13 by 2 BCS increments, and 3 by 3 BCS increments. Only 48% of owners were able to correctly estimate their dog's BW; of those that got it wrong, 33% underestimated and 19% overestimated the weight of their dog by greater than 10% of actual BW (Figure 1).

(a) (b)

Figure 1. Owner assessment of canine (**a**) bodyweight in kilograms and (**b**) body condition score on a five-point scale. Bars are divided in owner misperception types (correct estimation, overestimation and underestimation).

Table 2. The number and percentage of owners who correctly assessed, overestimated, and underestimated the body condition score (BCS) by word description and picture, bodyweight (BW) of their dog, and amount to feed following interpretation of wet and dry pet food labels.

	Sample Size (n=)	Correct n (%)	Overestimation n (%)	Underestimation n (%)
BCS (word)	174	96 (55%)	30 (17%)	48 (28%)
BCS (picture)	174	89 (51%)	19 (11%)	66 (38%)
BW Estimate	174	84 (48%)	33 (19%)	57 (33%)
Wet Pet Food Label	174	85 (49%)	13 (7%)	76 (44%)
Dry Pet Food Label	92	40 (44%)	13 (14%)	39 (42%)

Forty-nine percent of owners could correctly quantify how much wet food to feed by interpretation of the wet food label, 44% of owners underestimated, and 7% overestimated. For the dry food, 44% of owners correctly quantified, 42% underestimated, and 14% overestimated how much to feed by interpretation of the dry food label. Chi-square analysis demonstrated a significant positive association (p = 0.01) for owners correctly estimating their dog's BW and interpreting the wet pet food label, but there was no statistically significant association between the estimate of BW and dry food requirement (Table 3).

Table 3. Pearson's chi-square test for the assessment of correct interpretation of pet food labels and the correct estimation of bodyweight (BW).

Analysis	Chi-Square Value	Degrees of Freedom	p-Value
Correct Wet Label vs. Correct BW	6.579	1	0.010
Correct Dry Label vs. Correct BW	0.005	1	0.941

4. DISCUSSION

The Pet Food Manufacturers Association estimates that 88% of UK owners choose a commercially prepared pet food [2] and advise owners to follow feeding guidelines, adapting to the individual needs of their pet. Feeding guidelines are a vital piece of information on the pet food label as they provide recommended portion sizes based on size/weight of a pet. Therefore, owner ability to accurately assess BCS and BW and interpret pet food feeding guidelines is important for the correct provision of calorie requirements. The current study demonstrated that in a cohort of randomly selected dog owners, just over half could accurately assess their dog's BCS and of those that got it wrong, 78% underestimated it. These findings of misperception of body shape substantiate those previously reported and owner underestimation of BCS has been shown to be most prevalent in owners of obese dogs and is a risk factor of obesity [14,15,16]. The current study also concurs with previous findings in that owners struggle to correctly identify BCS even with the aid of pictures [17], so the use of these images on pet food labels will only be useful if owners are educated on how to interpret them correctly.

Overall, we found that 38% of dogs were either overweight or obese. In the most recently published canine studies, between 16 and 34% of dogs were found to be overweight [18,19] and between 33 and 41% obese globally [20], which is in line with our current findings. Owners need to be educated about modifying feeding rations if their pet's BCS is not within an ideal range. Based on these findings, investigating whether owners are aware of this is justified as part of a future study. Furthermore, less than half of owners were able to correctly estimate BW within 10% accuracy. Although there was a significant association between owner estimation of BW and interpretation of the wet pet food label, this was not the case for the dry label, and this is a concern given that most owners in the UK feed dry food to their dogs [1].

Owner motivation to read pet food labels may correlate with how motivated they are to read and acknowledge human food labels [21]. If owners neither recognise that their pets are underweight or overweight, nor realise the dangers of each, and if they are not used to looking at labels on food for their own consumption and have difficulty interpreting pet food labels, they may not in turn regulate the amount of food they provide to their pet appropriately. The ramification of inaccurate feeding is that owners could be either inadvertently providing excess calories to their pets, resulting in overweight and obesity, or providing insufficient calories, resulting in malnutrition. Balanced nutrition ensuring adequate intakes of energy, protein, minerals, and vitamins is essential for dogs to ensure health and longevity [22]. An interesting finding from this study was that many owners seemed to underestimate the amount that their dogs should be fed—for both dry and wet food. Given the high prevalence of overweight and obesity in the study population, this was a surprising finding. One can speculate the reasons for this; it may be that owners who underestimated how much to feed based on the pet food labels may be providing calories to their dogs by other means such as in the form of table scraps and treats; estimates suggest 53% of owners who feed table scraps do so daily [2]. Owners may be feeding a combination of dry and wet pet foods, thus incurring errors in quantity required or may have adjusted the quantity for their dog's energy level (up to 50% more for active dogs or up to 20% less for less active dogs) as stated on the wet pet food label even though it was not specifically stated on the questionnaire. Alternatively, owners are not measuring the food correctly, or the pets are not expending sufficient calories for the amount of food given.

Further causes of error in energy intake can result from varying energy allowances, recommended for maintenance of adult dogs; current label recommendations are an inexact science. Brands may use different energy maintenance requirement formulas to guide feeding instructions, causing variability in energy content, as figures ranging from less than 90 kcal $ME/kg^{0.75}$ (377 kJ) to approximately 200 kcal $ME/kg^{0.75}$ (810 kJ) have been published for adult maintenance [23]. Furthermore, the method by which the energy content of food is determined might be important, namely whether measured in feeding trials, or calculated and, if so, by what method, modified Atwater factors [24] or NRC guidelines [25]. Calculations based upon modified Atwater factors often underestimate the actual energy content of commercial foods of dogs [26], perhaps accounting for the current propensity for canine obesity. The cause of inaccurate feeding is complex and owners should be advised about appropriate levels of feeding and the modification required for underweight or overweight pets.

A significant association was found between wet food label interpretation and correct BW estimation. This may be due to canned pet food having less exact feeding guidelines (halves and thirds of a can used as guidelines) compared to dry food. The feeding of dry food may be complicated further since, even if owners do interpret the label correctly, they typically underestimate the amount of food they are providing when using feeding cups [27]. Hence, even if owners could interpret the label, the dogs may still be overfed. The same argument could be applied to the large numbers of owners, who also underestimated how much dry food to feed, as they too may ultimately end up overfeeding. For both reasons, namely an inability to interpret the dry food label and imprecision when using feeding cups, recommendations to weigh the appropriate amount of dry food for pets is important to accurately feed overweight dogs. Dry pet food labels may have been more difficult to interpret for a few reasons, one being that the dry pet food label lists broad ranges of weight categories and grams per feed. Grams per feed differences ranged from 85 to 225 g. Secondly, the open-ended, quantitative question for the dry food (blank with number of grams to be written by the owner) may have confused owners in comparison to the wet food label, where the options were listed out for them (specific categories of fractions of cans) to choose. This may help to explain a difference in the accuracy of comparing the wet and dry pet food labels.

Whilst the aim of our study was fulfilled and indicated that owners are inaccurate when feeding pets, the results should be interpreted with caution. A limitation of the study was that the participants were recruited from free services and may not be representative of the general population. Therefore, it would be worthwhile to pursue additional larger studies enlisting owners from a wider cross section of society and comparing the accuracy of BCS, BW, and/or feeding instruction among groups of owners with underweight, ideal and overweight dogs, or owners with varied demographics, which has been shown to influence pet obesity. It would have been interesting to ascertain whether owners were able to interpret labels once given the true weight of their pets. Additionally, the data from the two different time points (2014 and 2015) were not statistically assessed in this study, so a bias in assessments or any differences between the findings from the two collection periods cannot be ruled out. Although the owners were asked to interpret only two labels, a wet and dry pet food label, we did not establish what food the owners typically fed their pet, and additional labels were not investigated. The questionnaire could include the dog's typical feeding regimen, thus allowing a comparison between current and novel feeding. In this regard, in future studies, a range of different labels could be investigated from both dry and wet foods to tease out further information about the presentation of data and the ease, or otherwise, of interpretation. Finally, the expected daily caloric requirement for each animal was not calculated and compared to caloric density of the dry or wet pet food. Various formulae for maintenance energy requirements [23] are available, but limited information about factors that could alter these requirements, such as physical activity and home conditions, were not known, so the crude feeding guidelines on the wet and dry food labels were followed. As feeding guidelines for dry food are suggested in "grams" and for wet food as "cans", we designated greater than ±20% grams according to feeding guidelines for dry food and ±25% for wet food based on the feeding guidelines as correct. Other values could have been selected, but this seemed a sensible starting point by which to ascertain a general guideline of owners' ability to estimate these measurements (particularly without any other knowledge about other factors impacting on calorie requirements of these animals being available). Further, it would have been interesting to ask the owners to actually 'plate up' the food they would give in a day so that this could be measured.

This study has shown that many owners are not aware of their pet's BCS and BW and that many owners cannot accurately interpret pet food labels. Perhaps the current findings could be viewed as encouraging inasmuch as inaccurate feeding due to lack of recognition of body size and shape together with inability to interpret pet food labels are factors that can all be tackled by owner education. The current findings lend credence to the instigation of owner education about the monitoring of their pets' BCS and BW, adjusting feeding regimes accordingly, and the interpretation of pet food labels.

5. CONCLUSIONS

This study used a novel questionnaire to determine how well owners could assess BCS and BW and accurately interpret wet and dry pet food labels. Analysis of the questionnaire scores showed that around half of the sample population was unable to correctly identify any of these parameters. These figures clearly indicate that more needs to be done by veterinarians and the pet food industry to educate owners on the importance of a pet's weight, a pet's shape, and pet food label interpretation. Veterinary surgeons need to change owner behaviour by counselling them about the importance of appropriate BW and the deleterious effects of obesity and should aim to help owners with their feeding regimes, particularly for dry food, educating them throughout their dog's life so that they are fed to an ideal BCS for health and longevity.

ACKNOWLEDGMENTS

The study was funded by the University of Glasgow, and the authors would like to thank all participating pets and owners. Thank you to Donna MacKenzie and Lucy MacMillan for their help in collecting the data and to Tim Parkin for his statistical advice. The authors would also like to thank Dundee City Council, Liverpool City Council, Glasgow City Council, and East Renfrewshire Council for their cooperation at the microchipping van locations.

AUTHOR CONTRIBUTIONS

Philippa S. Yam conceived and designed the experiments; Gregory Naughton and Christina F. Butowski performed the experiments; Philippa S. Yam and Amanda L. Root analysed the data and wrote the paper.

CONFLICTS OF INTEREST

The authors declare no conflict of interest.

REFERENCES

1. Pet Food Manufacturer's Association: PFMA Annual Report 2014. Available online: http://www.pfma.org.uk/_assets/docs/Annual%20Report%202014%20Website%20Format%20-%20for%20printing%20A4.pdf (accessed on 10 April 2017).
2. Pet Food Manufacturers Association: Pet Data Report from TNS Data 2017. Available online: http://www.pfma.org.uk/_assets/docs/Data%20report/PFMA-Pet-Population- Report%202017_WEB.pdf (accessed on 10 April 2017).
3. Laflamme, D.P.; Abood, S.K.; Fascetti, A.J.; Fleeman, L.M.; Freeman, L.M.; Michel, K.E.; Bauer, C.; Kemp, B.L.; Doren, J.R.; Willoughby, K.N. Pet feeding practices of dog and cat owners in the united states and australia. J. Am. Vet. Med. Assoc. 2008, 232, 687–694.
4. Fediaf 2017 Code of Good Labelling Practice for Pet Food: Version of October 2011 European Pet Food Industry Federation (FEDIAF). Available online: http://www.fediaf.org (accessed on 10 April 2017).
5. Courcier, E.A.; Thomson, R.M.; Mellor, D.J.; Yam, P.S. An epidemiological study of environmental factors associated with canine obesity. J. Small Anim. Pract. 2010, 51, 362–367.
6. Lund, E.M.; Armstrong, P.J.; Kirk, C.A.; Klausner, J.S. Prevalence and risk factors for obesity in adult dogs from private us veterinary practices. Int. J. Appl. Res. Vet. Med. 2006, 4, 177–186.

7. McGreevy, P.D.; Thomson, P.C.; Pride, C.; Fawcett, A.; Grassi, T.; Jones, B. Prevalence of obesity in dogs examined by Australian veterinary practices and the risk factors involved. Vet. Rec. 2005, 156, 695–702.
8. German, A.J.; Ryan, V.H.; German, A.C.; Wood, I.S.; Trayhurn, P. Obesity, its associated disorders and the role of inflammatory adipokines in companion animals. Vet. J. 2010, 185, 4–9.
9. German, A.J.; Holden, S.L.; Wiseman-Orr, M.L.; Reid, J.; Nolan, A.M.; Biourge, V.; Morris, P.J.; Scott, E.M. Quality of life is reduced in obese dogs but improves after successful weight loss. Vet. J. 2012, 192, 428–434.
10. Kealy, R.D.; Lawler, D.F.; Ballam, J.M.; Mantz, S.L.; Biery, D.N.; Greeley, E.H.; Lust, G.; Segre, M.; Smith, G.K.; Stowe, H.D. Effects of diet restriction on life span and age-related changes in dogs. J. Am. Vet. Med. Assoc. 2002, 220, 1315–1320.
11. Yam, P.S.; Butowski, C.F.; Chitty, J.L.; Naughton, G.; Wiseman-Orr, M.L.; Parkin, T.; Reid, J. Impact of canine overweight and obesity on health-related quality of life. Prev. Vet. Med. 2016, 127, 64–69.
12. Linder, D.; Mueller, M. Pet obesity management: Beyond nutrition. Vet. Clin. North. Am. Small Anim. Pract. 2014, 44, 789–806.
13. Edney, A.T.; Smith, P.M. Study of obesity in dogs visiting veterinary practices in the United Kingdom. Vet. Rec. 1986, 118, 391–396.
14. Rohlf, V.I.; Toukhsati, S.; Coleman, G.J.; Bennett, P.C. Dog obesity: Can dog caregivers' (owners') feeding and exercise intentions and behaviors be predicted from attitudes? J. Appl. Anim. Welf. Sci. 2010, 13, 213–236.
15. Courcier, E.A.; Mellor, D.J.; Thomson, R.M.; Yam, P.S. A cross sectional study of the prevalence and risk factors for owner misperception of canine body shape in first opinion practice in Glasgow. Prev. Vet. Med. 2011, 102, 66–74.
16. White, G.A.; Hobson-West, P.; Cobb, K.; Craigon, J.; Hammond, R.; Millar, K.M. Canine obesity: Is there a difference between veterinarian and owner perception? J. Small Anim. Pract. 2011, 52, 622–626.
17. Eastland-Jones, R.C.; German, A.J.; Holden, S.L.; Biourge, V.; Pickavance, L.C. Owner misperception of canine body condition persists despite use of a body condition score chart. J. Nutr. Sci. 2014, 3, e45.
18. Diez, M.; Picavet, P.; Ricci, R.; Dequenne, M.; Renard, M.; Bongartz, A.; Farnir, F. Health screening to identify opportunities to improve preventive medicine in cats and dogs. Eur. J. Comp. Anim. Pract. 2016, 2, 54–62.
19. German, A.J.; Blackwell, E.; Evans, M.; Westgarth, C. Overweight dogs are more likely to display undesirable behaviours: Results of a large online survey of dog owners in the UK. J. Nutr. Sci. 2017, 6, e14.
20. Montoya-Alonso, J.A.; Bautista-Castaño, I.; Peña, C.; Suárez, L.; Juste, M.C.; Tvarijonaviciute, A. Prevalence of canine obesity, obesity-related metabolic dysfunction, and relationship with owner obesity in an obesogenic region of Spain. Front. Vet. Sci. 2017, 4, e59.
21. Lähteenmäki, L. Consumer interpretation of nutrition and other information on food and beverage labels. In Advances in Food and Beverage Labelling: Information and Regulations; Woodhead Publishing: Cambridge, England, UK, 2014; pp. 133–135.
22. FEDIAF 2016 Nutritional Guidelines for Complete and Complementary Pet Food for Cats and Dogs. Available online: http://www.fediaf.org (accessed on 8 May 2017).
23. Bermingham, E.N.; Thomas, D.G.; Cave, N.J.; Morris, P.J.; Butterwick, R.F.; German, A.J. Energy requirements of adult dogs: A meta-analysis. PLoS ONE 2014, 9, e109681.
24. Association of American Feed Control Officials (AAFCO). Dogs and Cats Nutrient Profiles; Official Publication: Washington, DC, USA, 2010; pp. 169–183.
25. NRC Energy. National Research Council. Nutrient Requirements of Dogs and Cats; National Academic Press: Washington, DC, USA, 2006; pp. 22–48.
26. Hill, R.C. Challenges in measuring energy expenditure in companion animals: A clinician's perspective. J. Nutr. 2006, 136, 1967–1972.

27. German, A.J.; Holden, S.L.; Mason, S.L.; Bryner, C.; Bouldoires, C.; Morris, P.J.; Deboise, M.; Biourge, V. Imprecision when using measuring cups to weigh out extruded dry kibbled food. J. Anim. Physiol. Anim. Nutr. 2011, 95, 368–373.

CHAPTER 10

Lens Capsule HSPG-Perlecan Regulates Lens Fibre Differentiation during Chick Embryo Development

Cristina Martín[1], María I. Alonso[1], Francisco Lamus[1], José A. Moro[1], A. De la Mano[1], José M. Fernandez[2], Alberto Caballero[1], Ángel Gato[1*]

[1]*Departamento de Anatomía y Radiología, Laboratorio de Desarrollo y Teratología del Sistema Nervioso. Instituto de Neurociencias de Castilla y León (INCYL); Facultad de Medicina, Universidad de Valladolid, Valladolid, España*

[2]*Departamento de Biología Celular, Histología y Farmacología; Facultad de Medicina, Universidad de Valladolid, Valladolid, Spain*

ABSTRACT

Lens fibre differentiation is a life-long process related with lens transparency, and is particularly intense during development, being related with an FGF-2 antero-posterior gradient at the equator level as the main growth factor involved which has been related with the basal membrane of the lens anlagen known as "Lens capsule". However the lens fibre differentiation induced by FGF2 depends, as in other biological systems, on the local bioavailability of FGF-2 regulated by their relationship with extracellular matrix molecules as Heparan Sulphate Proteoglycans. Here, we try to clarify how Perlecan (a heparan sulphate proteoglycan specific from basement membranes) is involved in lens fibre differentiation at earliest stages of eye development. Our results show that Perlecan, is a major component in the lens capsule during the earliest stages of lens development in chick embryos being present during lens plate induction, lens vesicle stage and the onset of lens fibre differentiation. In order to demonstrate a direct involvement of HSPG-Perlecan in lens fibre differentiation, we generate depleted lenses by HSPG-Perlecan synthesis disruption and specific enzymatic digestion. The HSPG-Perlecan depleted lens show a significant delay or abolition in the lens fibre differentiation which remains in an immature cells displaying DNA synthesis in the posterior epithelium and a decrease in FGF2 lens expression. These data support the hypothesis that lens capsule HSPG-Perlecan is a key molecule involved in lens fibre differentiation during development, probably by involvement in FGF-2 biodisponibility.

Keywords: Eye Development, Lens Fibre Differentiation, HSPG-Perlecan, FGF-2, Chick Embryo

1. INTRODUCTION

From early development, lens vesicle epithelial cells undergo a differentiation process towards primary lens fibre, and these cellular dynamics remain through- out life with cell replication in the anterior epithelium and migration through the equatorial (peripheral) area towards the posterior epithelium, in which cells differentiate in lens fibres [1] . This cellular behaviour associated with lens fibre differentiation is restricted later at foetal and adult periods, to the equatorial area. Despite the involvement of growth factors like IGF-I and TGF-β has been reported in lens fibre differentiation [2] [3] [4] , the FGF-2 remains as the strongest candidate in the regulation of lens fibre differentiation during development both in rodents and chick [5] [6] .

Currently, however, modulators of growth factor signalling in lens fibre differentiation induced by FGF-2 are seen to be influential and two kinds of mechanisms have been proposed. One is based on molecules able to repress the FGF2 activity such as "sprouty" and "sef" families of genes, which are strongly expressed in the lens anterior epithelium exerting a repressive action in FGF-2 lens cells signalling and maintain their epithelial state [7] [8] . The other mechanism proposed is that of regulation of FGF-2 bioavailability and receptor activation by the extracellular matrix proteoglycans, as described for Heparan Sulphate proteoglycans in many biological systems [9] [10] . In this context, there are previous reports of the presence and involvement of proteoglycans in several aspects of early lens development [11] [12] ; in addition, special attention has been given to the influence of the CSPG and HSPG of the lens capsule in lens fibre differentiation in birds and mammals [13] [14] , suggesting that HSPG from the lens capsule might be involved in lens fibre differentiation by FGF-2 activity regulation [15] . However, experimental support for this relation is necessary.

In this paper, we show the expression pattern of Perlecan A besal membrane HSPG in the lens capsule at the earliest stages of chick embryo lens development, and we provide experimental support of direct relation between lens capsule HSPG-Perlecan and lens fibre differentiation during chick eye development.

2. MATERIALS AND METHODS

2.1. Obtaining Chick Embryos and Experimental Approach

Fertile white hybrid hens eggs were incubated at 38°C in a humidified atmosphere to obtain chick embryos at different developmental stages, ranging from 13 to 25 H.H. [16] , in order to study the expression pattern of HSPG-Perlecan. A total of 20 embryos were incubated up to stage 17 H.H. (corresponding to that of the lens vesicle, prior to differentiation of lens fibres), at which a single dose of 24 µl b-D-xyloside (a sulphated proteoglycans synthesis disruptor) 4 mM (Sigma) was injected subgerminally with a Hamilton microsyringe [12] , and the eggs were reincubated for 24 hours until stage 23 H.H. (corresponding to the development of lens fibres). Controls (15 embryos) used in the experiment were treated with a sterile saline solution with the same procedure. All the animal handling and procedures were supervised by the research and welfare service from the Valladolid University.

A total of 30 embryos were treated with heparinase as follows: once embryos reached stage 17 H.H., a small opening was made in the shell to expose the right eye of the embryo, after which the vitelline membrane was cut with tungsten needle and 12 nl of a heparinase II solution from Flavobacterium heparinum (Sigma) (10 U of enzyme dissolved in 25 µl of PBS), was microinjected into the subectodermal space in the area of the lens equator. As controls, we microinjected 15 embryos with the same amount of heat-inactivated enzyme as for the enzyme treated. Microinjection was performed with a microinjector (Medical Systems Corp, Greenvale, NY 11548, U.S.A, PLI-100). Following this, the eggs were sealed and re-incubated for a period of 18 hours until stage 21 H.H.

In both cases (β-D-xyloside and heparinase II treatment) a single dose of 200 nl of BrdU (0.3% solution) was microinjected in the outflow of the heart just 1 hour before the end of the reincubation period. The embryos were then removed from the extraembryonic membranes and fixed for 1 hr in Carnoy's.

2.2. Immunohistochemistry

Histological sections (8 µ) from five different lens anlagen at each stage were washed in phosphate-buffered saline (PBS), pre-incubated with normal horse serum (1/20 in PBS) and incubated overnight at room temperature with a 1:1000 dilution of Anti Heparan Sulphate-Perlecan monoclonal antibody (Upstate Biotechnology). After two washings in PBS, the sections were reincubated for 30 minutes in fluorescein-conjugated goat anti-rat Ig-G FITC conjugate (Sigma) as a secondary antibody, mounted in Aquamount (Gurr) and examined under a Leica SPE laser confocal microscope. Control sections were prepared as described above but pre-immune serum was used as the primary antibody and no labelling was observed.

In others cases, lens anlagen histological sections (8 µm) were deparaffinised and BrdU was detected following standard procedures [17] . The sections were incubated in a solution containing a 1/100 dilution of monoclonal antibody to BrdU (Dako) for 30 minutes at room temperature. An avidin-extravidin system conjugated to peroxidase (mouse anti-rabbit 1/20 for 30 minutes and extravidin 1/20 for 10 minutes; Sigma) was used to detect the primary antibody and was developed with DAB. A Nikon microphot-FXA photomicroscope was employed to visualize and photograph the preparations.

In other cases, lens anlagen frozen sections (5 µm) were used to detect FGF2 following standard procedures. The sections were incubated in a solution containing a 1/1000 dilution of monoclonal antibody to FGF2 (anti-bovine FGF2 antibody from Sigma) for 30 minutes at room temperature. An ExtrAvidin stainig Kit antimouse system conjugated to peroxidase (from Sigma) was used to detect the primary antibody and was developed with DAB. A Nikon microphot-FXA photomicroscope was employed to visualize and photograph the preparations.

Finally, lens anlagen histological sections (8 µm) were deparaffinised, dehydrated and stained with Haematoxilin-Eosin technique following standard procedures. We used a Nikon microphot-FXA photomicroscope to visualize and photograph the preparations.

3. RESULTS

3.1. Immunohistochemical Detection of HSPG-Perlecan during Early Lens Development

In this study we have used a monoclonal antibody which recognizes Perlecan, a type of Heparan Sulphate Proteoglycan present in basement membranes [18] [19] [20] . As we show in Figure 1, the antibody specifically labels basement membranes in the eye anlagen, including the surface ectoderm, lens and optic cup, at different stages of development.

At stage14 H.H., we show (Figure 1(a)) intense immunolabelling in the basement membrane of both the lens placode (similar to surface ectoderm basement membrane labelling) and the optic cup, although in some case both basement membranes seem to be in direct contact. During lens placode invagination (stages 15 - 16 H.H.) (Figure 1(b) and Figure 1(c)) there is still intense immunolabelling on the basement surface of the lens epithelium, however during this period intensity on the basement surface of the optic cup decreases. At these stages it is possible to observe several areas in which both basement membranes (lens and optic cup) are in close contact. A similar disposition in terms of Perlecan immunomarking can be seen at stages 16 - 17 H.H. (Figure 1(d)); furthermore, at these stages it can be observed how basement membrane immunolabelling in the lens anterior epithelium is restored as the final step in lens isolation from the ectodermal surface.

From stages 18 to 25 H.H. (Figure 1(e)-(h)), the lens capsule is the structure with the most intense immunolabelling in the eye anlage, appearing as a strong, continuous and apparently uniform layer on the basement side of the lens. At the beginning of lens fibre differentiation we cannot appreciate any labelling in the anterior and posterior epithelium cells.

As can be observed in Figure 1(e) and Figure 1(f), at these stages of eye development (E18 - 20 H.H.) immunolabelling on the surface ectoderm and neural retina decrease considerably; however, in some sections a point of contact can be perceived between the neural retina at the margin of the optic cup and the basement membrane of the lens equator (Figure 1(h)), both labelled by the Perlecan antibody. Finally, we detect the "pecten" (Figure 1(g)) as another structure showing strong immunolabelling during chick eye development.

3.2. Disruption of Lens HSPG Alters Normal Lens Fiber Growth and Differentiation

With the aim to test the role of Heparan Sulphate-Perlecan proteoglycan on lens development, especially in the differentiation of lens fibres from epithelial cells, we carried out two experimental approaches: first, we used β-Dxyloside, a xylosiltransferase substrate which disrupts the assembly of sulphated proteoglycans including different forms of HSPG [21] ; secondly, we used Heparinase type II, which specifically digest Heparan Sulphate including Perlecan [22] . Both treatments were administered at stage 17 H.H., which corresponds with the lens vesicle stage, before initial lens fibre differentiation, and were maintained until the 23 H.H., when the histological structure of the lens anlagen (Figure 2(a)) shows a polarized lens with a thin anterior epithelium, a transitional equatorial (marginal) area and a thick posterior epithelium displaying the elongated primary fibres with a central nucleus, closing the inner cavity.

As seen in Figure 2(d), BrdU immunolabelling showed the nucleus with DNA synthesis activity (which at these stages of development could be assimilated to mitotic activity) in the anterior epithelium, in which the homogeneous presence of a layer of BrdU-positive nucleus reaching the equatorial area could be observed; however, in the posterior epithelium, there was a total absence of BrdU-positive nucleus. This mitotic pattern was consistent with the dynamic cellular behaviour described during lens development [23] with cellular replication in the anterior epithelium, migration through the equatorial zone and differentiation in the posterior epithelium. As we described before, the eyes of stage 23 H.H. controls showed intense immunolabelling with the anti-perlecan antibody in the lens capsule (Figure 1(f)). Finally, lens capsule immunodetection of FGF2 show the absence of labelling in the anterior epithelium and a ubiquitous label in the lens fibre which was more intense in the posterior part of the lens capsule (Figure 2(j)).

All the β-Dxyloside treated embryos showed a severe, specific disruption in lens anlagen development. The histological study shows the lens anlagen with an undifferentiated posterior epithelium with no lens fibre growth, resembling the lens vesicle stage (Figure 2(b)). In other cases (Figure 2(c)), some areas of the posterior epithelium show a partial lens fibre growth while others remain undifferentiated and the lens vesicle cavity remains uncollapsed. In the β-Dxyloside treated embryos, BrdU immunodetection (Figure 1(e)) shows a clear disruption in the mitotic behaviour of the lens

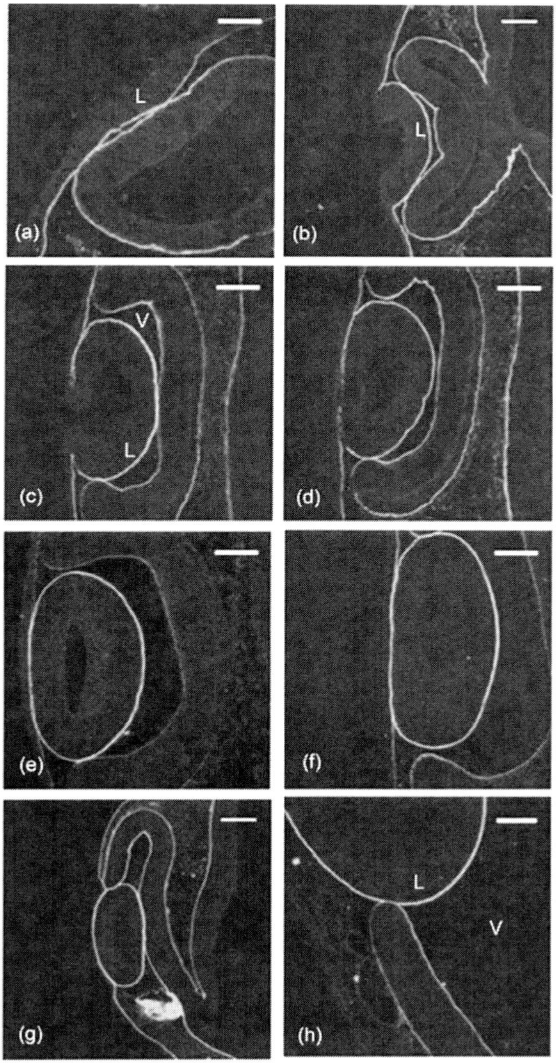

Figure 1. Immunohistochemistry micrographs of chick embryo lens primordium sections, processed with anti-Heparan-Sulphate-Perlecan antibody and a conjugate secondary antibody with fluorescein isothiocyanate (FITC). The images show the HSPG-Perlecan expression patterns during lens development. (a) (E14 H.H. lens placode). (b) (c) (E15-16 H.H. lens placode invagination). (d) (E16-17 H.H. lens vesicle). (e) (E18H.H.), (f) (E20 H.H.) and (g) (h) E25 H.H.) showing the lens fibre differentiation process. (a), (c), (d), (e), (f): Scale Bar = 50 μm; (b), (g): Scale Bar = 100 μm; H: Scale Bar = 30 μm. L: Lens. V: Vitreous.

anlagen cells, with an increased number of positive nuclei in the anterior epithelium which appear disposed in several layers and a significant number of positive nuclei in the posterior epithelium which seems to remain in an undifferentiated state. In order to test the efficiency of β-Dxyloside treatment in HSPG-Perlecan lens capsule disruption, we studied the expression of this proteoglycan by immunohistohemistry. Figure 2(g) shows how, in the β-Dxyloside treated lens, there was a large decrease in Perlecan immunolabelling compared with the controls (Figure 2(f)), displaying a non-homogeneous distribution and extensive zones with no Perlecan expression, mainly located in the posterior epithelium. Finally the FGF2 lens anlagen immunolabelling was dramatically reduced in the fibre cells and posterior part of the lens capsule in β-Dxyloside treated lens (Figure 2(k)) with respect to that of the control embryos (Figure 2(h)).

In order to confirm the specificity of our results, we made a selective enzymatic digestion of HSPG (Perlecan) in the developing lens with Heparinase II microinjection in the lens anlagen equator. Heparinase II treated embryos (Figure 3(b)) show a hystolological appearance of the lens anlagen with the same alterations as in the β-D-Xyloside treated embryos, namely, a reduced thickness in the posterior epithelium, with a dramatic decrease in the length of the lens fibre cytoplasm (with the lens vesicle cavity remaining completely uncollapsed) compared with the control (Figure 3(a)); occasionally the Heparinase II treated embryos presented a completely undifferentiated posterior epithelium with no signs of lens fibre cytoplasmic growth (Figure 3(c)). Compared with the control embryos (Figure 3(d)), BrdU immunodetection in the Heparinase II treated lens showed (Figure 3(e)) a greater number of BrdU positive nucleus in the anterior epithelium arranged in several layers and the abnormal presence of BrdU positive nucleus in the undifferentiated posterior epithelium. In order to test the effectiveness of Heparinase II enzymatic activity on HSPG-Perlecan in the chick embryo lens anlagen, we made an immunohistochemical study, which shows that the enzyme treated lens (Figure 3(g)) had greatly reduced HSPG-Perlecan in the lens capsule compared with the control embryos (Figure 3(f)), showing a non homogeneous labelling pattern with several areas of no expression of Perlecan, almost in the posterior capsule. Finally, as is showed in Figure 3(h), a dramatic decrease in FGF2 expression was detected, both in the lens fibres and in the posterior lens capsule, in the Heparinase II treated lens in comparison with the control embryos (Figure 3(i)).

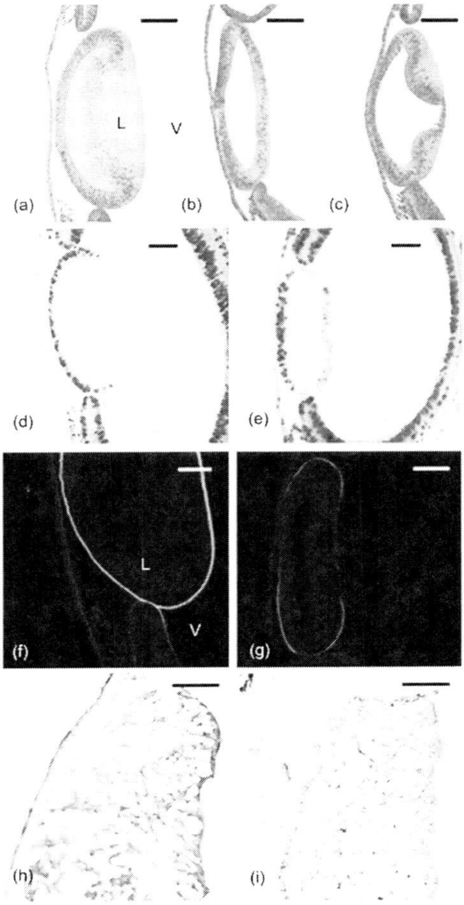

Figure 2. β-D-Xyloside treated lens anlagen. Micrographs of middle sections of chick lens anlagen of E23 H.H. embryos. Haematoxylin-Eosin stained sections of a control (a) and a β-D-Xyloside-treated embryo (b) (c) showing disruption in lens fibre differentiation. Cell replication detected by BrdU immunolabelling in a control (d) and a β-D-Xyloside-treated embryo (e); note the presence of labelled nucleus in the posterior epithelium. Anti-Heparansulphate- Perlecan antibody fluorescein-labelled immunohisto chemistry of a control (f) and a β-D-Xyloside-treated embryo (g) showing a clear decrease in HSPG- Perlecan expression in the lens capsule. Anti FGF2 lens capsule immunolabelling of a control (h) and a β-D-Xyloside-treated embryo (i) showing a dramatic lost of (a, b, c, d, e) and g: Scale Bar =50 μm. f: Scale Bar =30 μm. h and i: Scale Bar =100 μm. L: Lens. V: vitreous.

4. DISCUSSION

Our results demonstrate that Perlecan, a macromolecule specific for basement membranes, is a ubiquitous and permanent component of the lens capsule during the first stages of lens fibre differentiation. Our results also strongly suggest that Perlecan is involved in regulating lens cell behaviour, playing a key role in lens fibre differentiation, probably in association with lens capsule FGF-2 activity.

The lens capsule has been described as a specific extracellular matrix structure which isolates the lens from its surroundings, and has been shown to be a key structure in controlling lens cell behaviour, regulating the outside-inside traffic signals (Danysh [15] . Here we described the presence of Perlecan in chick embryo lens at earliest stages of development which include the beginning of the lens fibre differentiation. Although other proteoglycans have been shown to be present in the lens capsule, the main sulphated proteoglycan related with early lens development is HSPG [24] , which has been proposed to be related with the control of lens cell behaviour given its growth factor binding properties [25] [14] . Previous studies by [26] have reported the presence of Perlecan in chick developing lens from morula to 17 H.H. stage, so during lens plate induction and invagination to form the lens vesicle but this period does not include the beginning of the lens fibre differentiation; at these stages, Perlecan could be related with invagination morphogenetic mechanisms such as has been described in otic placode invagination [27] . Furthermore they describe a strong Perlecan expression in the lens without specific localization. Here we report a precise description of HSPG-Perlecan during lens vesicle period and also during lens fibre differentiation stages showing that Perlecan expression is restricted to the lens capsule.

The use of β-D xyloside or Heparinase II (specific for HSPG molecules) treatment in our experimental approach, with similar results, demonstrates that lens fibre differentiation is disrupted by the depletion of some kinds of lens HSPG. As we comment before, β-D xyloside disrupts the synthesis of sulphated proteoglycans [28] , including Heparan sulphate and Chondroitin Sulphate Proteoglycans which also have shown to play several important roles during early eye development [13] [29] [30] [31] [32] [33] , consequently, our results with β-D_xyloside treatment could be due, at least in part, to the disruption of other proteoglycans than HSPG. However, here we report experimental evidences that the disruption of β-D_xyloside treatment made at 17 H.H. stage induces similar lens development disruption than the treatment made with Heparinase II and is coincident with Perlecan depletion in the lens capsule, so we can conclude that our results are generated by HSPG-Perlecan lens depletion.

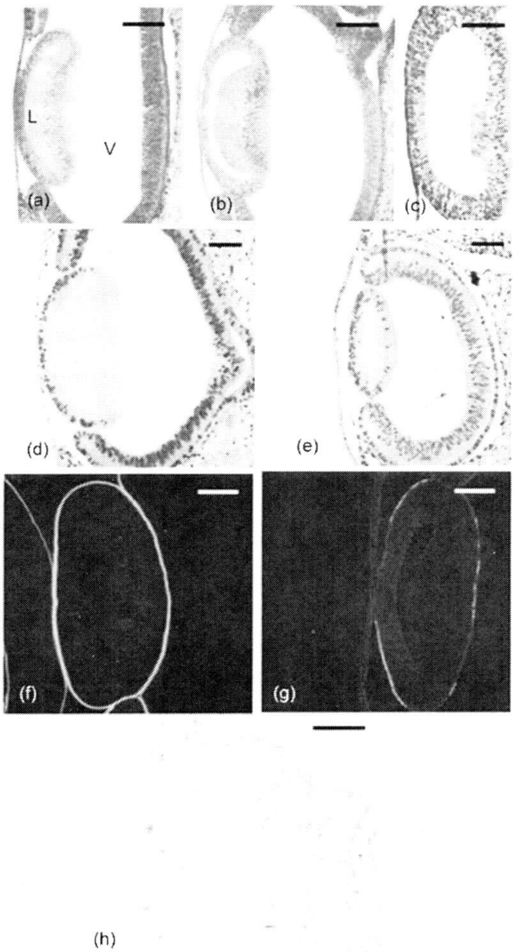

Figure 3. Heparinase II-treated lens anlagen: Micrographs of middle sections of chick lens anlagen of E23H.H. embryos. Haematoxylin-Eosin stained sections of a control (a) and a Heparinase II-treated embryo (b, c) showing disruption in lens fibre differentiation. Cell replication detected by BrdU immunolabelling in a control (d) and a Heparinase II-treated embryo (e); note the presence of labelled nucleus in the posterior epithelium. Anti-Heparan-Sulphate-Perlecan antibody fluorescein labelled immune-histochemistry of a control (f) and a Heparinase II-treated embryo (g) showing a clear decrease in HSPG-Perlecan expression in the lens capsule. Anti FGF2 lens capsule immunolabelling of a control (h) and a Heparinase II-treated embryo (h) showing a dramatic lost of FGF2. a, b and h:Scale Bar =100 μm. c, d, e, f, g: Scale Bar = 50 μm. L: Lens. V: Vitreous.

Some studies described the presence of non specific HSPG located in the lens capsule during development [14] [15] . This work highlights the key role of Perlecan (a complex macromolecule rich in heparan sulphate [34]), in lens fibre differentiation at early chick development. Supporting our results, an important role for Perlecan has been proposed by Rossi et al. [35] during late development in view of lens capsule degeneration in knock-out mice lacking the protein core of Perlecan. Perlecan has been also described as a key constituent of the lens capsule in adult cows [36] [37] .

Our results also show that the disruption of Perlecan in the lens capsule of chick embryos, affects the whole range of cellular dynamics in the developing lens, showing an increase in the number of BrdU positive cells in the anterior epithelium, the presence of undifferentiated cells (BrdU positive) in the posterior epithelium and a deficit in lens fibre differentiation from the morphological (cytoplasmic elongation) point of view. Although it has been suggested that the different components of lens cell dynamics (replication, migration and differentiation) may depend on different and specific factors [38] , our results suggest that all of them depend, at least initially, on common regulatory factors linked to the HSPG-Perlecan in lens capsule.

Given this important role of lens capsule Perlecan during development, an interesting question is the source of this macromolecule detected in the lens capsule. In this regard, Dong et al. [39] have demonstrated (by Perlecan m-RNA expression) that lens Perlecan, at the earliest stages of lens development, originates from the primary fibres in the posterior epithelium, and later Perlecan mRNA expression becomes restricted to lens fibres in the equatorial area, where all the cellular processes (replication, migration and differentiation) of lens differentiation take place. Our results show that the lens capsule is the principal structure containing Perlecan during early lens development, which strongly suggests that the primary lens fibres could play a key role in controlling new lens fibre differentiation, regulating the synthesis and "deposition" of Perlecan in the lens capsule.

Given the binding biological properties of HSPG containing molecules, as Perlecan, they are been proposed as molecules able to create a reservoir of growth factors in the surroundings of lens cells and in particular near to the region of lens fibre differentiation [25] . Our results show that Perlecan depleted lens, lost their FGF2 expression both in the lens capsule and in the fibre cells allowing us suppose that the relation of lens capsule Perlecan with lens fibre differentiation are also linked to the regulation of FGF2 concentration during lens developmet.

The lens is a biological structure in which two differentiated but interdependent cellular sub-populations coexist with transitional cellular behaviour (replication, migration and differentiation); the location is in the equatorial area and duration is life-long. A complex control involves many different molecular events in which FGF2 activity remains, to date, the main and initial regulatory mechanisms involved [36] [37] [40] [41] [43] . The traditional concept of FGF2 activity has been based on the gradient theory, which implies an increase in FGF2 concentration from anterior (surface) to posterior (deep) cellular populations, especially in the equatorial area [42] , Chamberlain [5] [43] [44] . However, in the last decade a new concept has developed in which the influence exerted by FGF2 on lens cellular behaviour is based on local gradients of FGF2 "activity". Recently, have been described new systems of complex regulation have in which the primary activity of and FGF-2 is activated and/or inhibited by secondary control mechanisms [45] [46] . These new theories propose that the undifferentiated stage of lens anterior epithelial cells is due to the expression of inhibitory genes such as Sprouty and Sef, which block the activity of the FGF-2 receptor [8] [39] [41] , making a "low FGF2 activity" area.

Besides, several studies support the hypothesis of the reservoir of extracellular matrix growth factors, which in the lens capsule could also have an important regulatory influence on lens cell behaviour [36] [47] . Our results show that a depletion in lens capsule Perlecan during early lens chick development induces a near complete loss of FGF2 in the lens (both citoplasmic and capsular) and leads to severe disruption in lens dynamic and fibre differentiation. Supporting these results, Perlecan-linked HSPG in the adult cow lens capsule has been shown to regulate FGF2 activity, both increasing the local concentration of this growth factor near the lens fibres as well as enhancing FGF2 activity, probably by an FGF2 transdimerization which increases the efficiency of transmembrane signalling [34] [36] [48] . Moreover, the biological activity of lens capsule Perlecan could be regulated by an equilibrium between the synthesis of Perlecan by the lens fibres [39] , which regulate the concentration of FGF2, and the activity of Metalloproteinases such as MMP2 in the lens capsule, which regulates the release of free FGF2 from this reservoir; this has been demonstrated in the adult cow lens [15] [36] .

Taken together, all these data strongly support the hypothesis that the extracellular matrix organised as lens capsule plays a key role in lens fibre differentiation during development. This influence seems to be due in a great extent to their content in Perlecan which probably acts as FGF2 concentration regulator.

ACKNOWLEDGEMENTS

The authors thank to Dr. David Rixham for language translation assistance. This work was supported by the following Grants: Ministerio de Educación y Ciencia (BFU207/6516), Instituto de Salud Carlos III (PI020961), Junta de Castilla y León (VA21A07, VA049/04, SAN673/VA15/08) and Federación Regional de Cajas de Ahorro.

CITE THIS PAPER

Martín, C., Alonso, M.I., Lamus, F., Moro, J.A., De la Mano, A., Fernandez, J.M., Caballero, A. and Gato, Á. (2017) Lens Capsule HSPG-Perlecan Regu- lates Lens Fibre Differentiation during Chick Embryo Development. Open Journal of Veterinary Medicine, 7, 9-22.

REFERENCES

1. Chow, R.L. and Lang, R.A. (2001) Early Eye Development in Vertebrates. Annual Review of Cell and Developmental Biology, 17, 255-296.
2. Bassas, L., Zelenka, P.S., Serrano, J. and de Pablo, F. (1987) Insulin and IGF Receptors Are Developmentally Regulated in the Chick Embryo Eye Lens. Experimental Cell Research, 168, 561-566.
3. Bassnett, S. and Beebe, D.C. (1990) Localization of Insulin-Like Growth Factor-1 Binding Sites in the Embryonic Chicken Eye. Investigative Ophthalmology & Visual Science, 31, 1637-1643.
4. de Iongh, R.U., Lovicu, F.J., Overbeek, P.A., Schneider, M.D., Joya, J., Hardeman, E.D. and McAvoy, J.W. (2001) Requirement for TGFbeta Receptor Signaling during Terminal Lens Fiber Differentiation. Development, 128, 3995-4010.
5. Chamberlain, C.G. and McAvoy, J.W. (1989) Induction of Lens Fibre Differentiation by Acidic and Basic Fibroblast Growth Factor (FGF). Growth Factors, 1, 125-134.
6. Wang, Q.A., McAvoy, J.W. and Lovicu, F.J. (2010) Growth Factor Signaling in Vitreous Humor-Induced Lens Fiber Differentiation. Investigative Ophthalmology & Visual Science, 51, 3599-3610.
7. Kim, H.J. and Bar-Sagi, D. (2004) Modulation of Signalling by Sprouty: A Developing Story. Nature Reviews Molecular Cell Biology, 5, 441-450.
8. Boros, J., Newitt, P., Wang, Q., McAvoy, J.W. and Lovicu, F.J. (2006) Sef and Sprouty Expression in the Developing Ocular Lens: Implications for Regulating Lens Cell Proliferation and Differentiation. Seminars in Cell & Developmental Biology, 17, 741-752.
9. Ruoslahti, E. and Yamaguchi, Y. (1991) Proteoglycans as Modulators of Growth Factor Activities. Cell, 64, 867-869.
10. Goetz, R. and Mohammadi, M. (2013) Exploring Mechanisms of FGF Signalling through the Lens of Structural Biology. Nature Reviews Molecular Cell Biology, 14, 166-180.

11. Hendrix, R.W. and Zwaan, J. (1974) Changes in the Glycoprotein Concentration of the Extracellular Matrix between Lens and Optic Vesicle Associated with Early Lens Differentiation. Differentiation, 2, 357-362.
12. Gato, A., Martin, C., Alonso, M.I., Martinez-Alvarez, C. and Moro, J.A. (2001) Chondroitin Sulphate Proteoglycan Is Involved in Lens Vesicle Morphogenesis in Chick Embryos. Experimental Eye Research, 73, 469-478.
13. Bard, J.B.L. and Abbott, A.S. (1979) Matrices Containing Glycosaminoglycans in the Developing Anterior Chambers of Chick and Xenopus Embryonic Eyes. Developmental Biology, 68, 472-486.
14. Halfter, W. and Schurer, B. (1994) A New Heparan Sulfate Proteoglycan in the Extracellular Matrix of the Developing Chick Embryo. Experimental Cell Research, 214, 285-296.
15. Danysh, B.P. and Duncan, M.K. (2009) The Lens Capsule. Experimental Eye Research, 88, 151-164.
16. Hamburger, V. and Hamilton, H.L. (1951) A Series of Normal Stages in the Development of the Chick Embryo. Journal of Morphology, 88, 49-92.
17. Martin, C., Alonso, M.I., Santiago, C., Moro, J.A., De la Mano, A., Carretero, R. and Gato, A. (2009) Early Embryonic Brain Development in rats Requires the Trophic Influence of Cerebrospinal Fluid. International Journal of Developmental Neuroscience, 27, 733-740.
18. Murdoch, A.D. and Iozzo, R.V. (1993) Perlecan: The Multidomain Heparan Sulphate Proteoglycan of Basement Membrane and Extracellular Matrix. Virchows Archiv A, 423, 237-242.
19. Timpl, R. (1993) Proteoglycans of Basement Membranes. Experientia, 49, 417-428.
20. Timpl, R. (1994) Proteoglycans of Basement Membranes. EXS, 70, 123-144.
21. Lugemwa, F.N. and Esko, J.D. (1991) Estradiol Beta-D-Xyloside, an Efficient Primer for Heparan Sulfate Biosynthesis. Journal of Biological Chemistry, 266, 6674-6677.
22. Govindraj, P., West, L., Smith, S. and Hassell, J.R. (2006) Modulation of FGF-2 Binding to Chondrocytes from the Developing Growth Plate by Perlecan. Matrix Biology, 25, 232-239.
23. Schulz, M.W., Chamberlain, C.G., de Iongh, R.U. and McAvoy, J.W. (1993) Acidic and Basic FGF in Ocular Media and Lens: Implications for Lens Polarity and Growth Patterns. Development, 118, 117-126.
24. Haloui, Z., Jeanny, J.C., Jonet, L., Courtois, Y. and Laurent, M. (1988) Immunochemical Analysis of Extracellular Matrix during Embryonic Lens Development of the Cat Fraser Mouse. Experimental Eye Research, 46, 463-474.
25. Lovicu, F.J. and McAvoy, J.W. (1993) Localization of Acidic Fibroblast Growth-Factor, Basic Fibroblast Growth-Factor, and Heparan-Sulfate Proteoglycan in Rat Lens—Implications for Lens Polarity and Growth-Patterns. Investigative Ophthalmology & Visual Science, 34, 3355-3365.
26. Soulintzi, N. and Zagris, N. (2007) Spatial and Temporal Expression of Perlecan in the Early Chick Embryo. Cells Tissues Organs, 186, 243-256.
27. Moro-Balbas, J.A., Gato, A., Alonso, M.I., Martin, P. and de la Mano, A. (2000) Basal Lamina Heparan Sulphate Proteoglycan Is Involved in Otic Placode Invagination in Chick Embryos. Anatomy and Embryology, 202, 333-343.
28. Schwartz, N.B., Galligani, L., Ho, P.L. and Dorfman, A. (1974) Stimulation of Synthesis of Free Chondroitin Sulfate Chains by Beta-D-Xylosides in Cultured Cells. Proceedings of the National Academy of Sciences of the United States of America, 71, 4047-4051.
29. Peterson, P.E., Pow, C.S., Wilson, D.B. and Hendrickx, A.G. (1995) Localisation of Glycoproteins and Glycosaminoglycans during Early Eye Development in the Macaque. Journal of Anatomy, 186, 31-42.
30. Ring, C., Lemmon, V. and Halfter, W. (1995) Two Chondroitin Sulfate Proteoglycans Differentially Expressed in the Developing Chick Visual System. Developmental Biology, 168, 11-27.

31. Yao, R.H., Alcala, J. and Maisel, H. (1996) Developmental Changes in Glycoconjugate Composition during Chick Lens Morphogenesis. Experimental Eye Research, 62, 419-431.
32. Van Rybroek, J.J. and Olson, M.D. (1981) Surface Coat Material Associated with the Cells of the Developing Lens Vesicle in the Chick Embryo. Anatomical Record, 201, 261-271.
33. Gato, A., Moro, J.A., Alonso, M.I., Pastor, J.F., Represa, J. and Barbosa, E. (1993) Chondroitin Sulfate Proteoglycan and Embryonic Brain Enlargement in the Chick. Anatomy and Embryology, 188, 101-106.
34. Knox, S., Merry, C., Stringer, S., Melrose, J. and Whitelock, J. (2002) Not All Perlecans Are Created Equa: Interactions with Fibroblast Growth Factor (FGF) 2 and FGF Receptors. Journal of Biological Chemistry, 277, 14657-14665.
35. Rossi, M., Morita, H., Sormunen, R., Airenne, S., Kreivi, M., Wang, L., Fukai, N., Olsen, B.R., Tryggvason, K. and Soininen, R. (2003) Heparan Sulfate Chains of Perlecan Are Indispensable in the Lens Capsule but Not in the Kidney. EMBO Journal, 22, 236-245.
36. Tholozan, F.M., Gribbon, C., Li, Z., Goldberg, M.W., Prescott, A.R., McKie, N. and Quinlan, R.A. (2007) FGF-2 Release from the Lens Capsule by MMP-2 Maintains Lens Epithelial Cell Viability. Molecular Biology of the Cell, 18, 4222-4231.
37. Tholozan, F.M. and Quinlan, R.A. (2007) Lens Cells: More than Meets the Eye. International Journal of Biochemistry & Cell Biology, 39, 1754-1759.
38. McAvoy, J.W., Stump, R.J.W., Ang, S., Chen, Y. and Lovicu, F.J. (2005) Growth Factor Regulation of Lens Development: Key Roles for FGF and Wnt in Promoting Lens Cell Differentiation and Polarity. Mechanisms of Development, 122, S99-S99.
39. Dong, S., Landfair, J., Balasubramani, M., Bier, M.E., Cole, G. and Halfter, W. (2002) Expression of Basal Lamina Protein mRNAs in the Early Embryonic Chick Eye. Journal of Comparative Neurology, 447, 261-273.
40. Le, A.C.N. and Musil, L.S. (2001) FGF Signaling in Chick Lens Development. Developmental Biology, 233, 394-411.
41. Lovicu, F.J. and McAvoy, J.W. (2005) Growth Factor Regulation of Lens Development. Developmental Biology, 280, 1-14.
42. Wride, M.A. (1996) Cellular and Molecular Features of Lens Differentiation: A Review of Recent Advances. Differentiation, 61, 77-93.
43. Chamberlain, C.G. and McAvoy, J.W. (1997) Fibre Differentiation and Polarity in the Mammalian Lens: A Key Role for FGF. Progress in Retinal and Eye Research, 16, 443-478.
44. McAvoy, J.W., Chamberlain, C.G., Deiongh, R.U., Richardson, N.A. and Lovicu, F.J. (1991) The Role of Fibroblast Growth-Factor in Eye Lens Development. Annals of the New York Academy of Sciences, 638, 256-274.
45. Iyengar, L., Patkunanathan, B., McAvoy, J.W. and Lovicu, F.J. (2009) Growth Factors Involved in Aqueous Humour-Induced Lens Cell Proliferation. Growth Factors, 27, 50-62.
46. Dawes, L.J., Sugiyama, Y., Lovicu, F.J., Harris, C.G., Shelley, E.J. and McAvoy, J.W. (2014) Interactions between Lens Epithelial and Fiber Cells Reveal an Intrinsic Self-Assembly Mechanism. Developmental Biology, 385, 291-303.
47. Mott, J.D. and Werb, Z. (2004) Regulation of Matrix Biologyogy by Matrix Metalloproteinases. Current Opinion in Cell Biology, 16, 558-564.
48. Goodger, S.J., Robinson, C.J., Murphy, K.J., Gasiunas, N., Harmer, N.J., Blundell, T.L., Pye, D.A. and Gallagher, J.T. (2008) Evidence That Heparin Saccharides Promote FGF2 Mitogenesis through Two Distinct Mechanisms. Journal of Biological Chemistry, 283, 13001-13008.

CHAPTER 11

Raw Meat-Based Diets in Dogs and Cats

Maria Fredriksson-Ahomaa [1,*], Tiina Heikkilä [1], Noora Pernu [1], Sara Kovanen [1], Anna Hielm-Björkman [2] and Rauni Kivistö [1]

[1] *Department of Food Hygiene and Environmental Health, Faculty of Veterinary Medicine, P.O.Box 66, 00014 University of Helsinki, Finland*

[2] *Department of Equine and Small Animal Medicine, Faculty of Veterinary Medicine, P.O.Box 66, 00014 University of Helsinki, Finland*

ABSTRACT

Feeding pets raw meat-based diets (RMBDs) is commonly practiced by many companion animal owners and has received increasing attention in recent years. It may be beneficial for the animals, but may also pose a health risk for both pets and their owners, as RMBDs may be contaminated by enteric pathogens—such as *Campylobacter, Salmonella,* and *Yersinia*—which are the most common zoonotic bacteria causing enteritis in humans. Little information exists on the prevalence of these pathogens in pet food, and thus one aim was to investigate the prevalence of *Campylobacter, Salmonella,* and *Yersinia* in commercial RMBDs from retail stores. Little evidence also exists on the significance of raw meat feeding on the shedding of *Campylobacter, Salmonella,* and enteropathogenic *Yersinia* in the feces of pets, and therefore, the second goal was to study the presence of these pathogens in dogs and cats fed RMBDs. Polymerase chain reaction (PCR) only sporadically detected *Campylobacter, Salmonella,* and enteropathogenic *Yersinia* in RMBDs.

These pathogens were not found by culturing, indicating a low contamination level in frozen RMBDs. They were also detected in the feces of dogs and cats, but the association with feeding RMBDs to them remained unclear.

Keywords: raw food; zoonotic enteric pathogens; PCR; dogs; cats

1. INTRODUCTION

Most dogs and cats are fed commercially produced dry (pellets/kibble) or wet (canned) food, but the popularity of raw-meat based diets (RMBDs) is rising [1,2]. Sometimes RMBDs are called BARF (Bones And Raw Food or Biologically Appropriate Raw Food) or RAP (Raw Animal Products) [3,4]. RMBDs can contain skeletal muscle, fat, internal organs, cartilage, and bones from farm animals (ruminants, pigs, and poultry), horses, game, and fish. RMBDs can be prepared at home or sold commercially. The most common form of commercial RMBDs are frozen, but freeze-dried products have recently also entered the market [3,5]. Some are intended to be nutritionally complete and balanced, but products intended only for supplemental feeding also exist [3].

Proponents of feeding RMBDs to their pets claim important health benefits for the diets, such as improvement of coat and skin, and reduction in dental diseases [3]. However, health risks, such as enteric pathogens in raw meat, an unbalanced diet and internal punctures caused by bones in the food, may also occur [3]. Raw meat and internal organs can easily be contaminated already during slaughter through feces and tonsils, which are frequently positive for *Campylobacter, Salmonella*, and *Yersinia* [6,7]. Raw poultry is most commonly contaminated by *Campylobacter* and *Salmonella*, and raw pork by *Yersinia* and *Salmonella* [6,8]. Therefore, pet food containing raw meat may also be contaminated with these bacterial pathogens. Still, only a limited number of studies are available on the prevalence of these enteric pathogens in RMBDs (Table 1).

Campylobacteriosis, salmonellosis, and yersiniosis are the most commonly reported notifiable gastroenteritis in the European Union (EU) [27]. The diseases are typically foodborne, especially through eating or handling undercooked or raw meat products. Contact with pets is also considered a risk factor for infection. These pathogens are widespread in nature and found in the gastrointestinal tract of wild and domestic mammals and birds. They have also been reported in dogs and cats with and without diarrhea [14,16,28]. However, the reported prevalence rates vary highly between the studies (Table 1). *Campylobacter* is the most commonly found zoonotic pathogen in the feces of both dogs and cats. They mainly carry *C. upsaliensis*, which very rarely causes campylobacteriosis in humans. *C. helveticus*, which is typically not associated with human disease, has frequently been isolated from cats (Table 1).

Table 1. The prevalence of *Campylobacter, Salmonella*, and *Yersinia enterocolitica* in feces or rectal swabs of dogs and cats and in commercial raw meat-based diets (RMBDs).

Sample	No.	Country	*Campylobacter*				*Salmonella*	*Yersinia Enterocolitica* [a]	Reference
			spp.	*Jejuni*	*Upsaliensis*	*Helveticus*			
Dogs	138	Canada					23%		[9]
	251	Canada	43%	6%	37%		1%		[10]
	1212	China						8%	[11]
	190	Finland	28%	9%	17%				[12]
	4325	Germany [b]						4%	[13]
	147	Ireland	43%	10%	29%				[14]
	171	Italy	17%	9%	5%				[15]
	90	New Zealand	36%	13%	23%	1%			[16]
	529	Norway	23%	3%	20%				[17]
	290	Spain	35%	14%	21%				[18]
	180	Sweden	37%	4%	29%	1%			[19]
	249	UK	38%	1%	38%				[20]
	126	UK					1%		[21]
	130	USA	1%				2%		[22]
	554	USA					5%		[23]
Cats	84	Finland	32%	8%	7%	17%			[12]
	2624	Germany [c]						0.3%	[13]
	35	Ireland	42%	9%	26%	6%			[14]
	102	Italy	15%	8%	6%	1%			[15]
	110	New Zealand	16%	5%	5%	7%			[16]
	301	Norway	18%	3%	13%				[17]
RMBDs	25	Canada	<4%				20%		[24]
	40	Canada	<3%				5%		[25]
	50	New Zealand	28%	22%					[16]
	196	USA					8%		[5]
	240	USA	<0.4%				7%		[26]

[a] pathogenic bioserotype or *ail* positive; [b] mainly Germany (77%), and 13 other European countries (33%); [c] mainly Germany (81%), and 10 other European countries (19%).

The goals of our study were to investigate the prevalence of *Campylobacter*, *Salmonella*, and enteropathogenic *Yersinia* in commercial RMBDs and to determine whether dogs and cats given daily RMBDs excrete these enteric bacteria in their feces.

2. MATERIALS AND METHODS

2.1. Samples and Sample Preparation

In total, 88 RMBDs originating from 12 producers were bought from retail shops in 2015 and 2016 and studied before expiration dates. All producers were from Finland and only domestic meat was used in the products, which were intended for dogs and/or cats. The most common raw meat in commercial RMBDs were beef (43%) and poultry (41%), followed by pork (27%) (Table 2). Twenty-one RMBDs contained more than one raw meat source. In addition to skeletal muscle, some of the diets included organs and bone/cartilage. All RMBDs were frozen and thawed at room temperature before screening with polymerase chain reaction (PCR) and culturing.

Table 2. Raw meat-based diets (RMBDs) from different producers studied in 2015 and 2016.

Producer	No. of RMBDs	Fed to		Number of Samples Containing						
		Dogs	Cats	Poultry	Pork	Beef	Sheep	Horse	Game	Fish
A	6	6	0	1	3	2	0	0	1	1
B	43	26	17	22	8	13	1	2	1	6
C	10	8	2	1	1	9	0	0	0	0
D	1	1	0	0	1	0	0	0	0	0
E	9	6	3	0	4	8	0	0	0	1
F	8	2	6	7	4	3	0	0	0	0
G	1	1	0	0	0	1	0	0	0	1
H	5	0	5	3	0	2	0	0	0	0
I	1	0	1	0	1	0	0	0	0	0
J	1	0	1	1	0	0	0	0	0	0
K	2	0	2	0	2	0	0	0	0	0
L	1	0	1	1	0	0	0	0	0	0
All	88	50	38	36	24	38	1	2	2	9

A total of 50 fecal samples from 29 dogs fed raw meat and 21 dogs not fed raw meat during 2013 and 2014. All fecal samples from dogs were stored at −70 °C until PCR screening and culturing. The dogs fed raw meat received only commercial or home-made RMBDs daily and the dogs not fed raw meat received only dry pellets daily several months before sampling. The dogs did not have signs of diarrhea.

A total of 75 fecal samples were collected from two indoor cats living in the same household, 47 samples of which were obtained from the older cat (born 19 April 2015) between 3 November 2015 and 28 November 2016 and 28 from the younger cat (born 22 January 2016) between 13 May 2016 and 28 November 2016. Both cats have received RMBDs daily and have lived indoor their whole lives. During the fecal sampling time from 3 November 2015 to 28 November 2016, 48 samples from 38 RMBDs fed to the cats were studied (Table 2). Poultry (61%) was the most common meat type in the diets, followed by pork (32%) and beef (29%).

Approximately a 10-g food sample was removed from the packaging and mixed thoroughly with the 90 mL of buffered peptone water (BPW, LAB M, Kerava, Finland) in the stomacher bag by hand or by a stomacher blender (LabBlender 400, London, UK) for 1 min. About a 1-g fecal sample was mixed with 9 mL BPW shortly with a vortex. All samples were incubated at 37 °C overnight (16–18 h) before DNA extraction.

2.2. PCR Screening of Pathogens

The presence of *Campylobacter, Salmonella*, and enteropathogenic *Yersinia* was screened by real-time PCR based on SYBRGreen [29]. DNA was isolated from 1 mL of the ON enrichment by ZR Fecal DNA MiniPred™ (Nordic BioSite Oy, Helsinki, Finland) according to the manufacturer's instructions. Two µL of the DNA was added to 18 µL of mastermix (iQ™ SYBRGreen Supermix, BioRad) containing primers specific for *Campylobacter* [30], *Salmonella* [31], and enteropathogenic *Yersinia* [32,33]. The threshold cycle (Ct) under 39 and a specific melting temperature indicated a positive result. Additionally, *ail*-positive samples were run separately with *ail* primers for *Y. enterocolitica* [32] and *Y. pseudotuberculosis* [33]. *Campylobacter* (*rrn*)-positive samples were separately studied with primers specific for *C. jejuni* (*map*) and *C. coli* (*ceu*) [34].

2.3. Isolation of Campylobacter, Salmonella, and Yersinia

The presence of *Salmonella* and *Yersinia* were studied by enrichment and selective agar plates [29]. For *Salmonella* isolation enrichment on semisolid Rappaport-Vassiliadis (MSRV, Labema) for 24 h at 42 °C was used before inoculating on selective xylose-lysine-deoxycholate (XLD, Labema) plates which were incubated for 24 h at 37 °C. Cold enrichment at 6 °C for at least three weeks was used for *Yersinia* isolation. After cold enrichment, 10 µL was inoculated on CIN plates which were incubated at 30 °C for 20–24 h. Up to four typical colonies on XLD and CIN plates were sub-cultured on blood agar plates and identified with API 20E (BioMerieux, France). Serotyping was done with commercial antisera (Denka Seikan, Japan). The presence of *Campylobacter* was studied by plating (1 g feces/1 mL 0.9% saline) on modified charcoal cefoperazone deoxycholate agar (mCCDA) with cefoperazone and amphotericin (Oxoid, Basingstoke, UK) and identification was done by gram staining and PCR according to Olkkola et al. [28].

2.4. Statistical Analyses

Statistical analyses were performed using the analytical software package SPSS® Statistics Version 24 (IBM Corp., Armonk, NY, USA). Fisher's exact test was used to analyze the relation between feeding RMBDs and dry pellets and the presence of *Campylobacter*, *Salmonella*, and enteropathogenic *Yersinia*.

3. RESULTS

Enteric pathogens were detected in 28% of the RMBDs, originating from 12 producers (Table 3). *Campylobacter* was the most frequent pathogen, detected by PCR in 15% of the RMBDs. However, all samples were negative for *C. jejuni* and *C. coli*. *Y. enterocolitica* carrying the *ail* gene was detected in 11% of the samples. *Salmonella* was detected in only 2% of the samples and, surprisingly, *Y. pseudotuberculosis* was also detected in two samples.

Table 3. Occurrence of zoonotic enteric bacteria in raw meat-based diets.

Producer	Number of Samples	Number of PCR-Positive Samples				
		All	*Campylobacter*	*Salmonella*	*Yersinia Enterocolitica*	*Yersinia Pseudotuberculosis*
A	6	1	0	0	1	0
B	43	14	8	1	6	0
C	10	0	0	0	0	0
D	1	0	0	0	0	0
E	9	2	2	0	0	0
F	8	2	0	0	1	1
G	1	1	0	0	1	0
H	5	2	1	0	1	0
I	1	0	0	0	0	0
J	1	1	1	0	0	0
K	2	1	1	0	0	0
L	1[a]	1	0	1	0	1
All	88	25	13	2	10	2

[a] One sample containing poultry from producer L was contaminated with *Salmonella* and *Y. pseudotuberculosis*.

Campylobacter was most frequently found in RMBDs containing beef (21%) and *Y. enterocolitica* in RMBDs containing pork (15%) (Table 4). The two *Y. pseudotuberculosis*-positive samples were both from raw poultry products, one of which was also *Salmonella* positive. All samples were culture negative.

Table 4. Occurrence of zoonotic enteric bacteria in different meats used in raw meat-based diets.

Meat	Number of Samples	Number of PCR-Positive Samples				
		All	*Campylobacter*	*Salmonella*	*Yersinia Enterocolitica*	*Yersinia Pseudotuberculosis*
Beef	38	9	8	0	1	0
Pork	24	8	4	0	4	0
Poultry	36	8	4	1	4	2
Sheep	1	0	0	0	0	0
Horse	2	1	0	1	1	0
Game	2	0	0	0	0	0
Fish	9	5	3	0	2	0

Campylobacter was detected by PCR in 55% of the tested fecal samples from 29 dogs fed raw meat and in 33% fecal samples from 21 dogs fed dry pellets (Table 5). *Campylobacter* was clearly, but not significantly (p = 0.158), more frequently detected in dogs fed RMBDs than in dogs fed dry pellets. Only a weak positive association (Phi = 0.216) was observed between dogs fed raw meat and the presence of *Campylobacter*. Of the dogs fed raw meat, two (7%) shed *Salmonella* and one (3%) shed *ail*-positive *Y. enterocolitica*.

Table 5. Detection of zoonotic enteric bacteria in fecal samples of dogs fed raw meat-based diets (RMBDs) or dry pellets.

Feeding	Number of Dogs	Number of Dogs Shedding					
		Campylobacter		*Salmonella*		*Y. Enterocolitica*	
RMBDs	29	16	(55%)	2	(7%)	1	(3%)
Dry pellet	21	7	(33%)	0		0	

Campylobacter was isolated from 20 dogs and most (90%) of them shed *C. upsaliensis*. *C. jejuni* was isolated from two dogs, and *Salmonella* and *ail*-positive *Y. enterocolitica* from one dog each. All these three pathogens were isolated only from dogs fed RMBDs.

Both cats were sampled between 10 May 2016 and 28 November 2016, and *Campylobacter* was detected in all (100%; 26/26) fecal samples from the old cat and in most (96%; 27/28) samples from the young cat by PCR and *C. helveticus* was isolated from both cats. Between 3 November 2015 and 3 May 2016, before the young cat came to the household, the old cat was *Campylobacter*-positive only once (5%; 1/21).

Y. enterocolitica was detected in two fecal samples from the old cat in January, and *Y. enterocolitica* of bioserotype 4/O:3 was isolated from both samples. In June 2016, during the same sampling day, *Salmonella* was detected from both cats by PCR, but not by culturing.

4. DISCUSSION

Pets are important members of many households, and several health benefits from pet ownership have been documented, including reduction in loneliness and depression. Despite the benefits, companion animals may be one potential source of zoonotic infections to the owners. Feeding RMBDs, a current trend among dog and cat owners, may transmit zoonotic meat-borne pathogenic bacteria to pets via contaminated RMBDs. However, only a few studies have been conducted on the prevalence of zoonotic enteropathogenic bacteria, such as *Salmonella*, *Campylobacter*, and *Yersinia*, in RMBDs (Table 1). In our study, *Campylobacter* was detected by PCR in 15% of the commercial RMBDs. However, no *C. jejuni*or *C. coli*, which are typical human pathogenic species, were identified, showing that these pathogens appear to be rare in pet food, at least in Finland where the prevalence of *Campylobacter* is below 10% in broilers at retail level (https://www.evira.fi/en/about-evira/news/2016/campylobacter-most-prevalent-in-poultry-in-late-summer/). *Y. enterocolitica*carrying the *ail*-gene was detected in 11% of the samples and more often in RMBDs containing pork. Fattening pigs at slaughter carry pathogenic *Y. enterocolitica* frequently in their tonsils, and raw pork is considered the main infection source [35]. *Salmonella* was rarely detected in our study and the occurrence (2% by PCR and <1% by culturing) was clearly lower than reported in Canada and USA (Table 1). One reason is probably the very low (<1%) prevalence of *Salmonella* in Finnish farm animals, including poultry [36]. All samples in our study were negative by culturing, indicating that the number of zoonotic enteric bacteria is low in domestic frozen RMBDs. The bacteria may also be unculturable due to prolonged stress during storage in the freezer.

Pets fed raw meat are suspected to shed meat-borne pathogens more often in their feces than pets not fed raw meat. In our study, dogs fed raw meat shed *Campylobacter* more frequently than dogs fed dry pellets; however, the difference was not statistically significant. Furthermore, the most common species isolated from dog feces was *C. upsaliensis*, and *C. jejuni*was rarely isolated. *Campylobacter* is a very common finding, and *C. upsaliensis* is the most common species found in dog feces also in several other studies (Table 1). Dogs and cats are shown to be the main reservoirs for *C. upsaliensis*[12,14,18,20]. *C. upsaliensis* appears to be commensal in dogs and cats, and no differences between the isolation rates in healthy and diarrhoeic animals have been reported [12,17]. However, their lifestyle may affect the shedding of pathogens in the feces. Dogs exposed to outdoor water sources have an increased risk of shedding *C. upsaliensis* [10]. In our study, the association between feeding raw meat and shedding *C. upsaliensis* remained unclear; however, *C. jejuni* was only isolated from dogs fed RMBDs. *Salmonella* and *Y. enterocolitica* were rarely detected in the dog feces in our study. Both pathogens were isolated from only one dog each; however, both positive dogs were fed RMBDs. In other studies, the isolation rates of *Salmonella* in dog feces varied between 1% and 23% (Table 1). Lentz et al. demonstrated that dogs fed raw meat were more likely to shed *Salmonella* in their feces than dogs not fed raw meat [25]. Joffe and Schlesinger demonstrated that dogs fed raw chicken shed *Salmonella* in their feces, and they also showed that raw chicken was frequently contaminated with *Salmonella* [37].

We also studied two indoor cats fed RMBDs on a daily basis the whole life, and were able to demonstrate that ever since the young cat (three months old) came to the household, both cats shed *Campylobacter* in the feces. All samples from the old cat and all but one from the young cat were *Campylobacter*-positive from 5 October 2016 until 28 November 2016, when we stopped the sampling. *C. helveticus* was isolated from the feces of both cats, which is the species mostly found in cats [12,14]. It is possible that the young cat brought the *Campylobacter* to the household and infected the older one. Younger animals (under one year) are more likely to shed *Campylobacter*, as they may be more susceptible to colonization than older individuals [20]. However, the old cat shedding *Campylobacter* in its feces for over five months until the end of the sampling was also an interesting observation in our study. In an earlier study, long-term colonization of *Campylobacter* spp. has been described in dogs and cats [38]. In our study, during the period that the cats shed *Campylobacter* in the feces, *Campylobacter*was also detected more frequently in the food fed to the cats; however, the link between eating raw meat and shedding *Campylobacter* remained unclear.

Salmonella and enteropathogenic *Yersinia* were rarely detected in the cat feces and RMBDs fed to them in our study. However, *Y. enterocolitica* belonging to bioserotype 4/O:3, which is the most common human pathogenic type typically transmitted by raw or undercooked pork, was surprisingly isolated twice from the old cat's feces before the young cat came to the household. The cat has only lived indoors and was fed RMBDs containing pork, so most possibly the transmission to the cat occurred through contaminated food. Raw pork is a potential source of *Y. enterocolitica* 4/O:3 infections in dogs and cats [39]. *Salmonella* was detected in both cats on the same sampling time, indicating that RMBDs fed to them could be the infection source.

This study shows that RMBDs can contain zoonotic pathogens that can be a health hazard both to pets and in-contact humans; however, a clear link between feeding RMBDs and infections in pets and pet owners still remains mostly unclear [5,24,26,37,40]. As feeding RMBDs gains popularity, the potential risks should be reviewed, and lists of precautions and typical disease symptoms should be made available to dog owners feeding their pets RMBDs. However, consumers should handle all dog food products carefully, being mindful of their potential risks to human and animal health. Several outbreaks of human *Salmonella* infections have been caused by contaminated dry dog foods [41,42,43]. One reason that dry-food outbreaks have been reported more often than RMBDs outbreaks might be because feeding RMBDs is still marginal compared to feeding dry food. A study of in-household simultaneous canine and human fecal pathogen testing combined with clinical symptoms is now under way (Hielm-Björkman, personal communication).

5. CONCLUSIONS

Zoonotic meat-borne bacteria—such as *Campylobacter, Salmonella*, and enteropathogenic *Yersinia*—were only sporadically detected in RMBDs by PCR. These pathogens were not found by culturing, indicating a low contamination level in frozen commercial RMBDs produced in Finland. *C. upsaliensis* was a common finding in dogs irrespective of feeding RMBDs or dry pelletst. *Salmonella* and enteropathogenic *Yersinia* were detected only in dogs fed RMBDs; however, the infection source and transmission routes remained unclear. *Y. enterocolitica* bioserotype 4/O:3 and *Salmonella* were probably transmitted from contaminated RMBDs to the indoor cats but not *C. helveticus*. However, the indoor cats shed *C. helveticus* in their feces for months. The practice of feeding raw meat to dogs and cats may increase the potential transmission risk of meat-borne pathogens to people. Pet owners, especially individuals at increased risk for infectious diseases (small children, old people, and immunocompromised individuals), should be aware of the safety risks of feeding RMBDs. Attention should particularly be paid to storage and handling of raw meat. RMBDs should be kept frozen until use.

ACKNOWLEDGMENTS

Maria Stark and Urszula Hirvi are gratefully acknowledged for their technical assistance.

AUTHOR CONTRIBUTIONS

The manuscript was written by Maria Fredriksson-Ahomaa who also supervised Tiina Heikkilä and Noora Pernu, and contributed to the studies of food samples. Tiina Heikkilä contributed to the cat study by collecting and testing the cat and food samples. Noora Pernu, Sara Kovanen, Anna Hielm-Björkman, and Rauni Kivistö contributed to the dog feeding study. Noora Pernu studied the prevalence of bacterial pathogens in dog samples by PCR and isolated the salmonella and yersinia isolates. Sara Kovanen and Rauni Kivistö isolated and characterized the campylobacter isolates and Anna Hielm-Björkman was responsible for the design and sampling of the dog feeding study. Rauni Kivistö also supervised Noora Pernu and Sara Kovanen.

CONFLICTS OF INTEREST

The authors declare no conflict of interest.

REFERENCES

1. Schlesinger, D.P.; Joffe, D.J. Raw food diets in companion animals: A critical review. Can. Vet. J. 2011, 52, 50–54.
2. Parr, J.M.; Remillard, R.L. Handling alternative dietary requests from pet owners. Vet. Clin. North Am. Small Anim. Pract. 2014, 44, 667–688.
3. Freeman, L.M.; Chandler, M.L.; Hamper, B.A.; Weeth, L.P. Current knowledge about the risks and benefits of raw meat-based diets for dogs and cats. J. Am. Vet. Med. Assoc. 2013, 243, 1549–1558.
4. Morgan, S.K.; Willis, S.; Shepherd, M.L. Survey of owner motivations and veterinary input of owners feeding diets containing raw animal products. Peer. J. 2017, 2017.
5. Nemser, S.M.; Doran, T.; Grabenstein, M.; McConnell, T.; McGrath, T.; Pamboukian, R.; Smith, A.C.; Achen, M.; Danzeisen, G.; Kim, S.; et al. Investigation of Listeria, Salmonella, and toxigenic Escherichia coli in various pet foods. Foodborne Pathog. Dis. 2014, 11, 706–709.
6. Wideman, N.; Bailey, M.; Bilgili, S.F.; Thippareddi, H.; Wang, L.; Bratcher, C.; Sanchez-Plata, M.; Singh, M. Evaluating best practices for Campylobacter and Salmonella reduction in poultry processing plants. Poult. Sci. 2016, 95, 306–315.
7. Fredriksson-Ahomaa, M.; Gerhardt, M.; Stolle, A. High bacterial contamination of pig tonsils at slaughter. Meat Sci.2009, 83, 334–336.
8. Fosse, J.; Seegers, H.; Magras, C. Foodborne zoonoses due to meat: A quantitative approach for a comparative risk assessment applied to pig slaughtering in Europe. Vet. Res. 2008, 39.
9. Leonard, E.K.; Pearl, D.L.; Janecko, N.; Finley, R.L.; Reid-Smith, R.J.; Scott Weese, J.; Peregrine, A.S. Risk factors for carriage of antimicrobial-resistant Salmonella spp and Escherichia coli in pet dogs from volunteer households in Ontario, Canada, in 2005 and 2006. Am. J. Vet. Res. 2015, 76, 959–968.

10. Procter, T.D.; Pearl, D.L.; Finley, R.L.; Leonard, E.K.; Janecko, N.; Reid-Smith, R.J.; Weese, J.S.; Peregrine, A.S.; Sargeant, J.M. A cross-sectional study examining Campylobacter and other zoonotic enteric pathogens in dogs that frequent dog parks in three cities in south-western Ontario and risk factors for shedding of Campylobacter spp. Zoonoses Public Health 2014, 61, 208–218.
11. Wang, X.; Liang, J.; Xi, J.; Yang, J.; Wang, M.; Tian, K.; Li, J.; Qiu, H.; Xiao, Y.; Duan, R.; et al. Canis. lupus familiarisinvolved in the transmission of pathogenic Yersinia spp. in China. Vet. Microbiol. 2014, 172, 339–344.
12. Rossi, M.; Hänninen, M.L.; Revez, J.; Hannula, M.; Zanoni, R.G. Occurrence and species level diagnostics of Campylobacter spp., enteric Helicobacter spp. and Anaerobiospirillum. spp. in healthy and diarrheic dogs and cats. Vet. Microbiol. 2008, 129, 304–314.
13. Stamm, I.; Hailer, M.; Depner, B.; Kopp, P.A.; Rau, J. Yersinia enterocolitica in diagnostic fecal samples from European dogs and cats: Identification by fourier transform infrared spectroscopy and matrix-assisted laser desorption ionization-time of flight mass spectrometry. J. Clin. Microbiol. 2013, 51, 887–893.
14. Acke, E.; McGill, K.; Golden, O.; Jones, B.R.; Fanning, S.; Whyte, P. Prevalence of thermophilic Campylobacter species in household cats and dogs in Ireland. Vet. Rec. 2009, 164, 44–47.
15. Giacomelli, M.; Follador, N.; Coppola, L.M.; Martini, M.; Piccirillo, A. Survey of Campylobacter spp. in owned and unowned dogs and cats in Northern Italy. Vet. J. 2015, 204, 333–337.
16. Bojanic, K.; Midwinter, A.C.; Marshall, J.C.; Rogers, L.E.; Biggs, P.J.; Acke, E. Isolation of Campylobacter spp. from client-owned dogs and cats, and retail raw meat pet food in the Manawatu, New Zealand. Zoonoses Public Health.
17. Sandberg, M.; Bergsjo, B.; Hofshagen, M.; Skjerve, E.; Kruse, H. Risk factors for Campylobacter infection in Norwegian cats and dogs. Prev. Vet. Med. 2002, 55, 241–253.
18. Carbonero, A.; Torralbo, A.; Borge, C.; García-Bocanegra, I.; Arenas, A.; Perea, A. Campylobacter spp., C. jejuni and C. upsaliensis infection-associated factors in healthy and ill dogs from clinics in Cordoba, Spain. Screening tests for antimicrobial susceptibility. Comp. Immunol. Microbiol. Infect. Dis. 2012, 35, 505–512.
19. Holmberg, M.; Rosendal, T.; Engvall, E.O.; Ohlson, A.; Lindberg, A. Prevalence of thermophilic Campylobacter species in Swedish dogs and characterization of C. jejuni isolates. Acta Vet. Scand. 2015, 57.
20. Parsons, B.N.; Porter, C.J.; Ryvar, R.; Stavisky, J.; Williams, N.J.; Pinchbeck, G.L.; Birtles, R.J.; Christley, R.M.; German, A.J.; Radford, A.D.; et al. Prevalence of Campylobacter spp. in a cross-sectional study of dogs attending veterinary practices in the UK and risk indicators associated with shedding. Vet. J. 2010, 184, 66–70.
21. Lowden, P.; Wallis, C.; Gee, N.; Hilton, A. Investigating the prevalence of Salmonella in dogs within the Midlands region of the United Kingdom. BMC Vet. Res. 2015, 11.
22. Hackett, T.; Lappin, M.R. Prevalence of enteric pathogens in dogs of north-central Colorado. J. Am. Anim. Hosp. Assoc.2003, 39, 52–56.
23. Leahy, A.M.; Cummings, K.J.; Rodriguez-Rivera, L.D.; Rankin, S.C.; Hamer, S.A. Evaluation of fecal Salmonellashedding among dogs at seven animal shelters across Texas. Zoonoses Public Health 2016, 63, 515–521.
24. Weese, J.S.; Rousseau, J.; Arroyo, L. Bacteriological evaluation of commercial canine and feline raw diets. Can. Vet. J.2005, 46, 513–516.
25. Lenz, J.; Joffe, D.; Kauffman, M.; Zhang, Y.; Lejeune, J. Perceptions, practices, and consequences associated with foodborne pathogens and the feeding of raw meat to dogs. Can. Vet. J. 2009, 50, 637–643.
26. Strohmeyer, R.A.; Morley, P.S.; Hyatt, D.R.; Dargatz, D.A.; Scorza, A.V.; Lappin, M.R. Evaluation of bacterial and protozoal contamination of commercially available raw meat diets for dogs. J. Am. Vet. Med. Assoc. 2006, 228, 537–542.

27. EFSA and ECDC. The European Union summary report on trends and sources of zoonoses, zoonotic agents and food-borne outbreaks in 2015. EFSA J. 2016; 14, p. 4634.
28. Olkkola, S.; Kovanen, S.; Roine, J.; Hänninen, M.; Hielm-Björkman, A.; Kivistö, R.; Chang, Y. Population genetics and antimicrobial susceptibility of canine Campylobacter isolates collected before and after a raw feeding experiment. PLoS ONE 2015, 10, e0132660.
29. Nousiainen, L.; Joutsen, S.; Lunden, J.; Hänninen, M.; Fredriksson-Ahomaa, M. Bacterial quality and safety of packaged fresh leafy vegetables at the retail level in Finland. Int. J. Food Microbiol. 2016, 232, 73–79.
30. Lund, M.; Nordentoft, S.; Pedersen, K.; Madsen, M. Detection of Campylobacter spp. in chicken fecal samples by real-time PCR. J. Clin. Microbiol. 2004, 42, 5125–5132.
31. Malorny, B.; Paccassoni, E.; Fach, P.; Bunge, C.; Martin, A.; Helmuth, R. Diagnostic real-time PCR for detection of Salmonella in food. Appl. Environ. Microbiol. 2004, 70, 7046–7052.
32. Thisted Lambertz, S.; Nilsson, C.; Hallanvuo, S.; Lindblad, M. Real-time PCR method for detection of pathogenic Yersinia enterocolitica in food. Appl. Environ. Microbiol. 2008, 74, 6060–6067.
33. Thisted Lambertz, S.; Nilsson, C.; Hallanvuo, S. TaqMan-based real-time PCR method for detection of Yersinia pseudotuberculosis in food. Appl. Environ. Microbiol. 2008, 74, 6465–6469.
34. Best, E.L.; Powell, E.J.; Swift, C.; Grant, K.A.; Frost, J.A. Applicability of a rapid duplex real-time PCR assay for speciation of Campylobacter jejuni and Campylobacter coli directly from culture plates. FEMS Microbiol. Lett. 2003, 229, 237–241.
35. Bottone, E.J. Yersinia enterocolitica: Revisitation of an enduring human pathogen. Clin. Microbiol. Newsl. 2015, 37, 1–8.
36. Maijala, R.; Ranta, J.; Seuna, E.; Peltola, J. The efficiency of the Finnish Salmonella control programme. Food Control2005, 16, 669–675.
37. Joffe, D.J.; Schlesinger, D.P. Preliminary assessment of the risk of Salmonella infection in dogs fed raw chicken diets. Can. Vet. J. 2002, 43, 441–442.
38. Parsons, B.N.; Williams, N.J.; Pinchbeck, G.L.; Christley, R.M.; Hart, C.A.; Gaskell, R.M.; Dawson, S. Prevalence and shedding patterns of Campylobacter spp. in longitudinal studies of kennelled dogs. Vet. J. 2011, 190, 249–254.
39. Fredriksson-Ahomaa, M.; Korte, T.; Korkeala, H. Transmission of Yersinia enterocolitica 4/O:3 to pets via contaminated pork. Lett. Appl. Microbiol. 2001, 32, 375–378.
40. Finley, R.; Raid-Smith, R.; Weese, J.S. Human health implications of Salmonella-contaminated natural pet treats and raw pet food. Clin. Infect. Dis. 2006, 42, 686–691.
41. Behravesh, C.B.; Ferraro, A.; Deasy, M. , III; Dato, V.; Moll, M.; Sandt, C.; Rea, N.K.; Rickert, R.; Marriott, C.; Warren, K.; et al. Human Salmonella infections linked to contaminated dry dog and cat food, 2006–2008. Pediatrics 2010, 126, 477–483.
42. Imanishi, M.; Rotstein, D.S.; Reimschuessel, R.; Schwensohn, C.A.; Woody, D.H.; Davis, S.W.; Hunt, A.D.; Arends, K.D.; Achen, M.; Cui, J.; et al. Public veterinary medicine: Public health outbreak of Salmonella enterica serotype Infantis infection in humans linked to dry dog food in the United States and Canada, 2012. J. Am. Vet. Med. Assoc.2014, 244, 545–553.
43. Li, X.; Bethune, L.A.; Jia, Y.; Lovell, R.A.; Proescholdt, T.A.; Benz, S.A.; Schell, T.C.; Kaplan, G.; McChesney, D.G. Surveillance of Salmonella prevalence in animal feeds and characterization of the Salmonella isolates by serotyping and antimicrobial susceptibility. Foodborne Pathog. Dis. 2012, 9, 692–698.

CHAPTER 12

Comparative Studies in Tick-Borne Diseases in Animals and Humans

Ulrike Munderloh

Department of Entomology, University of Minnesota, 1980 Folwell Avenue, St. Paul, MN 55108, USA

In temperate zones of the earth, ticks are the most important arthropod vectors of zoonotic pathogens affecting humans and domestic animals. Ticks transmit the widest range of pathogen types, from viruses to rickettsiae, other bacteria, and protozoa. Historically, research with tick-borne pathogens has lagged behind research with mosquitoes, owing to a combination of their much longer life-cycle spanning months even under laboratory conditions, the requirement for multiple blood meals, and lack of standardized rearing systems. This issue presents a collection of original and review articles of the most up-to-date research and perspectives in the field. It highlights the complex life cycle strategies tick-borne disease agents have evolved to take advantage of their extraordinarily divergent hosts in order to facilitate continued circulation in nature. Importantly, emerging tick-borne pathogens are commonly zoonotic; they are maintained in complex cycles involving wild reservoir hosts and ticks, and are transmitted to humans and domestic animals as incidental hosts. The development and application of innovative technologies has spurred advances that address long-standing questions about the behavior of pathogenic and non-pathogenic symbiotic bacteria in ticks, and enable modelling of arbovirus development in tick cell and organ cultures. Cell culture systems have long been indispensable tools in medical research, and are amenable to sophisticated molecular and cell-biological approaches to dissect the pathogenesis of obligate intracellular bacteria, many of which are associated with ticks. Cell lines from a number of tick species important in human and veterinary medicine are increasingly utilized to address these questions, as highlighted by three articles in this special issue. Unlike rickettsiae that are in direct contact with host cytoplasm, *Anaplasmatceae* are sheltered from host defenses inside a membrane-bound compartment, and have evolved mechanisms to accommodate their needs. The article by Crowder and colleagues [1] provides new insights into how *A. phagocytophilum* remodels that barrier to ensure protection, flow of nutrients, and availability of conduits for the secretion of effectors to control host cell functions. The intracellular niche occupied by *A. phagocytophilum* provides unique opportunities to uncover novel cell defense signaling pathways that cannot be identified using facultatively intracellular bacteria, as reviewed by Shaw and colleagues in this issue [2].

Whole genome sequencing of tick-borne bacteria has accelerated the study of microbial ecology and host adaptation to yield a detailed picture of how bacterial communities across species that utilize the same generalist vector form similar patterns of host utilization to maximize host availability. At the same time, detailed molecular analyses have begun to illuminate bacterial signatures that underlie pathogenicity and immunogenicity, guiding the development of therapeutics and vaccines. Nevertheless, a number of tick-borne pathogens have surprisingly sophisticated capabilities to evade the mammalian immune response, and to remain either undetected or cause overwhelming infections. Thus, many still pose challenges for diagnosis. Moreover, *Borrelia* spp. that infect a variety of vertebrates have evolved mechanisms that enable them to utilize species-specific features of complement pathways to infect their many hosts [3]. Clearly, tick-borne zoonotic bacteria communicate extensively with the infected vertebrates, responding with adaptive changes as well as by active manipulation of hosts.

Far less research has focused on the effects of the pathogens on their tick vector. It is generally assumed that midgut and salivary glands are target organs in the tick, as demonstrated using PCR-based detection of dissected tissues. From those limited investigations, it is clear that *A. phagocytophilum* recognizes these different target organs as distinct as evident from their apoptosis signaling and pathway responses (see Alberdi et al. [4] in this issue). This specificity is also reflected in the divergent regulation of apoptosis pathways in tick cell lines, which makes them useful models to study tick immune responses.

Hard and soft ticks and their associated pathogens are spread widely across the globe, yet recognition of their presence and disease incidence is linked to the amount of scientific scrutiny expended. Globally, regions where ticks serve as vectors of zoonotic pathogens will continue to yield records of the range expansion of known pathogens and of the discovery of newly recognized agents. Thus, the field of tick-borne zoonotic disease agents is continuously changing and far from static. This issue represents a snapshot of our current understanding, and is meant to stimulate discussion to advance all aspects of this fascinating subject.

CONFLICTS OF INTEREST

The authors declare no conflict of interest

REFERENCES

1. Crowder, C.D.; Langeroudi, A.G.; Estabragh, A.S.; Lewis, E.R.G.; Marcsisin, R.A.; Barbour, A.G. Pathogen and Host Response Dynamics in a Mouse Model of Borrelia hermsii Relapsing Fever. *Vet. Sci.* **2016**, *3*.
2. Shaw, D.K.; McClure, E.E.; Wang, X.; Pedra, J.H.F. Deviant Behavior: Tick-Borne Pathogens and Inflammasome Signaling. *Vet. Sci.* **2016**, *3*.
3. Kraiczy, P. Travelling between Two Worlds: Complement as a Gatekeeper for an Expanded Host Range of Lyme Disease Spirochetes. *Vet. Sci.* **2016**, *3*
4. Alberdi, P.; Espinosa, P.J.; Cabezas-Cruz, A.; de la Fuente, J. *Anaplasma phagocytophilum* Manipulates Host Cell Apoptosis by Different Mechanisms to Establish Infection. *Vet. Sci.* **2016**, *3*.

CHAPTER 13

Growth Performance and Serum Lipids Profile of Clarias gariepinus Catfish Following Experimental Dietary Exposure to Fumonisin B_1

Bolade Thomas Adeyemo[1,2*], Tiamiyu Lateef Oloyede[2,3], Ayuba Victoria Ogeh[2], Cheikyula Joseph Orkuma[2]

[1]*Department of Veterinary Pathology, Faculty of Veterinary Medicine, University of Abuja, Abuja, Nigeria*

[2]*Department of Fisheries and Aquaculture, College of Forestry and Fisheries, University of Agriculture Makurdi, Makurdi, Nigeria*

[3]*Department of Aquaculture and Fisheries, Faculty of Agriculture, University of Ilorin, Ilorin, Nigeria*

ABSTRACT

Fumonisin B_1 (FB_1) is produced by fungus of the genus Fusarium (Fusarium verticiloides and F. proliferatum), and occurs predominantly in maize. The consumption of feed contaminated with FB_1 has been reported to cause deleterious effects in some fish species. This study was designed to determine the effects of dietary FB_1 on growth and lipids profile of Clarias gariepinus. 450 juvenile catfish were stocked into 5 groups of tanks consisting of 3 tanks per group and fed one of five diets amended with FB_1 (0.0 mg; 10.0 mg; 20.0 mg; 40.0 mg and 80.0 mg FB_1/kg) for 56 days. At time point's day 7, 14, 28 and 56, five fish were sampled from each tank weighted, length measured and bled for of lipids profile determinations. Results show that there was a significant reduction ($P < 0.05$), in the mean body length of the fish fed diets amended with various amounts of FB_1 compared with those fed control diet; also, there was a significant reduction ($P < 0.05$) in the weight gain of fishes fed diets amended with FB_1 compared with the control. The specific growth rate and the feed conversion ratio at 56 days shows fish fed 0.0 mg FB_1/kg had the highest specific growth rate (0.39 ± 0.14%/day) and the lowest feed conversion ratio (0.59 ± 0.01) whereas, fish fed 80.0 mg FB_1/kg had the least specific growth rate (0.07% ± 0.01%/day) and the highest feed conversion ratio (1.95 ± 0.11). Dietary FB_1 caused significant increases ($P < 0.05$) in serum cholesterol, HDL-C; LDL-C; triglycerides and the sphinganine-sphingosine ratio. Dietary FB_1 at an inclusion rate ≥ 20 mg FB_1/kg of diet produced significant reduction in weight gain and hyperlipidemia marked by hypercholesterolemia, increased blood high-density lipid cholesterol, increased blood low-density lipid cholesterol, elevated blood triglycerides and elevated sphinganine-sphingosine ratio.

Keywords: Clarias gariepinus, Fumonisin B_1, Lipids Profile, Growth Performance, Cholesterol, Catfish

1. INTRODUCTION

Fumonisin B_1 (FB_1) is a mycotoxin produced by fungus of the genus Fusarium (Fusarium verticiloides and F. proliferatum), and occurs predominantly in maize and in products made from maize [1] . FB_1 inhibits sphinganine N-acyltransferase (ceramide synthase), the enzyme that catalyzes the production of sphingolipids. The reduction of sphingolipids by FB_1 alters the control of cellular growth and cell-cell interactions with a resultant elevation of sphingoid bases [2] . The accumulation of the sphingoid bases and their metabolites alters phosphatidic acid phosphatase and monoacylglycerol acyltransferase activity with a concomitant interruption of phospholipids and fatty acid biosynthesis [2] [3] . Phospholipids and fatty acids have been shown to play an important role in the structure and function of biological membranes [4] .

Whereas the consumption of feed and/or feed ingredients contaminated with FB_1 has been reported to cause deleterious effects such as diminution of weight gain, reduction of haematological parameters and immunosuppression, in some fish species [5] [6] , other fish species notably the warm water catfishes are reported not to be sensitive to dietary fumonisin B_1 [7] [8] . The Clariid fish Clarias gariepinus is the largest group of cultured fish species in Nigeria [9] . These fish species are preferred because not only they command a good commercial value [10] , but also they are generally considered to be hardy and easy to culture [10] [11] . The culture of clariid fishes involves the utilization of feeds containing up to 30% maize as a source of energy [12] , thereby inadvertently imposing the risk of the introduction of FB_1 in the diets formulated for feeding the fishes [13] [14] . Aside from the report of [15] , on the effects of fumonisin B_1 on growth and haematology of Clarias gariepinus catfish fingerlings, there is a paucity of information on the effects of dietary FB_1 on the lipids profile of clariid catfishes; this study therefore, was designed to determine the effects of dietary fumonisin B1 on growth and the lipids profile of Clarias gariepinus catfish following dietary FB_1 exposure.

2. MATERIALS AND METHODS

Research grade FB_1 used for the study was purchased from Sigma Aldrich (St Loius, M.O USA). Purity was ascertained by HPLC-fluorescence detection after derivatization with o-phthaldiadehyde (OPA, Sigma) to be greater than 98%. Other chemicals and reagents used for the study were purchased commercially at the highest degree of purity available.

2.1. Preparation of the FB_1 Stock Solution

Fumonisin B_1 stock solution was prepared by dissolving 1 gm fumonisin B_1 (Sigma chemicals St Loius, USA) with 1000 μl of acetonitrile-water (Vol.:Vol.) resulting in a 1 μL:1 mg solution of FB_1.

2.2. Preparation of the Basal Diets

The basal diet was formulated according to [16] , with slight adjustments, using the following ingredients (fish meal 19%, Soybean cake 37%, maize 32.25%, palm oil 1.0%, fish oil 6.0%, Starch binder 2.0%, vitamin/min- eral premix 0.5%, Bone meal 1.0%, salt 0.25%) to meet the nutritional requirements of juvenile Clariid fishes. Formulated diets were then subjected to proximate analysis to ascertain their nutritional status (Table 1).

Preparation of the Experimental FB_1 Diets

From the FB_1 stock solution, the volume of solution needed to produce the experimental diets for the various FB_1 inclusions were pipetted into 1000 ml beakers into which had been placed 200 ml warm distilled water. After careful stirring, weighted portions of the starch binders were added, followed by the addition of the weighed portions of the basal diets and finally, pelletizing using a bench extruder. The finished feeds were oven dried at 65°C for 5 hours, allowed to cool to room temperature and packed in cellophane bags, labelled (control diet, D0 = 0.0 mg FB_1/kg; diet D1 = 10.0 mg FB_1/kg; diet D2 = 20.0 mg FB_1/kg; diet D3 = 40.0 mg FB_1/kg and diet D4 80.0 mg FB_1/kg) and there after stored at 4°C until used.

2.3. Experimental Fish

450 juvenile fish were procured from a commercial hatchery, acclimatized for 15 days. The fishes were then weighted (151.64 ± 2.11 g), total length measured (27.00 ± 1.39 cm) and then randomly distributed into fifteen 1000 L capacity tanks which were divided into five groups consisting of three tanks each labelled A1, A2, A3 (for the group A tanks); B1, B2 and B3 (for the group B tanks); C1, C2, C3 (for the group C tanks); D1, D2, D3 (for the group D tanks and E1, E2 and E3 (for the group E tanks) respectively at a stock density of 30 fish per tank. Stocked fish were then fed diets D0 (Control, 0.0 mg FB_1/kg); D1 (10.0 mg FB_1/kg); D2 (20.0 mg FB_1/kg); D3 (40.0 mg FB_1/kg) and diet D4 (80.0 mg FB_1/kg) for 56 days. At time points 7, 14, 28, and 56 days, using a hand held net, 5 fish were randomly selected from each tank, weighed and total length measured. At these same time intervals, blood for serum lipids determinations were obtained from the fish by caudal veni-puncture (using a 23 G needle fitted on a 5 ml syringe).

2.3.1. Determination of Growth Performance

The effects of dietary fumonisin B_1 on growth performance was determined by the evaluation of the Specific Growth Rate (SGR) and the Feed Conversion Ratio (FCR) after 56 days of feeding. The specific growth rate was determined according to [17] , using the formulae SGR = 100 × [(InW_2 – InW_1)/t_2 – t_1]. Were, W_2 is weight attained after period of feeding on diet; W_1 is weight at commencement of the feeding experiment; t_2 – t_1 is the time period (in days) of feeding on the experimental diets. The feed conversion ratio was calculated according to [18] using the formulae FCR = Amount of feed eaten (dry weight) (g)/fish weight gain (g).

2.3.2. Collection of Blood Sample for Determination of Serum Lipid Profiles

Blood samples were collected from both the control and the treatment groups. The aspirated blood were then dispensed into clean glass tubes and allowed to clot and then spurn with a bench centrifuge at 3000 RPM for 5 minutes. Serum were separated from the clotted blood using a clean syringe fitted with a clean needle and dispensed into clean ependorph tubes and stored at 0°C until used.

1) Determination of serum lipid profile

Total serum cholesterol was estimated using automatic serum chemistry auto analyser and kit (AUTOPAK supplied by Beyer Diagnostics India) by the enzymatic (Cholesterol Esterase, Cholesterol Oxidase and Peroxidase method of [19]). Results obtained were expressed as $mg{\cdot}dL^{-1}$ of serum.

2) Determination of serum triglycerides

Serum Triglycerides estimation was carried out with the use of the automatic serum analyser and kit (AUTOPAK, Bayer Diagnostics India) by the enzymatic Lipoprotein lipase, Glycerol kinase, Glycerol-3-Phosphate Oxidase method of [20] . Results are expressed as $mg{\cdot}dL^{-1}$ of serum.

3) Determination of serum high density lipid (HDL)

The high density lipid in the serum was estimated using the automatic serum analyser and kits (AUTOPAK, Bayer Diagnostics India) by the phosphotungstate method. Results obtained were expressed as $mg{\cdot}dL^{-1}$ of serum.

Very Low Density Lipoprotein-Cholesterol (VLVD) in serum was estimated using the Friedewald formula [21] .

$$\text{VLVD-Chol} = \text{Serum triglycerides}/5$$

The results obtained are expressed as $mg{\cdot}dL^{-1}$ of serum.

4) Determination of serum low density lipid

The serum Low Density Lipoproteins was estimated mathematically as the difference between the serum total cholesterol and the sum of the Very Low Density Lipid cholesterol and the High Density Lipid cholesterol according to the Friedwald formula [21] .

$$\text{LDL-Chol}\ (\text{mg/dL serum}) = \text{Total Chol} - (\text{VLDL-Chol} + \text{HDL-Chol}).$$

2.3.3. Quantification of Biomarkers of FB_1 Exposure

The quantification of serum sphinganine (Sa) and sphingosine (So) in serum was done according to the method of [22] . Briefly, 500 μl of serum was de-proteinized with 2 ml of methanol (CH_3OH), and then centrifuged for 10 minutes at 1200 g and 100C. A 1.9 ml aliquot of the supernatant was collected and mixed with an equal volume of water (1.9 ml) and 1.2 ml NH_4OH 0.35 M, and was then extracted with 4 ml of $CHCL_3$. After thorough mixing, the phases were separated by centrifugation at 1200 g and 10°C for 10 minutes. Then a 3 ml aliquot of the chloroformic phase was transferred to a silica gel mini column to eliminate the interferents.

The 3 ml chloroformic extracts was cleaned by passing it through polypropylene columns containing 5 g of anhydrous Na_2SO_4 crystals packed on top of 0.2 g of silica gel 60 (15 - 40 μm). The mini columns were preconditioned with 3 ml $CHCL_3$ (at a flow rate of less than 2 ml per minute), the chloroformic extracts were eluted and the column washed with 1 ml $CHCL_3$, discarding the eluate. The sphingoid bases (Sa and So) were eluted with 4 ml of $CHCL_3$:CH_3OH:NH_4OH (50:50:2) solution, thus evaporating the solvent from this eluate. The residues obtained was then dissolved with 1 ml of a 0.125 M KOH solution in CH_3OH:$CHCL_3$ (4:1, vol.:vol.), and were later incubated at 37°C for 90 minutes. Then 1 ml of $CHCL_3$ was added to each sample, and the organic phase was washed with alkaline water (0.6 ml of NH_4OH 0.35 M in 250 ml of bi-distilled water). The chloroformic phases were recovered by centrifugation for 10 min at 1200 g and 10°C; before being dried.

The dried residues obtained from the sera were then dissolved with 250 μL of CH_3OH:H_2O (9:1). They were then derivatized with 50 μL of O'-pthaldialdehide (OPA) reagent, and were analysed by reverse-phase HPLC. The OPA was prepared by adding 10 ml 3% Boric acid (pH adjusted to 10.5 with KOH); to a solution prepared with 5 ml OPA dissolved in 100 μL CH_3OH and 5 μL of 2-mercaptoethanol. The Sa and So levels were then quantified by a Hewlett Packard 1100 HPLC equipped with fluorescence detector set at 335 nm and 440 nm wave lengths for excitation and emission respectively. The Sa and So levels were calculated by extrapolating the peak areas obtained for experimental samples, in a calibration curve constructed by injection of commercial standards of Sa and So (44.4; 177.7 and 355.3 ng/ml).

3. STATISTICAL EVALUATION

Results were expressed as the mean ± Standard Deviation of the mean of each group (n = 30), and analysed by a one way analysis of variance (ANOVA). Variant means were then separated by the Turkey-Kramer Post hoc test (SPSS version 20). Differences were considered significant at $P < 0.05$.

4. RESULTS

4.1. Effects on Growth Performance

The nutrient composition and proximate analyses of the control and fumonisin B_1 amended diets as well as the FB_1 contents of the diets are as presented in Table 1. It shows the presence of fumonisin B_1 did not affect the percentage crude protein, metabolizable energy and digestible energy levels of the diets.

Table 2 depicts the growth performance of juvenile Clarias gariepinus catfish fed varied amounts of FB_1. It shows there were no significant difference ($P > 0.05$) in both the length and weight of the fishes fed the control and experimental diets at the start of the experiments. Table 2 also show that after 56 days of dietary FB_1 exposures, there were significant difference ($P < 0.05$), in the mean body length of the fishes fed the control diets when compared with those of fishes fed the diets amended with various amounts of FB_1. Table 2 shows the fishes fed the control diets had the highest gain in body length (47.63 ± 1.09 cm representing a 69.50% ± 1.20% gain in body length), while fishes fed the diets containing 40 mg FB_1/kg had the least gain in body length (34.02 ± 0.48 cm which represent 17.31% ± 0.48% gain in body length). Although, there was a significant difference ($P < 0.05$) in the mean final body length in a comparison of fishes fed the control diets and the experimental diets; Turkey post hoc evaluation reveals, there were no variation ($P > 0.05$) in the mean final length of fishes fed the diets amended with the various amounts of FB_1.

One way analysis of variance (ANOVA) reveals there was a significant difference ($P < 0.05$) in the percentage gain in total length in a comparison of fishes fed the control fish and those fed the diets amended with fumonisin B1. Turkey post hoc reveals there were significant differences ($P < 0.05$) in the percentage gain in length amongst the fishes fed the diets amended with fumonisin B1. One way analysis of variance (ANOVA) of the data obtained for the mean final weight gain shows there was a significant difference ($P < 0.05$) in the comparison of the weight gain of fishes fed the control diets and those fed the diets amended with fumonisin B1; Turkey post hoc test shows there are no variations ($P > 0.05$) in the weight gain of fishes fed the diets amended with 10.0 mg FB_1/kg and 20.0 mg FB_1/kg; these were however significantly different ($P < 0.05$) from those of fishes fed diets containing 40.0 mg FB_1/kg and 80.0 mg FB_1/kg. Also, there was no significant variation ($P > 0.05$) in the mean gain in weight of fishes fed diets containing 40.0 and 80.0 mg FB_1/kg (Table 2).

Table 1. Nutrient composition, fumonisin B_1 content and proximate analysis of formulated diets.

	FUMONISIN B_1 AMENDED DIETS				
Parameter	D0	D1	D2	D3	D4
Fish meal	19.00	19.00	19.00	19.00	19.00
SoyBean cake	37.00	37.00	37.00	37.00	37.00
Maize	32.25	32.23	32.23	32.25	32.25
Palm oil	1.00	1.00	1.00	1.00	1.00
Fish oil	6.00	6.00	6.00	6.00	6.00
Vit/min Premix	0.50	0.50	0.50	0.50	0.50
Bone meal	1.00	1.00	1.00	1.00	1.00
Salt (NaCl)	0.23	0.25	0.25	0.25	0.25
Fumonisin B1	0.00	10.00	20.00	40.00	80.00
Starch Binder	2.00	2.00	2.00	2.00	2.00
Crude Protein	40.01	40.00	40.04	40.01	40.02
Gross Energy	19.77	20.00	20.01	19.86	20.00
Digestible Energy	12.00	12.01	11.98	12.01	12.10

D0 = 0.0 mg FB_1/kg, D1 = 10.0 mg FB_1/kg; D2 = 20.0 mg FB_1/kg; D3 = 40.0 mg FB_1/kg; D4 = 80.0 mg FB_1/kg.

Table 2. Growth performance of Clarias gariepinus juveniles fed varied levels of fumonisin B_1 (FB_1) fifty six days.

	FUMONISIN B_1 AMENDED DIETS				
Parameter	D0	D1	D2	D3	D4
Mean initial body length (cm)	28.10 ± 0.10^a	27.00 ± 1.39^a	28.60 ± 1.41^a	29.00 ± 0.93^a	27.70 ± 0.66^a
Mean final body length (cm)	47.63 ± 1.09^a	35.24 ± 0.88^b	34.87 ± 0.41^b	34.02 ± 0.48^b	36.38 ± 1.75^b
Mean gain in length (%)	69.50 ± 1.20^a	30.52 ± 0.33^b	21.92 ± 0.09^c	17.31 ± 0.48^d	31.33 ± 0.93^b
Mean initial weight (g)	153.09 ± 1.37^a	151.64 ± 2.11^a	152.07 ± 1.33^a	152.61 ± 0.93^a	151.78 ± 1.27^a
Mean final weight (g)	190.01 ± 2.03^a	168.14 ± 0.01^b	166.30 ± 0.12^b	164.21 ± 0.81^c	163.41 ± 0.14^c
Mean gain in weight (%)	24.12 ± 1.31^a	10.88 ± 1.27^b	8.56 ± 0.01^c	7.60 ± 0.07^d	7.66 ± 0.12^d
SGR (%/day)	0.39 ± 0.14^a	0.19 ± 0.08^b	0.16 ± 0.02^b	0.13 ± 0.01^b	0.07 ± 0.01^c
FCR	0.59 ± 0.01^a	0.98 ± 0.01^b	1.01 ± 0.00^b	1.24 ± 0.01^c	1.95 ± 0.11^c

Rows with different superscript are significantly different ($P < 0.05$). D0 = 0.0 mg FB_1/kg; D1 = 10.0 mg FB_1/kg; D2 = 20.0 mg FB_1/kg; D3 = 40.0 mg FB_1/kg; D4 = 80.0 mg FB_1/kg.

Evaluation of the data obtained for the specific growth rate (SGR) and the feed conversion ratio (FCR) at 56 days of dietary exposure to fumonisin B1 shows fishes fed the control feed (0.0 mg FB_1/kg) had the highest specific growth rate (0.39% ± 0.14%/day) and the lowest feed conversion ratio (0.59 ± 0.01) whereas, fishes fed 80.0 mg FB_1/kg had the least specific growth rate (0.07% ± 0.01%/day) and the highest feed conversion ratio (1.95 ± 0.11). One way analysis of variance (ANOVA) reveals significant variation ($P < 0.05$) in the SGR of fishes fed the control diets compared with fish fed the diets containing various amounts of fumonisin B_1. Also, Turkey post hoc shows whereas there were no significant difference ($P > 0.05$) in the SGR of fishes fed diets containing 10.0, 20.0 and 40 mg FB_1/kg, there was a significant difference ($P < 0.05$) in the SGR of fishes fed diets amended with fumonisin $B_1 \leq 40.0$ mg FB_1/kg of the diet compared with fishfed 80.0 mg FB_1/kg of the diet (Table 2).

4.2. Effects of FB_1 on Serum Lipids

The results obtained for the effects of dietary exposure to fumonisin B_1 on the serum total cholesterol are as shown in Table 3. It shows at 7 days of feeding on the diet, serum cholesterol increased from 92.20 ± 1.00 $mg \cdot dL^{-1}$ in fish fed diet D0(the control diet) to 107.29 ± 2.02 $mg \cdot dL^{-1}$ in fish fed diet D4 (80.0 mg FB_1/kg).One way ANOVA shows significant variations ($P < 0.05$) in the serum cholesterol levels of the fish fed the control diet compared with those fed diets amended with FB_1; further, Turkey post hoc showed whereas the serum cholesterol level of fish fed diet D2 (20.0 mg FB_1/kg) and D3 (40.0 mg FB_1/kg) are not significantly different ($P > 0.05$) they were however, significantly different from those of fish fed diet D1 (10.0 mg FB_1/kg) and diet D4 (80.0 mg FB_1/kg). At 14 day dietary exposure to fumonisin B_1, the serum cholesterol level ranged from 94.89 ± 3.48 $mg \cdot dL^{-1}$ in fish fed diet DF1 to 129.19 ± 3.41 $mg \cdot dL^{-1}$ in fish fed diet D4. There was no difference ($P > 0.05$) in the serum cholesterol concentration of fish fed diet D0 and the fish fed diet D1, D3 and D4; the serum cholesterol of fish fed D2 however, differed significantly ($P < 0.05$) from those of fish fed the other diets. At 28 days of fumonisin B1 exposure, the serum cholesterol levels ranged from 95.41 ± 1.11 $mg \cdot dL^{-1}$ (in fish fed the control diet) to 130.72 ± 2.40 $mg \cdot dL^{-1}$ in fish fed diet D4 (80.0 mg FB_1/kg). Turkey post hoc showed that except for fish fed diet D2, the serum cholesterol levels were not significantly different ($P > 0.05$) in the fish fed the diets amended with FB_1. At day 56 of the feeding trial, the serum cholesterol level of fish fed diet D2 (20.0 mg FB_1/kg) was not significantly different ($P > 0.05$) from the serum cholesterol of fish fed diet D3 (40.0 mg FB_1/kg) but were however, significantly different ($P < 0.05$) from those of fish fed the control diet (0.0 mg FB_1/kg), diet D1 (10.0 mg FB_1/kg) and diet D4 (80.0 mg FB_1/kg).

Table 4 shows the serum high density lipid cholesterol (HDL-C) concentration of juvenile Clarias gariepinus exposed to varied amounts of dietary FB_1 for 56 days. It shows although there were marginal increases in the serum HDL-C levels in fish exposed to dietary FB_1, except for fish fed diet containing 10.0 mg FB_1/kg, at 7 days of feeding, the serum HDL-C level

Table 3. Serum total cholesterol concentration (mg·dL^{-1}) of juvenile Clarias gariepinus fed diets amended with varied amounts of fumonisin $B_1$for fifty six days.

	FUMONISIN B_1 AMENDED DIETS				
Days feeding	D0	D1	D2	D3	D4
7	92.20 ± 1.00^{a}	98.07 ± 0.13^{b}	101.24 ± 1.09^{c}	103.31 ± 1.07^{c}	107.29 ± 2.02^{d}
14	95.19 ± 1.31^{a}	94.89 ± 3.48^{a}	112.81 ± 1.09^{b}	128.99 ± 1.78^{c}	129.19 ± 3.41^{c}
28	95.41 ± 1.11^{a}	96.69 ± 1.09^{a}	113.78 ± 1.06^{b}	128.43 ± 1.12^{c}	130.77 ± 2.40^{c}
56	93.56 ± 0.66^{a}	94.27 ± 1.96^{a}	135.44 ± 2.37^{b}	135.99 ± 1.08^{b}	155.41 ± 6.13^{c}

Rows with different superscript are significantly different ($P < 0.05$). D0 = 0.0 mg FB_1/kg; D1 = 10.0 mg FB_1/kg; D2 = 20.0 mg FB_1/kg; D3 = 40.0 mg FB_1/kg; D4 = 80.0 mg FB_1/kg.

Table 4. Serum high density lipid cholesterol concentration (mg·dL^{-1}) of juvenile Clarias gariepinus fed diets amended with varied amounts of fumonisin B_1 for fifty six days.

Days feeding	D0	D1	D2	D3	D4
7	10.01 ± 1.21^{a}	12.07 ± 1.00^{b}	10.23 ± 1.37^{a}	10.24 ± 1.41^{a}	10.19 ± 1.67^{a}
14	7.91 ± 0.83^{a}	9.37 ± 0.30^{bc}	9.81 ± 1.00^{c}	8.99 ± 1.00^{b}	9.10 ± 0.41^{b}
28	7.79 ± 0.77^{a}	10.94 ± 0.38^{b}	10.17 ± 1.21^{c}	9.29 ± 1.47^{d}	9.96 ± 0.41^{cd}
56	7.03 ± 0.01^{a}	9.48 ± 1.87^{b}	9.99 ± 0.67^{b}	9.71 ± 0.99^{b}	10.00 ± 0.01^{b}

Rows with different superscript are significantly different ($P < 0.05$). D0 = 0.0 mg FB_1/kg; D1 = 10.0 mg FB_1/kg; D2 = 20.0 mg FB_1/kg; D3 = 40.0 mg FB_1/kg; D4 = 80.0 mg FB_1/kg.

of the control fish was not significantly different ($P > 0.05$) from those of the fish fed the diets amended with varied amounts of FB_1. At 14 days and as was at day 28 and 56 post dietary exposure to FB_1, the serum HDL-C levels in the exposed fish was significantly increased ($P < 0.05$) compared with those of the control fish; Turkey post hoc evaluation revealed the variation in the serum HDL-C of the fish fed the diets amended with FB_1 at days 14, 28 and 56 to be significant ($P < 0.05$).

Results obtained from the dietary exposure of juvenile Clarias gariepinus catfish to FB_1 on serum low density lipid cholesterol (LDL-C) are shown in Table 5. It shows at 7 days, except for fishes fed diets D4 (80.0 mg FB_1/kg), there was reduction of the LDL-C from 70.63 ± 0.98 mg·dL^{-1} (in fish fed 0.0 mg FB_1/kg) to 63.80 ± 0.84 mg·dL^{-1} (in fish fed diet containing 10.0 mg FB_1/kg). Table 5, also shows that at 7 days of feeding, there was no significant difference ($P > 0.05$) in the serum concentration of the LDL-C in fish fed diets containing between 10.0 and 40.0 mg FB_1/kg these values where however significantly ($P < 0.05$) different when compared with those of fish fed the control diet (0.0 mg FB_1/Kg) and diet D4 (80.0 mg FB_1/kg). At 56 days of dietary exposure, the serum LDL-C ranged from 64.59 ± 0.37 mg·dL^{-1} (in fish fed 10.0 mg FB_1/kg) to 118.99 ± 0.64 mg·dL^{-1} (in fish fed diet containing 80.0 mg FB_1/kg). ANOVA shows there was significant difference ($P < 0.05$) in the LDL cholesterol of fish exposed to dietary FB_1 for 56 days compared with the control; Turkey post hoc further reveals except for fish fed diets with FB_1 inclusion rates ≥ 40 mg at 28 days of feeding, there were no significant difference in the serum LDL cholesterol concentration of fish fed the diet amended with the varied amounts of FB_1 for 56 days.

Dietary exposure to FB_1 caused significant increases in the serum triglyceride concentrations of the exposed fish (Table 6). At seven days of the feeding trial, the serum triglyceride levels ranged between 82.79 ± 0.34 $mg{\cdot}dL^{-1}$ to 134.90 ± 0.72 $mg{\cdot}dL^{-1}$. One way analysis of variance (ANOVA) showed there were significant variations ($P < 0.05$) in the serum triglycerides concentrations obtained for the fish fed the diets amended with FB_1 when compared with those of fish fed the control diets (Table 6). Further, Turkey-Kramer post hoc evaluation showed significant ($P < 0.05$) variation in the serum triglyceride concentration for the fish fed the diets amended with varied amounts of FB_1. At day 56, the serum triglyceride levels of the fish fed the control diet differed significantly ($P < 0.05$) compared to values obtained for fish fed the diets amended with FB_1.

Table 5. Serum low density lipid cholesterol concentration ($mg{\cdot}dL^{-1}$) of juvenile Clarias gariepinus fed diets amended with varied amounts of fumonisin B_1 for fifty six days.

	FUMONISIN B_1 AMENDED DIETS				
Days feeding	D0	D1	D2	D3	D4
7	70.63 ± 0.98^a	63.80 ± 0.84^b	64.49 ± 0.31^b	66.09 ± 1.00^b	70.56 ± 1.02^a
14	71.24 ± 0.32^a	64.02 ± 0.19^b	77.45 ± 0.56^b	93.84 ± 0.75^b	93.81 ± 1.01^b
28	71.85 ± 0.17^a	62.87 ± 0.45^b	77.69 ± 0.92^b	93.61 ± 0.17^c	94.93 ± 1.76^c
56	**70.31 ± 0.64^a**	**64.59 ± 0.37^b**	**99.76 ± 0.55^b**	**100.19 ± 0.04^b**	**118.99 ± 0.64^b**

Rows with different superscript are significantly different ($P < 0.05$). D0 = 0.0 mg FB_1/kg; D1 = 10.0 mg FB_1/kg; D2 = 20.0 mg FB_1/kg; D3 = 40.0 mg FB_1/kg; D4 = 80.0 mg FB_1/kg.

Table 6. Serum Triglycerides concentration ($mg{\cdot}dL^{-1}$) of juvenile Clarias gariepinus fed diets amended with varied amounts of fumonisin B_1 for fifty six days.

	FUMONISIN B_1 AMENDED DIETS				
Days feeding	D0	D1	D2	D3	D4
7	82.79 ± 0.34^a	111.02 ± 0.49^b	132.61 ± 0.39^c	134.90 ± 0.72^d	132.72 ± 0.33^c
14	80.22 ± 0.81^a	107.51 ± 0.54^b	128.13 ± 0.97^c	130.81 ± 0.19^d	131.40 ± 0.21^e
28	78.84 ± 0.37^a	114.42 ± 0.76^b	129.59 ± 0.83^c	127.67 ± 0.26^d	129.42 ± 0.37^c
56	81.17 ± 0.71^a	100.99 ± 0.28^b	128.44 ± 0.51^c	130.44 ± 0.90^d	132.08 ± 0.54^e

Rows with different superscript are significantly different ($P < 0.05$). D0 = 0.0 mg FB_1/kg; D1 = 10.0 mg FB_1/kg; D2 = 20.0 mg FB_1/kg; D3 = 40.0 mg FB_1/kg; D4 = 80.0 mg FB_1/kg.

Table 7 shows the results obtained for the serum sphinganine (Sa) and serum sphingosine (So) following dietary exposure of juvenile Clarias gariepinus catfish to diets amended amended with varied amounts of FB_1 for 56 days. At 7 days of dietary exposure, the sphiganine-sphingosine ratio of the fish fed the control diet was significantly different ($P < 0.05$) from those of fish fed the diets amended with FB_1. At this time point, the Sa-So ratio ranged from 0.45 ± 0.01 (in fish fed the control diet) to 1.71 ± 0.03 (in fish fed diet D4). Further, Turkey- Kramer post hoc analysis shows that whereas the Sa-So ratio varied significantly in a comparison of fishes fed diets amended with 10.0 and 20.0 mg FB_1/kg, the Sa:So values obtained for fish fed diets greater than 20 mg FB_1/kg were not significantly ($P > 0.05$) different from one another. At day 14 and 56, the Sa-So ratio of fed the control diets were significantly different from those of fish fed the diets amended with FB_1. Turkey-Kramer post hoc further shows that the Sa-So ratio of fish fed the FB_1 amended diets differed significantly ($P < 0.05$) from one another.

Results for the assessment of the fish culture water quality parameters are within the range prescribed for fresh water fishes and are as follows: the pH of the culture water ranges between 7.0 and 7.4; alkalinity ranged from 0.55 mmol/L to 0.70 mmol/L; the nitrates concentration of the fish culture water ranged from 0.014 mg/L to 0.048 mg/L while the nitrites concentration in the fish culture water ranged from 5.70 mg/l to 10.40 mg/L. Also, the dissolved oxygen concentration of the fish culture water ranged between 5.17 mg/L and 7.05 mg/L while the temperature of the culture water ranged from 27.5°C to 30.8°C throughout the duration of the experiment [23] [24] .

5. DISCUSSION

5.1. Effects on Growth Performance

In the present study, dietary FB_1 significantly reduced growth performance parameters measured; similar effects have been reported in other fish species (Ictalurus puntatus, [25] , Oreochromis niloticus, [26] [27]). The mean body length was reduced in the treated fish in a nonspecific manner as the FB_1 contents of the diets increased; except for fish fed 80.0 mg FB_1/kg, the percentage length gain decreased along the gradient of FB_1 inclusion. The mean weight gain and the percentage weight gain decreased along the gradient of concentration of FB_1 inclusion in the diets. Previous studies of the toxicity of FB_1 in other fish species have shown that age, weight and the duration of exposure modulates the susceptibility to FB_1. [7] , reported dosages of up to 20 mg FB_1/kg reduced weight gain in 1.2

g Channel catfish (Ictalurus punctatus) compared with the controls while the thresh hold for toxicity in larger fish (6.1 g) was higher (40.0 mg FB_1/kg); the same authors further noted that, 2 years old channel catfish exposed to FB_1 for 98 days gained less weight only at concentrations above 80.0 mg FB_1/kg. In Nile Tilapia (Oreocrhomis niloticus), dietary exposure of up to 40 mg FB_1/kg or higher was reported to reduce weight gain only after 56 days of dietary FB_1 exposure [26] . The loss in growth performance induced by dietary exposure to FB_1 in Clarias gariepinus is expected as FB_1 has been demonstrated to interfere with cellular growth and cell-cell interactions [4] . The feed conversion ratio (FCR) of the fish exposed to FB_1 was significantly increased when compared to fish fed the control feed, indicating these groups of fish needed to consume more feed to produce the same weight gain as the control fish similar to the findings of [27] in broiler chicks.

Table 7. Serum sphinganine-sphingonine ratio of juvenile Clarias gariepinus fed diets amended with varied amounts of fumonisin B_1 for fifty six days.

	FUMONISIN B_1 AMENDED DIETS				
Days feeding	D0	D1	D2	D3	D4
7	0.45 ± 0.01^a	1.57 ± 0.12^b	1.70 ± 0.19^c	1.68 ± 0.01^c	1.71 ± 0.03^c
14	0.39 ± 0.01^a	1.66 ± 0.03^b	1.60 ± 0.09^c	1.70 ± 0.03^d	1.66 ± 0.01^b
28	0.41 ± 0.02^a	1.87 ± 0.01^b	2.03 ± 0.01^c	2.12 ± 0.01^d	2.16 ± 0.13^d
56	0.41 ± 0.01^a	1.79 ± 0.02^b	1.68 ± 0.11^c	1.69 ± 0.02^c	1.88 ± 0.12^d

Rows with different superscript are significantly different ($P < 0.05$). D0 = 0.0 mg FB_1/kg; D1 = 10.0 mg FB_1/kg; D2 = 20.0 mg FB_1/kg; D3 = 40.0 mg FB_1/kg; D4 = 80.0 mg FB_1/kg.

5.2. Effects on Lipids Profile

Dietary fumonisin B1 exposure caused a significant increase in the sphinganine-sphingosine (Sa:So) ratio starting from day 7 of the feeding trial. In addition to the nutritional parameters commonly employed as markers of toxicity in FB1 exposures; the increase in the Sa-So ratio has also been used as toxicity indicator in diverse experimental animal models [28] . The increased serum Sa-So ratio in the exposed Clarias gariepinus catfish juveniles was expected as FB_1 is structurally similar to sphinganine and sphingosine [29] - [31] and therefore competitively inhibits the enzyme ceramide synthase resulting in the accumulation of free sphinganine and to a lesser extent, sphingosine leading to an increase in the Sa-So ratio in exposed fish.

Serum total cholesterol increased significantly throughout the duration of the dietary exposure. The increase in serum total cholesterol occurred along the gradient of concentration of the FB_1 content of the diets as well as on the duration of dietary FB_1 exposure with the highest increase (66.11%) occurring in fish fed diet containing 80.0 mg FB_1/kg at 56 days of dietary exposure. The increase in serum cholesterol in the juvenile Clarias gariepinus was an expected result as hypercholesterolemia has been reported to be an early finding in a number of exposed animals [13] [27] , including fishes [5] . Hypercholesterolemia in FB_1toxicosis has been attributed to a disruption of Cytochrome P450 enzymes activity [4] and may also be a result of hepatic dysfunction initiated by the hepatonecrosis caused by the ingested FB_1 in the diets [32] .

High density lipoproteins are responsible for the reverse transport of cholesterol from the peripheral cells to the liver where cholesterol is transformed to bile acids which are excreted into the biliary tract. Elevated HDL- cholesterol concentrations particularly in conjunction with an elevated triglycerides increases cardiovascular risk [33] . LDL-C transports cholesterol to the arteries where they can be deposited on the arterial proteoglycans leading to the formation of plaques which have been shown to narrow the arterial lumen [4] predisposing to such diverse pathophysiologies as cardiovascular diseases (CVD) and atherosclerosis. The increase of LDL cholesterol in this study may be indicative of an increased risk of cardiovascular disease in the dietary FB_1 exposed catfish.

When compared with the fish fed the control diet, the serum triglycerides concentration of fish fed diets amended with FB_1 increased in a nonspecific manner. Fish under condition of stress have been shown to mobilize triglycerides and protein to fulfill an increased demand for energy not only to cope with the detrimental condition imposed by the toxicant, but also, to meet the energy required to sustain increased physical activity, biotransformation and the excretion of the noxious substance [34] [35] . Results obtained from this study shows that FB_1 imposed a nonspecific stress on the Clarias gariepinus juvenile catfishes which caused an increase in the serum concentration of triglycerides.

6. CONCLUSION

In conclusion, the study revealed that dietary exposure to fumonisin B_1 at an inclusion rate greater than or equal to 20 mg FB_1/kg of diet produced significant loss in weight gain and hyperlipidemia marked by hypercholesterolemia, increased blood high density lipid cholesterol, increased blood low density lipid cholesterol, elevated blood triglycerides and elevated sphinganine-sphingosine ratio. It is therefore recommended that for increased profitability in Clarias gariepinus catfish culture, dietary fumonisin B1 content should not exceed 20.0 mg FB_1/kg.

CITE THIS PAPER

Bolade Thomas Adeyemo, Tiamiyu Lateef Oloyede,, Ayuba Victoria Ogeh, Cheikyula Joseph Orkuma (2016) Growth Performance and Serum Lipids Profile of *Clarias gariepinus* Catfish Following Experimental Dietary Exposure to Fumonisin B_1. *Open Journal of Veterinary Medicine*,**06**,127-138.

REFERENCES

1. Walter, F.O. and Maracas, W.F.O. (2001) Discovery and Occurrence of the Fumonsins Historical Perspectives. Environmental Health, 109, 239-243.
2. Young S.A., Minaj, G., Denny, P.W. and Smith, K. (2012) Sphingolipids and Ceramide Homeostasis. Biochemistry Research International, 10, 248-351.
3. Soriano, J.M., Gonzalez, L. and Catal, A.I. (2005) Mechanism of Action of Sphingolipid and Their Metabolites in the Toxicity of FB1. Progress in Lipids Research, 44, 345-356.
4. Casteel, S.W., Turk, J.R. and Rottinghaus, G.E. (1994) Chronic Effects of Dietary Fumonisin on the Heart and Pulmonary Vasculature of Swine. Fundamental Applied Toxicology, 23, 518-524.
5. Pepeljnjak, S., Petrinec, Z., Kovacic, S. and Segvic, M. (2002) Screening Toxicity Study in Young Arp (Cyprinuscarpio) on Feed Amended with Fumonisin B1. Mycopathologia, 156, 139-145.
6. Santos, G.A., Rodrigues, I., Naechrer, K. and Encarnacao, P. (2010) Mycotoxins in Aquaculture: Occurrence in Feed Components and Impact on Animal Performance. Aquaculture Europe, 35, 6-10.
7. Lumlertdacha, S., Lovell, R.T., Shelby, R.A., Lenz, S.D. and Kemppainen, B.W. (1995) Growth, Haemetology and Histopathology of Channel Catfish (Ictalurus punctatus), Fed Toxins from, Fusarium monilforme. Aquaculture, 130, 201-218.
8. Kori-Siahpere, O., Ikomi, R.B. and Ogbe, M.G. (2011) Biochemical Response of the African Catfish Clarias gariepinus (Burchell, 1822) to Sub-Lethal Concentrations of Potassium Permanganate. Annals of Biological Research, 2, 1-10.
9. Assmann, G. (1990) At What Levels of Total Low or High-Density Lipoprotein Cholesterol Should Diet/Drug Therapy Be Initiated? European Guideline. The American Journal of Cardiology, 65, 11-15.
10. Kojima, M., Masui, T., Nemoto, K. and Degawa, M. (2004) Lead Nitrate Induced Development of Hypercholesterolemia in Rats, Sterol Independent Gene Regulation of Hepatic Enzymes Responsible for Cholesterol Homeostasis. Toxicology Letters, 154, 35-44.
11. Tardieus, D., Tran, S.T., Auvergue, A., Babile, R., Benard, G., Bailly, J.D. and Guerre, P. (2006) Effects of Fumonisins on Liver and Kidney Sphinganine and the Shinganine to Sphingosine Ratio during Chronic Exposure in Ducks. Chemico-Biological Interactions, 160, 51-60.
12. Riley, R.T. and Voss, K.A. (2006) Differential Sensitivity of Rat Kidney and Liver to Fumonisin Toxicity: Organ-Specific Difference in Toxin Accumulation and Sphingoid Base Metabolism. Toxicological Science, 92, 235-345.

13. Haschek, W.M., Voss, K.A. and Beasley, V.R. (2006) Selected Mycotoxins Affecting Animal and Human Health. In: Haschek, W.M., Roussex, C.G. and Wallig, M.A., Eds., Handbook of Toxicologic Pathology, 2nd Edition, Academic Press, New York, 645-698.
14. Domijan, A.M. and Abramov, A.Y. (2011) Fumonisin B1 Inhibits Mitochondrial Respiration and Deregulates Calcium Homeostasis—Implication to Mechanism of Cell Toxicity. The International Journal of Biochemistry & Cell Biology, 43, 897-904.
15. Rauber, R.H., Dilkin, P., Mallman, A.O., Marchioro, A., Mallman, C.A., Borsoi, A. and Nascimento, V.P. (2012) Individual and Combined Effects of Salmonella typhimurium Lipopolysaccharide and Fumonisin B1 in Broiler Chickens. Poultry Science, 91, 2785-2791.
16. Petrinec, Z., Pepeljnjak, S., Kovacic, S. and Krznavic, A. (2004) Fumonisin B1 Causes Multiple Lesions in Common Carp (Cyprinus carpio). Deutsche Tierarztliche Wochenschrift, 111, 358-363.
17. Tuan, N.A., Manning, B.B., Lovell, R.T. and Rottinghaus, G.E. (2003) Responses of Nile Tilapia (Oreochromis niloticus) Fed Diets Containing Different Concentrations of Moniliformin of Fumonisin B1. Aquaculture, 217, 515-528.
18. Yildirim, M., Manning, B.B., Lovell, R.T. and Grizzler, J.M. (2000) Toxicity of Moniliformin and Fumonisin B1 Fed Singly and in Combination in Diets for Young Catfish Ictalurus punctatus. Journal of the World Aquaculture Society, 31, 599-608.
19. Swann, L. (2000) A Fish Farmer's Guide to Understanding Water Quality. Purdue University, West Lafayette.
20. Boyd, C.E. and Tucker, C.S. (1992) Water Quality and Pond Soil Analysis for Aquaculture. Agricultural Experiment Station, Auburn University, Auburn, 183.
21. Shephard, G.S. and van der Westhuizen, L. (1998) Liquid Chromatographic Determination of the Sphinganine/Sphin-gosine Ratio in Serum. Journal of Chromatography B: Biomedical Sciences and Applications, 710, 219-222.
22. Friedewald, W.T., Levy, R.I. and Fredrickson, D.S. (1972) Estimation of the Concentration of Low-Density Lipoprotein Cholesterol in Plasma, without Use of the Preparative Ultracentrifuge. Clinical Chemistry, 18, 499-502.
23. McGowan, M.W., Artiss, J.O., Strandbergh, D.R. and Zak, B. (1983) A Peroxidase-Coupled Method for the Colorimetric Determination of Serum Triglycerides. Clinical Chemistry, 29, 538-542.
24. Allain, C.C., Poon, L.S., Chen, C.S., Richmond, W. and Fu, P.C. (1974) Enzymatic Determination of Total Serum Cholesterol. Clinical Chemistry, 20, 470-475.
25. Sepahdar, A., Ebrahimzadeh, H.A., Sharifpour, I., Khorsaivi, A., Motallebi, A.A., Mohseni, M., Kakoolaki, S., Pourali, H.R. and Hallajian, A. (2009) Effects of Different Dietary Levels of AFB1 on Survival Rate and Growth Factors of Beluga (Huso huso). Iranian Journal of Fisheries Sciences, 9, 141-150.
26. Arnason, T., Bjornsson, B., Steinarsson, A. and Oddgeirsson, M. (2009) Effects of Temperature and Body Weight on Growth Rate and Feed Conversion Ratio in Turbot (Scohthalmus maximus). Aquaculture, 295, 218-225.
27. Ayinla, O.A. (2007) Analysis of Feeds and Fertilizers for Sustainable Development in Nigeria. NIOMR Technical Paper No. 83, 13 p.
28. Francis, A.G., Adewole, A.M., Oginni, O., Mercy, F., Ayodeji, O., Bada, M. and Akele, O. (2010) Growth Performance, Haematology and Serum Biochemistry of African Catfish (Clarias gariepinus) Fingerlings Fed Graded Levels of Dietary Fumonisin B1. Mycotoxin Research, 26, 221-227.
29. Monbaliu, S., Van Poucke, C., Detaverneir, C., Dumoulin, F., Van de Velde, M., Schoeters, E., Van Dyek, S., Averkieva, O., Van Peteghem, C. and De Saeger, S. (2010) Occurrence of Mycotoxins in Feed as Analyzed by a Multi-Mycotoxin LC-MS/MS Method. Journal of Agricultural and Food Chemistry, 58, 66-71.
30. Voss, K.A., Smith, G.W. and Hashcek, W.M. (2007) Fumonisins: Toxicokinetics, Mechanism of Action and Toxicity. Animal Feed Science and Technology, 137, 299-325.

31. Tiamiyu, L.O., Solomon, G.S. and Oketa, E.J. (2006) Effects of Different Boiling Periods of Soybean (Glycine max (L) Merill) on Growth, Performance of Tilapia (Oreochromis niloticus) Fingerlings. Journal of Aquatic Sciences, 21, 15-18.
32. Adeyemo, B.T. and Umeakuana, P.U. (2011) Some Clinico-Pathologic Changes in Clarias gariepinus and Heterobranchus longifilis Catfishes Following Ethylene Glycol Toxicosis. Vom Journal of Veterinary Science, 6, 23-25.
33. Adeogun, O.A., Ogunbadeji, H.K., Ayinla, O.A., Oresugun, A., Oguntade, O.R., Alhaji, T. and William, S.B. (2007) Urban Aquaculture: Producer Perception and Practices in Lagos State, Nigeria. Middle East Journal of Scientific Research, 2, 21-27.
34. FMARD (Federal Ministry of Agriculture and Rural Development of Nigeria) (2003) Aquaculture Development. Presidential Forum on the Fisheries Development Subsector, Federal Department of Fisheries, Federal Ministry of Agriculture and Rural Development, October 2003, 68.
35. Griessler, K. and Encarnacao, P. (2009) Fumonisins—Mycotoxins of Increasing Importance in Fish. Aquaculture Asia Magazine, XIV, 24-26.

CHAPTER 14

Exercise-Induced Cardiac Remodeling: Lessons from Humans, Horses, and Dogs

Rob Shave [1,*], Glyn Howatson [2,3], Dave Dickson [4] and Lesley Young [5]

[1]*Cardiff Centre for Exercise and Health, Cardiff Metropolitan University, Cardiff CF23 6XD, UK*

[2]*Faculty of Health and Life Sciences, Northumbria University, Newcastle upon Tyne TN1 8ST, UK*

[3]*Water Research Group, North-West University, Potchefstroom 2520, South Africa*

[4]*Heart Vets, Porthcawl, Wales CF36 5LD, UK*

[5]*Specialist equine Cardiology Services, Moulton, Suffolk CB8 8SD, UK*

ABSTRACT

Physical activity is dependent upon the cardiovascular system adequately delivering blood to meet the metabolic and thermoregulatory demands of exercise. Animals who regularly exercise therefore require a well-adapted heart to support this delivery. The purpose of this review is to examine cardiac structure, and the potential for exercise-induced cardiac remodeling, in animals that regularly engage in strenuous activity. Specifically, we draw upon the literature that has studied the "athlete's heart" in humans, horses, and dogs, to enable the reader to compare and contrast cardiac remodeling in these three athletic species. The available literature provides compelling evidence for exercise-induced cardiac remodeling in all three species. However, more work is required to understand the influence of species/breed specific genetics and exercise-related hemodynamics, in order to fully understand the impact of exercise on cardiac structure.

Keywords: echocardiography; exercise physiology; cardiac remodeling

1. INTRODUCTION

Across mammalian species, heart size increases with body size [1]; proportionate with an increased metabolic demand and the requirement to deliver a large blood volume. Early work from Stahl [2] showed that a mammal's heart size typically represents ~0.6% of their body mass (Figure 1). However, a closer inspection of individual mammalian species shows that there is considerable variance, and that in some animals, heart size represents as much as 1.7% of their body mass [3]. This variability appears to be related to the type of physical activity that the animals habitually engage in. Specifically, animals that run for prolonged periods of time, either for hunting (e.g., wolves) or sports performance (e.g., Thoroughbred horses and Greyhounds), have markedly larger hearts than those who are less physically active [1] (Figure 2). This exercise-induced cardiac remodeling is also clearly observed in human endurance athletes (Figure 2) [4], and has been termed the athlete's heart.

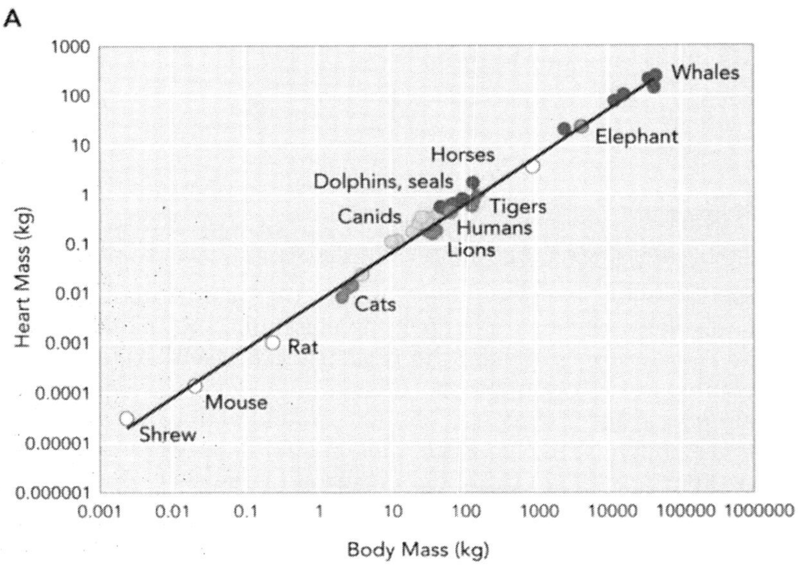

Figure 1. Relationship between heart mass and body mass across the mammalian species. Taken with permission from Williams et al. [1].

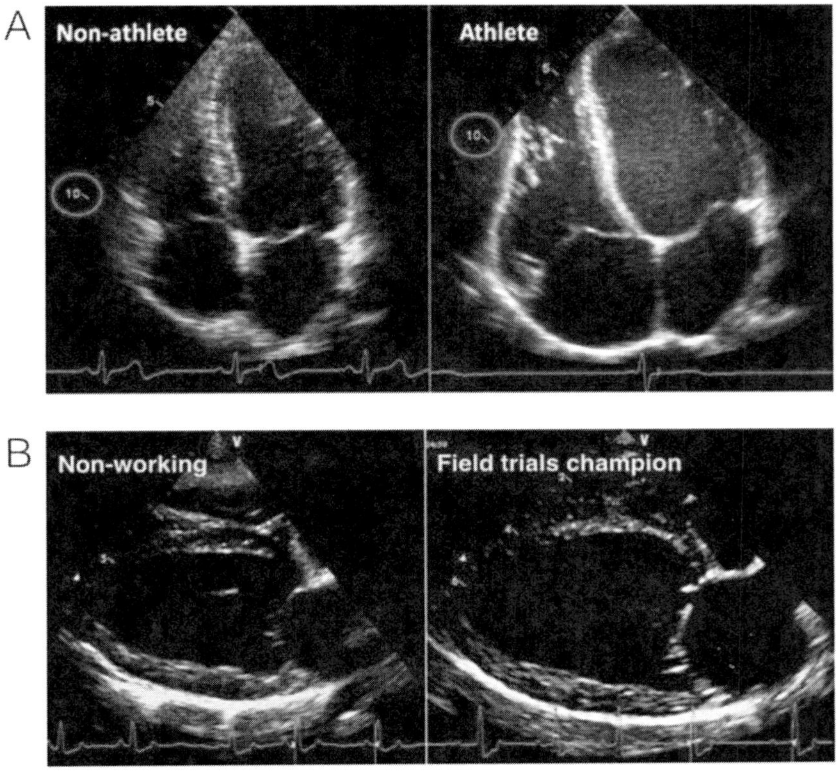

Figure 2. (**A**) Morphological changes in an athlete's heart compared with a non-athlete's heart. The normal heart of a 24-year-old librarian (**left**) is compared with that of a 23-year-old professional cyclist (**right**). Note the difference in scale such that the non-athlete's ventricle is ~10 cm in length (circled in red), whereas the 10-cm mark reaches only halfway along the ventricle of the athlete. Note that all cardiac chambers are enlarged and that increases in wall thickness are relative to dilation. Taken with permission from La Gerche et al. [4]; (**B**) Morphological changes in a working dog's (English Springer Spaniel) heart. The non-working dog (66 months old, 19 kg) (**left**) is compared with that of a Field trials champion (58 months old, 15.3 kg) (**right**). End diastolic volume in the non-working dog is 47 mL and 58 mL in the field trials champion.

Remodeling of the heart in both humans and animals often reflects a maladaptation in response to disease or underlying pathology. Enlargement of the ventricular cavities, or increases in myocardial wall thickness, occur in response to an enhanced volume load or an increase in wall stress, respectively. Thus, the altered hemodynamics associated with cardiac disease, such as valvular regurgitation or aortic stenosis, result in adverse pathologic cardiac remodeling. However, physiologic changes to hemodynamic loading of the heart are also unavoidable during exercise. Dependent on the specific exercise stimulus, the heart may acutely experience an increased volume load, an enhanced "pressure" afterload, or indeed a combination of both. Therefore, when exercise is regularly repeated over a prolonged period of time in a structured manner, akin to that prescribed to competitive athletes (human or animal), there are clear stimuli for cardiac adaptation and remodeling.

The purpose of this review is to examine the physiologic adaptations associated with the athlete's heart, drawing upon human, canine, and equine literature. Although there is crossover, and the potential for clinical confusion between pathologic and physiologic cardiac remodeling [5,6] this topic will not be addressed here. For more information regarding the differentiation of pathophysiologic remodeling, or on the cardiac pathology responsible for sudden cardiac death in sport, readers are directed to previous reviews [7,8,9,10,11]. Remodeling in human athletes is well recognized; however, in animal athletes the concept has received less attention. Accordingly, the present review will initially discus the characteristics of the human athlete's heart and the stimuli responsible for athletic remodeling. We will then examine the current evidence for exercise-induced cardiac remodeling in domesticated animal athletes, namely racehorses and dogs involved in athletic pursuits.

1.1. Acute Exercise and Cardiac Loading

Aerobic exercise is associated with an increased metabolic and thermoregulatory demand. For exercise to continue for any period of time beyond a few seconds, these competing demands must be met by an enhanced delivery of blood, both to the working muscle and the capillary beds responsible for heat exchange. This unavoidable need for blood flow to increase during exercise, explains the close relationship between cardiac output ($\dot{Q}$ —*the product of heart rate and stroke volume*) and the volume of oxygen consumed per minute ($\dot{V}O_2$) (See Figure 3) [12]. In humans, during dynamic (isotonic) aerobic exercise, heart rate and stroke volume increase, alongside a modest reduction in overall systemic vascular resistance; the net result being a large increase in cardiac output. In contrast, during resistance or static (isometric) exercise, a more modest increase in cardiac output is observed that is predominantly driven by an increase in HR, but which is also accompanied by a more pronounced rise in systemic blood pressure than observed during dynamic aerobic exercise. Hence, the hemodynamic load imposed on the heart is dependent upon the exercise intensity, duration, and modality. Importantly, it should be noted that exercise is very rarely either dynamic or static in isolation, rather, the majority of sports fall somewhere along a continuum that involves both elements. In an attempt to better characterize the relative load of individual sports, Mitchell and colleagues [13] have proposed an extended matrix (Figure 4), which provides a far better representation of the "mixed" cardiovascular load that athletes hearts are exposed to. Despite this, much of the available literature regarding cardiac remodeling in human athletes due to exercise has been presented within the context of either an exercise-induced pressure or volume stimulus.

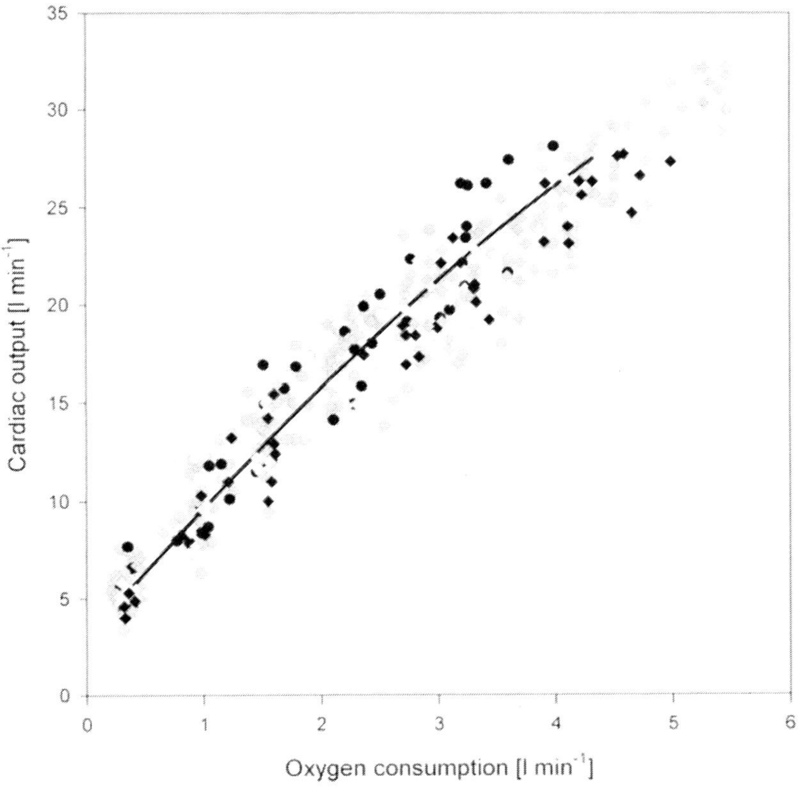

Figure 3. Simulated and experimental relationship between cardiac output and $\dot{V}O_2$. Taken with permission from Liguzinski and Korzeniewski [12].

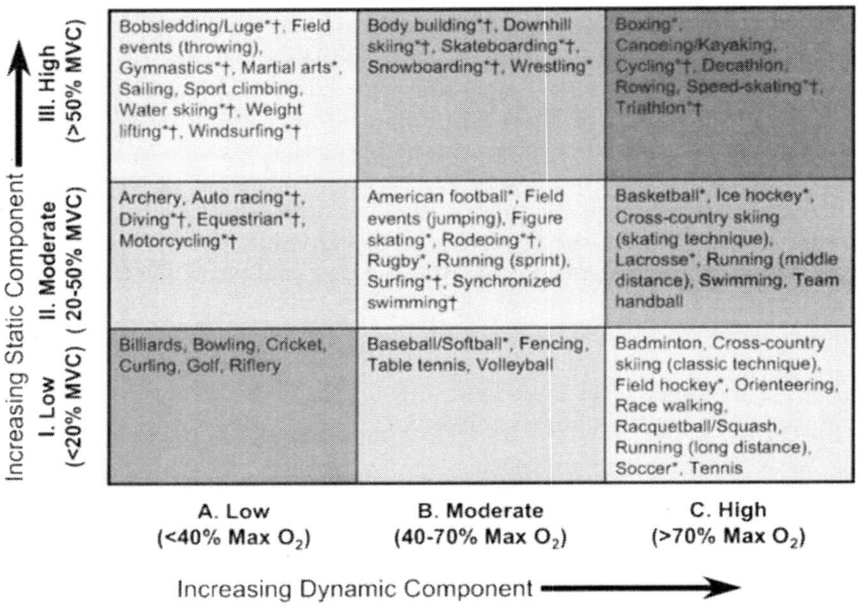

Figure 4. Sports classification model highlighting the combination of static and dynamic characteristics of differing sporting activities. Taken with permission from Mitchell et al. [13].

1.2. Athlete's Heart: Evidence from Humans

Cardiac remodeling in response to athletic training is a well-described phenomenon in humans [14,15]. As early as the 1890s, using auscultation and percussion, physicians in Sweden and the USA demonstrated increased cardiac dimensions in elite cross-country skiers [16] and rowers, respectively [17]. These data were followed by the first reports of athletic bradycardia in runners from the Boston Marathon [18]. Subsequently, with improved imaging technologies (e.g., cardiac ultrasound and magnetic resonance imaging), many studies have gone on to characterize the structural and functional adaptations observed in the human heart in response to athletic training. Since the mid-1970s, the field was driven by a dichotomous characterization of the athlete's heart. Specifically, it has been suggested that endurance athletes present with cardiac adaptations, induced by an increased "volume" load, while power-based athletes have a cardiac phenotype shaped by a higher "pressure" load. Simplistically, endurance athletes have large eccentrically remodeled hearts—*large ventricular volumes, modest wall thickening, and a low relative wall thickness*—associated with reductions in resting heart rate, while power athletes present with concentric remodeling—*thick ventricular walls, relatively small ventricular volumes, and a high relative wall thickness with minimal change in heart rate.* The latter is postulated to be induced by spikes in systemic blood pressure, and hence ventricular wall stress associated with repetitious strength/power-based activities. This dichotomous differentiation between endurance and power athletes was initially made by Morganroth [19], and has subsequently been termed the "Morganroth hypothesis". Until recently, this view has been widely accepted in the sports cardiology literature. However, whilst this hypothesis is attractive from a physiological basis, it is likely an over-simplistic representation of cardiac remodeling in response to athletic training.

The data for the original hypothesis was derived from early studies, using limited one-dimensional imaging techniques [19]. As imaging modalities have advanced, authors have theoretically [20] and empirically [21] argued against concentric remodeling in response to resistance-based exercise training. These authors cite inadequate approaches to cardiac scaling, insensitive imaging modalities, and the potential influence of anabolic agents in some of the populations studied, as reasons why early work supported concentric hypertrophy in these athletes. Haykowsky and colleagues [22] have provided further physiologic rationale as to why resistance athletes may not experience concentric remodeling. Individuals completing intense resistance training (e.g., power lifting) often perform a Valsalva manoeuvre during individual efforts. This respiratory manoeuver equalizes transmural pressures and so the wall stress, or afterload, faced by the left ventricle (LV), is not significantly greater than at rest [22]. Thus, if athletes do perform a Valsalva during resistance activities, the heart is not exposed to the adaptive stimulus. Although Haykowsky's data are compelling, it should be noted that not all resistance activities are performed with a Valsalva manoeuvre. Often, the nature of the athletic activity precludes its use, particularly when the activity is dynamic in nature. Accordingly, it is likely that in all but the most short-lived controlled resistance activities, athletes will experience repeated increases in cardiac afterload, to varying extents.

Presenting longitudinal data from American college football athletes, Weiner and colleagues [23] confirmed the presence of concentric remodeling in response to a prolonged period of resistance-based training. It should be noted, however, that these athletes also developed modest hypertension alongside the athletic training. As such, the increased LV afterload was not confined to the periods of training, and the observed increase in ventricular wall thickness cannot be solely attributed to the athletic stimulus. In contrast, Spence and colleagues [21], using cardiac MRI images, failed to show concentric remodeling in response to six months of resistance exercise training. It should be noted, however, that these individuals were naïve to training before the start of the study, and the training stimulus was limited to three one-hour sessions per week. Therefore, these results cannot be directly compared to the literature derived from competitive athletes who complete a far greater volume of training, and have done so for a considerably longer period of time.

A number of authors have adopted meta-analytical techniques for exploring the Morganroth hypothesis. The most recent of which confirmed evidence of eccentric remodeling in endurance-trained athletes [24], but also provided some support for cardiac remodeling in resistance-trained athletes. Although there were markedly fewer resistance-training studies included in the analyses, both the inter-ventricular septal wall and posterior wall thicknesses were greater in resistance-trained athletes than controls, and were similar to the values seen in endurance athletes. Ventricular cavity dimensions, but not volumes, were also greater in resistance athletes than the controls, but were lower than in the endurance athletes. In a previous meta-analysis, Pluim and colleagues [25] also provided support for concentric hypertrophy in resistance-trained athletes, based on an increased relative wall thickness (RWT, ratio of wall to chamber dimension). However, mean RWT data from Utomi et al. [24] were 0.40 for endurance athletes and 0.41 in resistance athletes, both within clinical norms and not significantly different. The authors therefore suggested that, "*as opposed to dichotomous cardiac structural responses to endurance and resistance training, it might be argued that both athlete groups present with a similar qualitative cardiac adaptation on a continuum, with greater cardiac dimensions in endurance athletes reflecting a greater overall training volume*". Fundamental to the Morganroth hypothesis is the assumption of a dichotomous hemodynamic load in sports; that is endurance and resistance athletes exclusively experience a volume or pressure overload, respectively. As outlined above, this is not the case for the vast majority of athletes; most athletes engage in a combination of resistance and endurance activities, and thus experience a combination of static and dynamic cardiovascular loads [13]. In all likelihood, as suggested by Utomi and colleagues [24], a continuum exists where the LV adapts in response to the many and varied acute and chronic hemodynamic challenges, induced by exercise training.

1.3. Cardiac Adaptations to Exercise Training beyond the Left Ventricle

To date, the vast majority of work examining the athlete's heart has focused on the LV. This is likely a consequence of the important role the LV plays in generating the cardiac output required to meet the demands of exercise, but is also probably related to the relative ease associated with imaging the left side of the heart. Notwithstanding, an increasing body of work has been completed which examines the potential of the right ventricle, atria, and aorta to remodel, in response to training in human athletes.

Given that exercise requires the stroke volume to increase in both the left and right ventricles, and that there is clear evidence of LV remodeling in endurance athletes, it should not be surprising that the RV also remodels. If there was not a proportionate increase in RV size, then a non-sustainable mismatch would occur in what is largely a closed-loop system. Using MRI images, Scharhag and colleagues [26] have provided evidence of the balanced bi-ventricular remodeling in elite endurance athletes (Figure 5). Although, in absolute terms, the resting afterload faced by the RV is considerably less than the LV, the rise in pulmonary artery pressure with exercise is proportionally greater than the rise in systemic arterial pressure [27]. This increase in RV afterload is related to both the inelastic properties of the pulmonary artery, and a relative lack of vasodilation in the pulmonary circulation. Accordingly, similar to the LV, there is a considerable stimulus for RV enlargement in athletes completing large volumes of endurance training. At present, less is known regarding RV remodeling in resistance-trained athletes and this requires further study [24].

Once again, drawing on the results from meta-analyses, it is apparent that the aorta [28] and left atrium respond to athletic training [29]. Using data from 23 previous studies, Iskandar and colleagues [28] showed that the aortic root is on average 3.2 mm lager, and the aortic valve annulus 1.6 mm greater, in athletes when compared to controls. The difference, however, appeared more marked in endurance-, than resistance-trained, athletes. The authors rightly recognize that they cannot rule out the potential confounding effects of body size on their findings, as larger individuals would unsurprisingly have larger aortas. In a similar fashion, the same authors [29] pooled data from 54 studies, including more than seven thousand athletes (endurance, strength, and combined athletes), in order to examine athletic remodeling of the left atria. Their analysis showed that the weighted mean left atrial diameter was 4.1 mm greater in athletes, in comparison to controls, and that LA volume indexed to body surface area was 7 mL/m^2 greater in athletes. Similar to the LV and the aorta, endurance athletes appear to show the greatest adaptation in the LA, but there is also some evidence of remodeling in resistance- and combination-trained athletes.

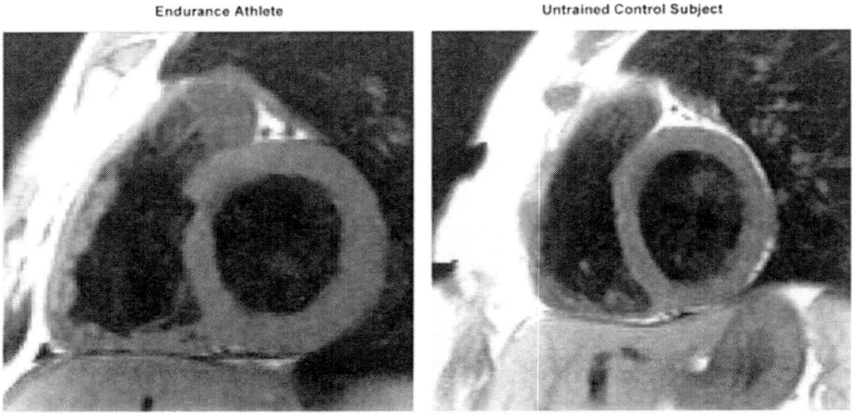

(a) (b)

Figure 5. End-diastolic T1-weighted short-axis slice from an endurance athlete (**a**) and an untrained control subject (**b**). Compared with the heart of the control subject, the endurance athlete's heart is characterized by an enlarged volume and a greater myocardial mass of both ventricles, while the proportions of the left and right heart are the same as in the untrained control subject. Taken with permission from Scharhag et al. [26].

1.4. Cardiac Mechanics

Recently, emerging technology has enabled the assessment of the mechanics that underpin LV function (e.g., deformation of the myocardium in the longitudinal, radial, and circumferential planes, and counter-directional rotation of the ventricular base and apex). Cardiac mechanics are intimately connected to the underlying myocardial architecture, which as outlined above, remodel significantly in response to exercise training. Authors have therefore sought to examine the effect of exercise training on LV mechanics. This is an emerging field, yet early data suggest that highly fit individuals have lower resting mechanics [30,31]. Similar to the bradycardia induced by endurance training, it is likely that a lower degree of myocardial deformation at rest facilitates a greater reserve to draw upon during exercise. Such a reserve would aid diastolic filling, and augment SV and cardiac output during periods of increased demand. Recent work has shown that this adaptation in ventricular mechanics is likely to be dependent upon the duration of training experience. During the initial phase of athletic training, LV rotational mechanics appear to increase, possibly related to an enhanced blood volume, which is then followed by a normalization, or even reduction, in the mechanics once the myocardium itself has remodeled [32].

2. SUMMARY

Since the 1800's, a significant body of work has been completed showing marked cardiac adaptations in the human heart in response to athletic training. These adaptations appear to be modulated by the relative volume and pressure load faced by the heart. This, in turn, is modulated by the type of activity undertaken by the athletes, which cannot be equated as simply either pressure or volume. As imaging modalities continue to improve, it is likely that we will gain a better understanding of the relationship between cardiac structure, mechanics, and overall function. Importantly, as this field advances, it will be important to move away from the pervasive dichotomous contextualisation of exercise training, to increase the number and quality of longitudinal studies, and to attempt to examine the influence of cardiac remodeling on cardiac function during both endurance and resistance exercise.

2.1. Cardiac Remodelling in Non-Human Mammalian Athletes

Many mammals are capable of impressive athletic performance, underpinned by highly adapted physiology [3]. Although the top speed (~70 mph) achieved by the cheetah [33] is extremely impressive, the combination of speed and endurance demonstrated by the pronghorn antelope (50 mph for several miles) [3], or the African Wild dog (41 mph for 10–60 min) [34], is equally, if not more impressive, given the physiology required to sustain such activity. These wild animals have evolved to fill a specific niche; however, through domestication, selective breeding, and tailored training practices, humans have artificially selected animal breeds capable of exceptional athletic performance. For example the Quarter Horse reaches maximal speeds of ~57 mph [35] and the Alaskan sled dog, during the Iditarod 1000+, can sustain a $\dot{V}O_2$ in excess of 100 mL/kg/min, for more than 10 h a day [3]. Such performance is underpinned by a highly adapted physiology and the ability to support both the metabolic and thermoregulatory demands of such activity. As discussed in the introduction above, the cardiovascular system is intimately linked to athletic performance, and it therefore follows that the heart of horses and dogs bred, and trained, specifically for performance, will have adapted and physiologically remodelled to meet the specific demands of the exercise imposed upon them.

2.2. The Heart of Athletic Horses

The level of endurance performance commonly observed in horses is only possible with a highly efficient physiology, primed to take in and transport large volumes of oxygen required for aerobic metabolism. The heart rate range of the horse is large, from ~26 $b \cdot min^{-1}$ at rest, to ~240 $b \cdot min^{-1}$ during intense exercise. This large HR reserve is coupled with the ability to double the packed-cell volume through splenic contraction [36]. When considered as part of the Fick equation:

$$\dot{V}O2max = \text{maximal cardiac output} \times \text{maximal a–VO2difference}$$

these adaptations are of huge benefit in optimising and aiding oxygen transport and extraction, respectively. Although the maximum heart rate is important in the determination of maximal cardiac output, it does not appear to differ between age-matched horses, and is not affected by exercise training [37]. As such, stroke volume in the equine athlete is probably the biggest determinant of $\dot{V}O_{2max}$.

Similar to the early work in human athletes, reports of large hearts in equine athletes date back several hundred years. Following the death of Eclipse in 1789, a horse that was unbeaten over 26 races, his heart was extracted and reportedly weighed, with a measurement of ~6.3 kg [3]. Similarly, at autopsy, the heart of the highly successful American racehorse Secretariat, was estimated to be ~10 kg. In comparison to normal horses, whose hearts weigh ~3.9 kg, hearts of these sizes would be capable of generating extremely large cardiac outputs. Examining the link between heart size and aerobic capacity in horses, Young and colleagues [38] have shown strong positive relationships between various measures of heart size and $\dot{V}O_{2max}$, a key determinant of endurance performance (see Figure 6). Based on the heart mass of Secretariat, it has been estimated that his maximal cardiac output would have been in excess of 500 $L \cdot min^{-1}$, which could have supported a $\dot{V}O_{2max}$ of approximately 240 $mL \cdot kg \cdot min^{-1}$ [3]. Unsurprisingly, there is also a relationship between race performance and heart size; Young et al. [39] have demonstrated a significant linear correlation between the British Horseracing Board Official rating or Timeform rating in National Hunt racing, and heart size, as assessed by ultrasound. This relationship has subsequently been confirmed in other horses, but specifically in horses where performance is dependent upon aerobic capacity, and thus cardiac output [40].

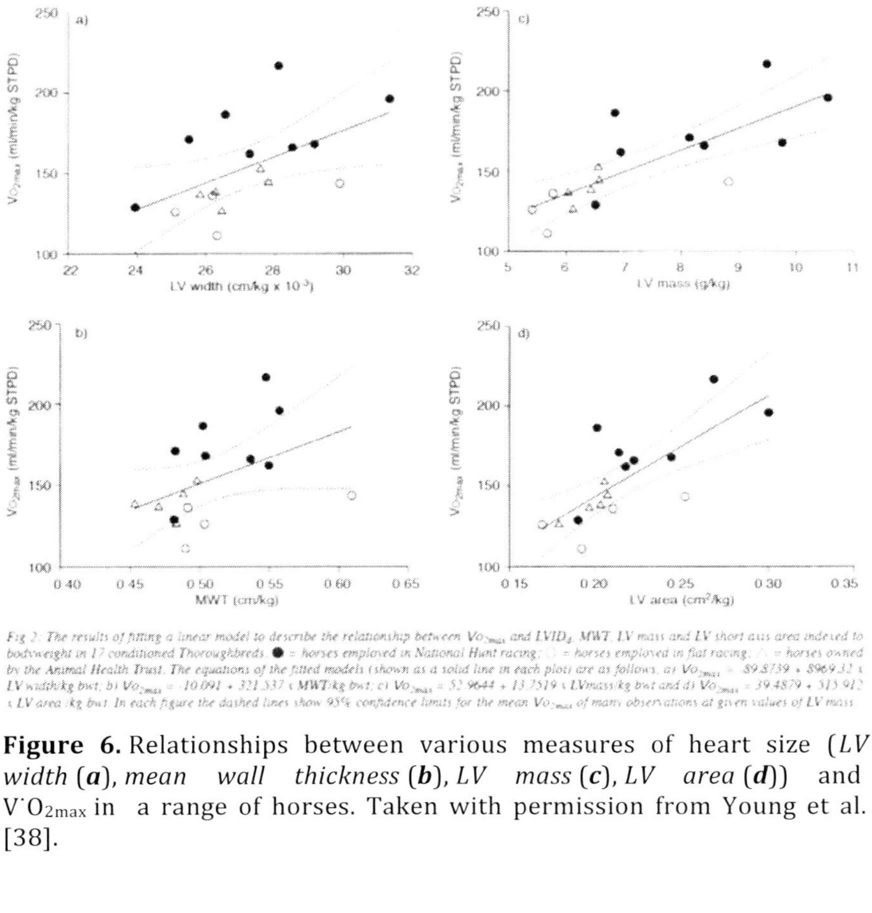

Figure 6. Relationships between various measures of heart size (*LV width* (***a***), *mean wall thickness* (***b***), *LV mass* (***c***), *LV area* (***d***)) and $\dot{V}O_{2max}$ in a range of horses. Taken with permission from Young et al. [38].

Although previous cross-sectional work in horses has shown that heart size is related to athletic performance, this does not prove a "cause and effect" relationship between athletic training and cardiac remodeling. Despite this limitation, there is evidence to support adaptive cardiac remodeling in response to exercise training, much the same as that previously described in humans. For example, autopsy studies have confirmed that trained Thoroughbreds have larger hearts than their untrained counterparts [41]. This has been further supported by longitudinal echocardiographic studies showing differences in cardiac structure between horses of differing training status [42,43]. Buhl and colleagues [43] examined the impact of approximately 18 months of exercise-training on cardiac structure and function, in 103 Standardbred horses. From the start to the end of the training period, internal ventricular diameters in diastole, LV mass, and mean ventricular wall thicknesses, all increased. Relative wall thickness, however, did not, suggesting that the Standardbred horse's ventricle undergoes eccentric remodeling, similar to that observed in human endurance runners. This contrasts with data from a previous longitudinal study examining Thoroughbred horses [42], which in addition to the increases in cardiac dimensions, documented an increase in RWT from 0.40 to 0.45, suggesting that these equine athletes, in comparison to the Standardbreds, experience a mixed ventricular remodeling. It is possible that Thoroughbred horses undergo a more demanding training regime, or that they experience greater elevations in systemic pressures during exercise, than Standardbreds. Of note, while mean systemic pressures in elite horses have been documented in excess of 200 mmHg [3], it has been shown that Thoroughbreds experience higher pulmonary capillary and wedge pressures, than Standardbreds at similar relative exercise intensities [44], and so it is entirely possible that the exercise-related hemodynamics on the left side of the heart are also markedly different.

Similar to human endurance athletes, right ventricular adaptations have also been observed in Thoroughbreds, following intense training. Lightfoot and colleagues [45] showed increases in right ventricular dimensions following a period of intense physical training. Such remodeling is likely related to the volume load experienced by the right side of the heart, induced by frequent elevations in venous return, and increases in total blood volume [46]. However, due to the difficulties in imaging the right ventricular free wall, the authors were unable to assess whether RV wall thickness also increased. Given the marked elevations in pulmonary artery pressures, and the increased blood viscosity related to the transient elevations in packed cell volume previously noted, it is likely that RV wall thickness would also increase, in order to minimize wall stress.

A finding repeatedly observed in the equine literature is that of an increased incidence of valvular incompetence in highly trained horses. The most common cardiac murmur is tricuspid valve regurgitation [47,48], the prevalence and severity of which is associated with both age and years of training [49]. In a prospective study completed by Lightfoot and colleagues, a simultaneous increase in RV diameters was concurrently observed with an increased incidence of tricuspid regurgitation, assessed using color flow Doppler (ranging from 10% of horses before training, to 23% after the initial 8–12 weeks of training, to 36% of all animals at full race fitness). Although it is impossible to suggest cause and effect, it appears likely that the RV remodeling experienced by the horses, alters the ventricular hemodynamics and hence propensity for regurgitation in this population of Thoroughbreds.

In summary, it is clear that the horse's heart is well adapted for endurance exercise and appears plastic in response to intense athletic training. Importantly though, the quadrupedal nature of locomotion, the coupling of respiration to stride frequency, the propensity to markedly increase oxygen carrying capacity through splenic constriction, and the pronounced increases in pulmonary artery pressure, all mean that the physiologic load placed upon the heart during exercise is very different to that observed in humans. As such, whilst we can learn a great deal from the horse in relation to athletic cardiac remodeling, it is very likely that species-specific phenotypic adaptations are present and responses to training may well differ. Future work using advanced imaging techniques, is required to develop a better understanding of the athletic heart syndrome in horses. Such work is warranted in order to better understand the physiology of equine athletes, but more importantly, to help address the causes of exercise-associated sudden cardiac death, which have been associated with the extreme physiology of horseracing [50,51].

2.3. The Athletic Heart in Dogs

In the wild, canids hunt for prolonged periods of time, and as such, are well adapted to meet the demands of aerobic exercise. Since their domestication at least 15,000 years ago [52], dogs have undergone artificial selection, resulting in the development of phenotypic traits suited for specific activities (e.g., terriers for digging, spaniels and retrievers for hunting and retrieving games, and sheep and cattle dogs for herd and flock management). It is likely, therefore, that through this process, the cardiovascular system of dogs has been directly or indirectly selected to facilitate the activities that the animals have been bred for. Surprisingly, however, relatively little work has been completed examining the cardiovascular structure and function in different breeds of dogs, and even less has been completed that considers the acute (within animal) or chronic (across generations) effect of exercise training. For example, generic echocardiographic guidelines for "dogs" are still often used when assessing cardiac structure and function for pathology [53,54]. If athletic remodeling does occur in dogs, the use of such generic guidelines could result in clinical confusion and inappropriate diagnosis [55].

Although relatively little work exists explicitly examining the effects of exercise on the canine heart, a small number of informative studies have been completed and provide the basis for our current understanding of exercise-induced cardiac remodeling in dogs. Similar to athletic humans, it has been shown that athletic training results in a reduction in resting HR in dogs [56,57,58,59,60]. Furthermore, the heart rate response to a standardized atropine challenge [57] or a fixed exercise workload [61,62], is also reduced following periods of exercise training. These lower resting, and sub-maximal heart rates, coupled with a maximum HR of ~300 b·min^{-1} [63], provide a large cardiac reserve in trained dogs that would help to meet the metabolic and thermoregulatory demands of exercise.

In contrast to the HR response to exercise training, there has been some conflict regarding the potential of structural cardiac remodeling in dogs. In early studies, increases in wall thickness and LV mass were consistently reported in response to exercise training [61,64], however, whether internal LV dimensions or volumes increased, was debated [61]. These early discrepancies likely arose due to the wide range of methodologies adopted, and importantly the different breeds of dogs studied. Notwithstanding, from a comparative perspective, it would be extremely interesting and informative if dogs do not remodel in a similar fashion to humans, in response to a similar endurance exercise stimulus.

Given the athletic prowess of the Greyhound, it is not surprising that this breed has featured heavily in the studies that have been completed examining the canine heart. Greyhounds are capable of running at almost 50 mph, and a single [65] animal has been recorded with a $\dot{V}O_{2max}$ of approximately 270 mL·kg·min^{-1} [3]. In comparison to breeds of a similar size, greyhounds have larger internal LV diameters, LV wall thicknesses, and higher heart to body weight ratios [65,66,67]. Similar findings have recently been confirmed in smaller racing dogs, such as Salukis and Whippets [68].

Previously, it has been debated whether the larger hearts in greyhounds are determined genetically, or reflect adaptations to training [67,69]. However, when training greyhounds were compared to non-training animals, their ventricular diameters, wall thicknesses, and cardiac output were shown to be larger [69,70], suggesting at least some degree of athletic-induced remodeling. Such remodeling of the ventricle should not be unexpected, especially as mean arterial blood pressure in the trained greyhound has been shown to be significantly higher than in mongrel dogs, or untrained greyhounds [69,71]. Similarly, athletic training has also been shown to result in an increased blood volume [59,72], which likely acts as a stimulus for remodeling of the ventricular cavity. Evidence therefore suggests that the large heart of the greyhound is genetically conferred, but it seems likely that athletic remodeling in response to both a pressure and volume load, also influences its cardiac morphology.

Another breed of dog capable of prodigious endurance performance is the Alaskan sled dog [73,74]. In the most comprehensive study examining exercise-induced cardiac remodeling in dogs, Stepien and colleagues [60] longitudinally assessed cardiac structure and function in 77 Alaskan sled dogs, before and after five months of endurance training. The training consisted of running ~20 km each day, pulling a sled and musher which was estimated to equate to ~30%–40% $\dot{V}O_{2max}$. Training was associated with significant increases in ventricular wall thicknesses (9%) and cavity dimensions (4%). The study included both veteran (*having completed at least one previous season of training*) and rookie (*in their first season*) dogs, and although ventricular wall thickness increased in both, following training, the remodeling was greater in the rookies. This finding likely relates to both the difference in biological age between the two groups, and also the retention of cardiac hypertrophy in the veteran dogs since the previous season.

The proportionally greater increase in ventricular wall thickness observed in the sled dogs following training, resulted in an increased relative wall thickness, which the authors suggest was attributable to a combined volume and pressure load associated with sled pulling exercise. While there is undoubtedly a brief period of resistance exercise associated with overcoming the initial inertia of the sled and musher, once moving, it would seem unlikely that the exercise undertaken is truly "resistive" in nature. As the authors themselves suggest, the exercise intensity equated to approximately 30%–40% $\dot{V}O_{2max}$. We therefore question whether the mixed ventricular remodeling observed is due to the "pressure load" induced by pulling the sled. Clearly, for wall thickness to increase to a greater extent than the internal cavity, a pressure load on the heart is required. However, instead of the exercise itself being resistance in nature, it is possible that the exercise *per se* is associated with a large pressure load in these animals. In chronically instrumented sled dogs, Van Citters and Franklin showed that systolic pressures could exceed 300 mmHg during aerobic exercise [75]. These pressures far exceed those seen in humans undertaking endurance exercise [76]. Accordingly, the pressure load described by Morganroth [19], to be associated with resistance exercise in humans, may be obligatory in sled dogs during endurance activity. This may therefore explain the increases in absolute and relative ventricular wall thickness, observed by Stepien and colleagues. Further work is needed to understand the blood pressure response to exercise in dogs, especially across different breeds. However, it is possible that the high systolic pressures observed in the sled dogs may be related to quadrupedal hemodynamics, or to differences in thermoregulation between dogs and humans (e.g., panting vs. sweating). These potential differences in canine hemodynamics may help to resolve why increases in ventricular wall thicknesses are a consistent finding in athletic dogs capable of significant endurance performance.

To summarize, while some work has been completed examining cardiac adaptation to exercise training in a small number of dog breeds, in general, there is a paucity of data regarding athletic remodeling in the canine world. As dogs have been bred for a variety of different tasks, it is highly likely that the cardiac phenotype will differ between breeds and how this phenotype responds to different training stimuli, will also possibly vary. Until more work is completed, when examining different breeds and different training interventions, care should be taken when trying to extrapolate findings from one breed to another.

3. CONCLUSIONS

In contrast to the human field, cardiac remodeling induced by athletic training has only received limited attention in the veterinary literature. As such, the subtleties of cardiac structure and function, which have been examined in great detail in human athletes, are yet to be fully discerned in our equine or canine equivalents. While clear data do exist regarding cardiac remodeling in human athletes, it should be noted that the response to training is heterogeneous. It is likely that this heterogeneous response is partially explained by differences in exercise stimulus, but genetic and epigenetic factors are also likely involved [77]. In contrast, the artificial selection imposed by man on both horses and dogs for athletic performance, likely means that these athletes are far more genetically similar than their human counterparts. Therefore, the study of cardiac remodeling in response to athletic training in these animals may provide greater insight into the potential for athletic activity to remodel the heart, independent of confounding variables. It should, however, be remembered that any adaptation observed in response to athletic training is superimposed on a species-specific, or indeed breed-specific, phenotype, which has been shaped by evolution and/or selective breeding. Care should therefore be taken when inferring between species or breeds in relation to exercise-induced cardiac remodeling.

AUTHOR CONTRIBUTIONS

All authors contributed to the drafting and editing of this manuscript.

CONFLICTS OF INTEREST

The authors declare no conflict of interest.

REFERENCES

1. Williams, T.M.; Bengtson, P.; Steller, D.L.; Croll, D.A.; Davis, R.W. The healthy heart: Lessons from nature's elite athletes. Physiology 2015, 30, 349–357.
2. Stahl, W.R. Organ weights in primates and other mammals. Science 1965, 150, 1039–1042.
3. Poole, D.C.; Erickson, H.H. Highly athletic terrestrial mammals: Horses and dogs. Compr. Physiol. 2011, 1, 1–37.
4. La Gerche, A.; Roberts, T.; Claessen, G. The response of the pulmonary circulation and right ventricle to exercise: Exercise-induced right ventricular dysfunction and structural remodeling in endurance athletes (2013 grover conference series). Pulm. Circ. 2014, 4, 407–416.
5. Maron, B.J.; Pelliccia, A. The heart of trained athletes: Cardiac remodeling and the risks of sports, including sudden death. Circulation 2006, 114, 1633–1644.
6. D'Silva, A.; Sharma, S. Exercise, the athlete's heart, and sudden cardiac death. Phys. Sportsmed. 2014, 42, 100–113.
7. Firoozi, S.; Sharma, S.; McKenna, W.J. Risk of competitive sport in young athletes with heart disease. Heart 2003, 89, 710–714.
8. Basso, C.; Corrado, D.; Thiene, G. Cardiovascular causes of sudden death in young individuals including athletes. Cardiol. Rev. 1999, 7, 127–135.
9. Maron, B.J. Distinguishing hypertrophic cardiomyopathy from athlete's heart physiological remodelling: Clinical significance, diagnostic strategies and implications for preparticipation screening. Br. J. Sports Med. 2009, 43, 649–656.
10. Pelliccia, A.; Maron, M.S.; Maron, B.J. Assessment of left ventricular hypertrophy in a trained athlete: Differential diagnosis of physiologic athlete's heart from pathologic hypertrophy. Prog. Cardiovasc. Dis. 2012, 54, 387–396.
11. Kim, J.H.; Baggish, A.L. Differentiating exercise-induced cardiac adaptations from cardiac pathology: The "grey zone" of clinical uncertainty. Can. J. Cardiol. 2016, 32, 429–437.
12. Liguzinski, P.; Korzeniewski, B. Oxygen delivery by blood determines the maximal VO2 and work rate during whole body exercise in humans: In silico studies. Am. J. Physiol. Heart Circ. Physiol. 2007, 293, H343–H353.
13. Mitchell, J.H.; Haskell, W.; Snell, P.; Van Camp, S.P. Task force 8: Classification of sports. J. Am. Coll. Cardiol. 2005, 45, 1364–1367.
14. George, K.P.; Wolfe, L.A.; Burggraf, G.W. The "athletic heart syndrome". A critical review. Sports Med. 1991, 11, 300–330.
15. Weiner, R.B.; Baggish, A.L. Exercise-induced cardiac remodelling: The need for assessment of regional myocardial function. J. Physiol. 2012, 590, 2829–2830.
16. Henschen, S. Eine medizinische sportstudie. Mitt. Med. Klin. Upsala 1899, 2, 15–18.
17. Darling, M.D. The effects of training. A study of the harvard university crews. Boston Med. Surg. J. 1899, 229–233.
18. White, P.D. The pulse after a marathon race. JAMA 1918, 71, 1047–1048.
19. Morganroth, J.; Maron, B.J.; Henry, W.L.; Epstein, S.E. Comparative left ventricular dimensions in trained athletes. Ann. Intern. Med. 1975, 82, 521–524.
20. Naylor, L.H.; George, K.; O'Driscoll, G.; Green, D.J. The athlete's heart: A contemporary appraisal of the 'morganroth hypothesis'. Sports Med. 2008, 38, 69–90.

21. Spence, A.L.; Naylor, L.H.; Carter, H.H.; Buck, C.L.; Dembo, L.; Murray, C.P.; Watson, P.; Oxborough, D.; George, K.P.; Green, D.J. A prospective randomised longitudinal mri study of left ventricular adaptation to endurance and resistance exercise training in humans. J. Physiol. 2011, 589, 5443–5452.
22. Haykowsky, M.; Taylor, D.; Teo, K.; Quinney, A.; Humen, D. Left ventricular wall stress during leg-press exercise performed with a brief valsalva maneuver. Chest 2001, 119, 150–154.
23. Weiner, R.B.; Wang, F.; Isaacs, S.K.; Malhotra, R.; Berkstresser, B.; Kim, J.H.; Hutter, A.M., Jr.; Picard, M.H.; Wang, T.J.; Baggish, A.L. Blood pressure and left ventricular hypertrophy during american-style football participation. Circulation 2013, 128, 524–531.
24. Utomi, V.; Oxborough, D.; Whyte, G.P.; Somauroo, J.; Sharma, S.; Shave, R.; Atkinson, G.; George, K. Systematic review and meta-analysis of training mode, imaging modality and body size influences on the morphology and function of the male athlete's heart. Heart 2013, 99, 1727–1733.
25. Pluim, B.M.; Zwinderman, A.H.; van der Laarse, A.; van der Wall, E.E. The athlete's heart. A meta-analysis of cardiac structure and function. Circulation 2000, 101, 336–344.
26. Scharhag, J.; Schneider, G.; Urhausen, A.; Rochette, V.; Kramann, B.; Kindermann, W. Athlete's heart: Right and left ventricular mass and function in male endurance athletes and untrained individuals determined by magnetic resonance imaging. J. Am. Coll. Cardiol. 2002, 40, 1856–1863.
27. La Gerche, A.; Heidbuchel, H.; Burns, A.T.; Mooney, D.J.; Taylor, A.J.; Pfluger, H.B.; Inder, W.J.; Macisaac, A.I.; Prior, D.L. Disproportionate exercise load and remodeling of the athlete's right ventricle. Med. Sci. Sports Exerc. 2011, 43, 974–981.
28. Iskandar, A.; Thompson, P.D. A meta-analysis of aortic root size in elite athletes. Circulation 2013, 127, 791–798.
29. Iskandar, A.; Mujtaba, M.T.; Thompson, P.D. Left atrium size in elite athletes. JACC Cardiovasc. Imaging 2015, 8, 753–762.
30. Nottin, S.; Doucende, G.; Schuster, I.; Tanguy, S.; Dauzat, M.; Obert, P. Alteration in left ventricular strains and torsional mechanics after ultralong duration exercise in athletes. Circ. Cardiovasc. Imaging 2009, 2, 323–330.
31. Stohr, E.J.; McDonnell, B.; Thompson, J.; Stone, K.; Bull, T.; Houston, R.; Cockcroft, J.; Shave, R. Left ventricular mechanics in humans with high aerobic fitness: Adaptation independent of structural remodelling, arterial haemodynamics and heart rate. J. Physiol. 2012, 590, 2107–2119.
32. Weiner, R.B.; DeLuca, J.R.; Wang, F.; Lin, J.; Wasfy, M.M.; Berkstresser, B.; Stohr, E.; Shave, R.; Lewis, G.D.; Hutter, A.M., Jr.; et al. Exercise-induced left ventricular remodeling among competitive athletes: A phasic phenomenon. Circ. Cardiovasc. Imaging 2015, 8.
33. Sharp, N.C. Timed running speed of a cheetah (acinonyx jubatus). J. Zool. 1997, 241, 493–494.
34. Nowak, R.M. Dogs, Wolves, Coyotes, Jackals, and Foxes: Canidae. In Walker's Carnivores of the World; Johns Hopkins University Press: Baltimore, MD, USA, 2005; p. 112.
35. Nielsen, B.D.; Turner, K.K.; Ventura, B.A.; Woodward, A.D.; O'Connor, C.I. Racing speeds of quarter horses, thoroughbreds and arabians. Equine Vet. J. Suppl. 2006, 128–132.
36. McKeever, K.H.; Hinchcliff, K.W.; Reed, S.M.; Robertson, J.T. Plasma constituents during incremental treadmill exercise in intact and splenectomised horses. Equine Vet. J. 1993, 25, 233–236.
37. Betros, C.L.; McKeever, K.H.; Kearns, C.F.; Malinowski, K. Effects of ageing and training on maximal heart rate and VO2max. Equine Vet. J. Suppl. 2002, 34, 100–105.

38. Young, L.E.; Marlin, D.J.; Deaton, C.; Brown-Feltner, H.; Roberts, C.A.; Wood, J.L. Heart size estimated by echocardiography correlates with maximal oxygen uptake. Equine Vet. J. Suppl. 2002, 34, 467–471.
39. Young, L.E.; Rogers, K.; Wood, J.L. Left ventricular size and systolic function in thoroughbred racehorses and their relationships to race performance. J. Appl. Physiol. (1985) 2005, 99, 1278–1285.
40. Sleeper, M.M.; Durando, M.M.; Holbrook, T.C.; Payton, M.E.; Birks, E.K. Comparison of echocardiographic measurements in elite and nonelite arabian endurance horses. Am. J. Vet. Res. 2014, 75, 893–898.
41. Kubo, K.S.T.; Sugimoto, O. Relationship between training and heart in thoroughbred racehorse. Exp. Rep. Eq. Hlth. Lab. 1974, 11, 89–93.
42. Young, L.E. Cardiac responses to training in 2-year-old thoroughbreds: An echocardiographic study. Equine Vet. J. Suppl. 1999, 30, 195–198.
43. Buhl, R.; Ersboll, A.K. Echocardiographic evaluation of changes in left ventricular size and valvular regurgitation associated with physical training during and after maturity in standardbred trotters. J. Am. Vet. Med. Assoc. 2012, 240, 205–212.
44. Hackett, R.P.; Ducharme, N.G.; Gleed, R.D.; Mitchell, L.; Soderholm, L.V.; Erickson, B.K.; Erb, H.N. Do thoroughbred and standardbred horses have similar increases in pulmonary vascular pressures during exertion? Can. J. Vet. Res.2003, 67, 291–296.
45. Lightfoot, G.; Jose-Cunilleras, E.; Rogers, K.; Newton, J.R.; Young, L.E. An echocardiographic and auscultation study of right heart responses to training in young national hunt thoroughbred horses. Equine Vet. J. Suppl. 2006, 36, 153–158.
46. Hinchcliff, K.W.; McKeever, K.H.; Schmall, L.M.; Kohn, C.W.; Muir, W.W., 3rd. Renal and systemic hemodynamic responses to sustained submaximal exertion in horses. Am. J. Physiol. 1990, 258, R1177–R1183.
47. Patteson, M.W.; Cripps, P.J. A survey of cardiac auscultatory findings in horses. Equine Vet. J. 1993, 25, 409–415.
48. Kriz, N.G.; Hodgson, D.R.; Rose, R.J. Prevalence and clinical importance of heart murmurs in racehorses. J. Am. Vet. Med. Assoc. 2000, 216, 1441–1445.
49. Young, L.E.; Wood, J.L. Effect of age and training on murmurs of atrioventricular valvular regurgitation in young thoroughbreds. Equine Vet. J. 2000, 32, 195–199.
50. Navas de Solis, C. Exercising arrhythmias and sudden cardiac death in horses: Review of the literature and comparative aspects. Equine Vet. J. 2016, 48, 406–413.
51. Lyle, C.H.; Uzal, F.A.; McGorum, B.C.; Aida, H.; Blissitt, K.J.; Case, J.T.; Charles, J.T.; Gardner, I.; Horadagoda, N.; Kusano, K.; et al. Sudden death in racing thoroughbred horses: An international multicentre study of post mortem findings. Equine Vet. J. 2011, 43, 324–331.
52. Driscoll, C.A.; Macdonald, D.W.; O'Brien, S.J. From wild animals to domestic pets, an evolutionary view of domestication. Proc. Natl. Acad. Sci. USA 2009, 106, 9971–9978.
53. Brown, D.J.; Rush, J.E.; MacGregor, J.; Ross, J.N., Jr.; Brewer, B.; Rand, W.M. M-mode echocardiographic ratio indices in normal dogs, cats, and horses: A novel quantitative method. J. Vet. Intern. Med. 2003, 17, 653–662.
54. Cornell, C.C.; Kittleson, M.D.; Della Torre, P.; Haggstrom, J.; Lombard, C.W.; Pedersen, H.D.; Vollmar, A.; Wey, A. Allometric scaling of m-mode cardiac measurements in normal adult dogs. J. Vet. Intern. Med. 2004, 18, 311–321.
55. Dickson, D.; Shave, R.; Rishniw, M.; Harris, J.; Patteson, M. Reference intervals for transthoracic echocardiography in the english springer spaniel: A prospective, longitudinal study. J. Small Anim. Pract. 2016, 57, 520–528.
56. Tipton, C.M.; Carey, R.A.; Eastin, W.C.; Erickson, H.H. A submaximal test for dogs: Evaluation of effects of training, detraining, and cage confinement. J. Appl. Physiol. 1974, 37, 271–275.
57. Barnard, R.J.; Duncan, H.W.; Baldwin, K.M.; Grimditch, G.; Buckberg, G.D. Effects of intensive exercise training on myocardial performance and coronary blood flow. J. Appl. Physiol. Respir. Environ. Exerc. Physiol. 1980, 49, 444–449.

58. Ritzer, T.F.; Bove, A.A.; Carey, R.A. Left ventricular performance characteristics in trained and sedentary dogs. J. Appl. Physiol. Respir. Environ. Exerc. Physiol. 1980, 48, 130–138.
59. Mackintosh, I.C.; Dormehl, I.C.; van Gelder, A.L.; du Plessis, M. Blood volume, heart rate, and left ventricular ejection fraction changes in dogs before and after exercise during endurance training. Am. J. Vet. Res. 1983, 44, 1960–1962.
60. Stepien, R.L.; Hinchcliff, K.W.; Constable, P.D.; Olson, J. Effect of endurance training on cardiac morphology in alaskan sled dogs. J. Appl. Physiol. 1998, 85, 1368–1375.
61. Wyatt, H.L.; Mitchell, J.H. Influences of physical training on the heart of dogs. Circ. Res. 1974, 35, 883–889.
62. Carey, R.A.; Ritzer, T.F.; Bove, A.A. Effect of endurance training on myocardial myosin adenosine triphosphatase activity of the dog. Med. Sci. Sports 1979, 11, 308–312.
63. Vatner, S.F.; Franklin, D.; Higgins, C.B.; Patrick, T.; Braunwald, E. Left ventricular response to severe exertion in untethered dogs. J. Clin. Investig. 1972, 51, 3052–3060.
64. Riedhammer, H.H.; Rafflenbeul, W.; Weihe, W.H.; Krayenbuhl, H.P. Left ventricle contractile function in trained dogs with cardial hypertrophy. Basic Res. Cardiol. 1976, 71, 297–308.
65. Page, A.; Edmunds, G.; Atwell, R.B. Echocardiographic values in the greyhound. Aust. Vet. J. 1993, 70, 361–364.
66. Schneider, H.P.; Truex, R.C.; Knowles, J.O. Comparative observations of the hearts of mongrel and greyhound dogs. Anat. Rec. 1964, 149, 173–179.
67. Carew, T.E.; Covell, J.W. Left ventricular function in exercise-induced hypertrophy in dogs. Am. J. Cardiol. 1978, 42, 82–88.
68. Seckerdieck, M.; Holler, P.; Smets, P.; Wess, G. Simpson's method of discs in salukis and whippets: Echocardiographic reference intervals for end-diastolic and end-systolic left ventricular volumes. J. Vet. Cardiol. 2015, 17, 271–281.
69. Pape, L.A.; Price, J.M.; Alpert, J.S.; Rippe, J.M. Hemodynamics and left ventricular function: A comparison between adult racing greyhounds and greyhounds completely untrained from birth. Basic Res. Cardiol. 1986, 81, 417–424.
70. Lonsdale, R.A.; Labuc, R.H.; Robertson, I.D. Echocardiographic parameters in training compared with non-training greyhounds. Vet. Radiol. Ultrasound 1998, 39, 325–330.
71. Cox, R.H.; Peterson, L.H.; Detweiler, D.K. Comparison of arterial hemodynamics in the mongrel dog and the racing greyhound. Am. J. Physiol. 1976, 230, 211–218.
72. McKeever, K.H.; Schurg, W.A.; Convertino, V.A. Exercise training-induced hypervolemia in greyhounds: Role of water intake and renal mechanisms. Am. J. Physiol. 1985, 248, R422–R425.
73. Hinchcliff, K.W.; Reinhart, G.A.; Burr, J.R.; Schreier, C.J.; Swenson, R.A. Metabolizable energy intake and sustained energy expenditure of alaskan sled dogs during heavy exertion in the cold. Am. J. Vet. Res. 1997, 58, 1457–1462.
74. Hinchcliff, K.W.; Constable, P.D.; Farris, J.W.; Schmidt, K.E.; Hamlin, R.L. Electrocardiographic characteristics of endurance-trained alaskan sled dogs. J. Am. Vet. Med. Assoc. 1997, 211, 1138–1141.
75. Van Citters, R.L.; Franklin, D.L. Cardiovascular performance of alaska sled dogs during exercise. Circ. Res. 1969, 24, 33–42.
76. Calbet, J.A.; Jensen-Urstad, M.; van Hall, G.; Holmberg, H.C.; Rosdahl, H.; Saltin, B. Maximal muscular vascular conductances during whole body upright exercise in humans. J. Physiol. 2004, 558, 319–331.
77. Bouchard, C.; Rankinen, T.; Timmons, J.A. Genomics and genetics in the biology of adaptation to exercise. Compr. Physiol. 2011, 1, 1603–1648.

CHAPTER 15

Stimulation of Phagocytosis and Production of Antibodies against Canine Herpesvirus Type 1 by Pidotimod (Adimod™)

Edgar G. Valdivia Lara[1], Andrés Ángeles Aldana[1], Jesica I. Ángeles Solis[2], Patricia Miranda Castro[3], Juan C. Del Río García[4], Guillermo Valdivia Anda[1*]

[1]*Laboratory of Microbial Pathogenicity, Multidisciplinary Research Unit, Faculty of Higher Education Cuautitlán, National Autonomous University of Mexico, Cuautitlán Izcalli, Mexico*

[2]*Laboratory DIVET™, Cuautitlan Izcalli, Mexico*

[3]*Biotechnology Laboratory, UNAM, Cuautitlan Izcalli, Mexico*

[4]*Laboratory of Mycotoxins, Multidisciplinary Research Unit, Faculty of Higher Education Cuautitlan, National Autonomous University of Mexico, Cuautitlan Izcalli, Mexico*

ABSTRACT

Neutrophils are the most important circulating phagocytes. Circulating monocytes and precursors of tissue macrophages also have the ability to phagocytize. Pidotimod (ADIMOD™) exerts immunostimulatory and immunoregulatory effects through the stimulation and regulation of cellular immune responses by lymphocytes Canine herpesvirus (CHV) mainly affect puppies between the first and second weeks of age, causing high morbidity in the litter. To date, there is only one commercial vaccine in Europe to prevent disease. In this work, inactivated CHV cultures were inoculated in rabbits, adsorbed and not adsorbed to chitosan nanoparticles. Phagocytosis in the presence or absence of specific antibodies was measured. Response of virus neutralizing antibodies was also evaluated. Adimod™ enhanced the nonspecific and specific phagocytotic response. The association of the virus to the nanoparticles increased the phagocytic ability of blood cells; however, Adimod™ alone had a greater effect on phagocytic activity and generated a stronger immune response that corresponded to the increased phagocytosis ($p < 0.05$). Moreover, the level of neutralizing antibodies was higher and increased more rapidly when Adimod™ was used.

Keywords: Phagocytosis, Canine Herpesvirus, Pidotimod Stimulation

1. INTRODUCTION

Antibodies are potent determinants of the humoral immune response. Although they are generated through interactions between B and T cells, antibodies trigger their cytotoxic effects by interacting with complement and innate effector cells; thus, they provide a functional link between the adaptive and innate immune systems [1] .

Neutrophils are the most important circulating phagocytes as they provide the first line of defense against invading particles, especially bacteria [2] ; however, circulating monocytes and precursors of tissue macrophages also have the ability to phagocytize [3] .

Pidotimod [(R)-3-(S)-(5-oxo-2-pyrrolidinyl)-carbonyl-thiazolidine-4-carboxylic acid] [4] (ADIMOD™) exerts immunostimulatory and immunoregulatory effects on T lymphocytes, enhances and stimulates the immunological mechanisms involved in the humoral and cellular immunity mediated by T lymphocytes and simultaneously stimulates macrophage migration, which is an essential aspect of the cellular immune response that enables adequate phagocytosis and clearance of infectious microbes [4] . Pidotimod is almost completely absorbed when administered orally, and urine recovery reaches nearly 93% of the unchanged product. Furthermore, it is widely distributed throughout the body and mainly eliminated in the urine and feces, and its bioavailability reaches approximately 45% and is dose-dependent. Pidotimod undergoes only slight hepatic metabolism; therefore, it is excreted unchanged in the urine and no metabolites are known. Its plasma protein binding is very poor, which is advantageous to prevent interactions with other medications. Pidotimod has a half-life of approximately 4 h [5] .

The mechanism of action of Pidotimod remains unknown but is considered to occur through the stimulation and regulation of cellular immune responses by lymphocytes, mainly T lymphocytes, with a concurrent stimulatory effect on macrophage migration. It should be noted that macrophages are involved in the cellular immune response; therefore, the display of Pidotimod is important for the mobilization of macrophages to allow them phagocytize and obtain the proper clearance of infectious microbes.

Pidotimod also presents a protective effect; although he did not have direct bactericidal or bacteriostatic activity, it globally stimulates the immune system, primarily macrophages.

To date, the role of Pidotimod in veterinary medicine, especially in small species, is unknown; therefore, the aim of this study was to evaluate its use as a vaccine adjuvant against Canine Herpesvirus (CHV).

2. METHODS

Were used sixty rabbits of the New Zealand breed with an approximate weight of 1700 g. The rabbits were obtained from the Animal Isolation Unit (Bioterio) of the Faculty of Higher Studies Cuautitlán, UNAM. Clinical and hematological evaluations were performed to ensure the health of the animals and set the initial parameters. Every procedure was conducted according to the use and care of experimental animals protocols mentioned in the Official Mexican Norm 062 (NOM062).

The virus was previously isolated from a 3-day-old dead puppy [6] ; Canine Herpesvirus (CHV) was replicated in MDCK cells at a concentration of 10^5 viral particles per ml and was inactivated by UV light exposure, as briefly described; the virus culture was placed in a 54 mm diameter glass Petri dish, capless at a distance of 20 cm from a mercury germicidal lamp (Osram HNS 10W), emitting mainly at 254 nm wavelength light. After 30 min an absorbed dose of approximately of 104 J/m^2(3.6 quantum/nucleotide) occurred, this UV dose has been described to cause inactivation of alphaviruses and herpesvirus.

Chitosan nanoparticles (NANO) were prepared at the UNAM-FESC Biotechnology Laboratory, and the particles presented an approximate molecular size of 1000 μm with a weight of approximately 1.5×10^6 Da [7] . A 1:10 dilution of 10^5/ml inactivated CHV was prepared using a chitosan solution. All preparations were photographed using transmission electron microscopy (TEM).

The experimental design was established as shown in Table 1.

All animals were bled from the jugular vein at 0, 7, 15, 30, 45 and 60 days. A complete blood count, CHV antibody detection in the serum and an in vitro phagocytosis test were performed.

An in vitro test to evaluate specific and nonspecific phagocytic activity was modified from a previously described protocol [1] . Blood samples were collected in plastic syringes containing 30 U of heparin per ml of blood. A 5% lysing solution was added to the sample, followed by centrifugation at 150 g, at 6°C, for 8 min to separate the leukocyte-rich plasma. The leukocytes were washed by adding Hank's balanced salt solution (HBSS) and centrifuging at 1118 g for 8 min. The cell pellet was re-suspended in Medium 199 (In Vitro™). The cell concentration was adjusted to 2.0 × 10^6 leukocytes per ml.

The phagocytosis assay was performed as follows: a suspension of peripheral blood leukocytes at a concentration between 10^3 and 10^4 per ml was prepared, and 0.5 ml of the suspension was added into 12-well Nunc™ plates. The chitosan- virus suspension (0.5 ml) was added to six wells in each plate, and 0.5 ml of the chitosan suspension was placed in the other six wells. Every well was filled with 0.5 ml of MEM medium containing 5% fetal bovine serum (In Vitro™) and 0.5 ml of serum from each rabbit. The plates were incubated overnight at 37°C. The content of each well was transferred to an Eppendorf tube and centrifuged at 1118 g for 8 min. at 6°C, the pellet was washed three times with PBS, and the pellet was re-suspended in 0.5 ml of MEM medium. For leukocyte counting, four smears from each well were fixed with cold acetone for 10 min. Then, 15 μL of fluorescein conjugated anti-CHV (VMRD™) were added. Finally, 100 cells were counted and the number of labeled cells per ml was calculated.

Table 1. Experimental design to determine the effect of Pidotimod (Adimod™) on phagocytosis in rabbits.

Treatment	Group 1	Group 2	Group 3	Group 4	Group 5
CHVi 1 ml sc	Yes				Yes
CHVi + NANO 1 ml sc		yes			
NANO 1 ml sc			Yes		
Adimod• oral seven days 2 ml				Yes	yes
Inactivated MDCK cells 1 ml sc			Yes	Yes	

Twelve animals in each group (n = 60) CHVi, canine herpesvirus inactivated; NANO, chitosan nanoparticles; sc, subcutaneous.

Total amount of serum antibodies against CHV was determined with a previously described seroneutralization test using 10^3 virus particles per well [8] [9] . The sera were inactivated at 57°C for 30 min, followed by a centrifugation step at 27,000 g for 10 min in a microcentrifuge. Because the virus used has a very low cytolytic effect, we employ a combination of routine test SN and observe cytopathic effect of the virus by the detection of cell infection by an immunoperoxidase technique, using antibodies against canine herpes virus of dog and anti canine IgG subsequently antibodies conjugated to peroxidase (Sigma™) developed by diaminobenzidine (Sigma™) on wells with cells [9] . Neutralizing antibodies were tested in all serum samples. A plaque formation test was used for viral titration: ten-fold dilutions of the virus culture (1:10, 1:100, 1:1000, 1:10,000) were made, and 0.5 ml of each was added to each well of a 12-well NUNC™ plate containing a monolayer of Madin-Darby canine kidney MDCK cells (In vitro™, p117) at 90% confluence in minimal essential medium (MEM) (In vitro™) with 5% newborn calf serum (NCS) (In vitro™). The cells were incubated for 1 h at 35°C to allow for viral absorption, with mixing every 15 min. Finally, MEM was added slowly with 1% newborn calf serum and 0.6% agarose (Bioline™) at 45°C to allow for solidification, and then incubated at 35°C for 3 days. A drop of formaldehyde (JTBaker™) was then added to each well so that the agarose could be subsequently removed. The wells were washed carefully with phosphate buffered saline (PBS) and fixed with acetone/methanol (ICR™) for 10 min. Subsequently, 0.5 ml of a 1:80 dilution of positive anti-CHV-1 control rabbit serum was added to each well and incubated for 30 min at 37°C. The wells were washed three times with PBS, and anti-rabbit IgG peroxidase (Sigma Aldrich™) was added and incubated for a further 30 min at 37°C. The plate was washed three times with PBS, and a 0.01% diaminobenzidine solution (Sigma Aldrich™) with hydrogen peroxide (ICR™) was added and held at 15 min at 37°C The plate was washed again with PBS and stained with Harris hematoxylin (Sigma Aldrich ™) for 15 s, and the plates were examined microscopically for the presence of plaque forming units, termed peroxidase-positive spotlights (PPS), which were counted only in wells containing 20 - 100 plaques.

The neutralization assay was performed by incubating equal amounts of virus (50 PPS) with rabbit sera at different dilutions (1:4, 1:8, 1:16, 1:32, and 1:64; pretreated for 30 min at 56°C) for 30 min at 37°C. The mixture (0.5 ml) was inoculated to NUNC™ plates containing a monolayer of MDCK cells at 90% confluence. Viral adsorption was allowed for 1 h at 35°C, while redistributing the inoculum every 15 min, and MEM supplemented with 1% calf neonate serum and 0.6% agarose at 45°C was added slowly, allowing for solidification. The plate was incubated at 35°C for three days, and then a drop of 40% formaldehyde was added to each well and the agarose layer was removed the next day. The titer was calculated as the inverse of the highest dilution in which a reduction of at least 50% of the number of plaques was observed. A serum sample was considered negative if the 1:8 dilution of serum did not show a decrease of 50% or more in plaque formation.

The results were analysed by the Mann-Whitney U-test used Statistic© software The numbers of samples are stated in the figure legends. The level of significance was set at $p < 0.05$. These is nonparametric test for the null hypothesis that there is no difference in phagocytosis between the two experiments. Assuming that we do not know if the experiments behave in a normal distribution, both experiments are independent and determine the difference between the two.

3. RESULTS AND DISCUSSION

The average counts of blood leukocytes in the 12 animals were not significantly different ($p > 0.05$) over time (results not shown). Therefore, the doses used for the immunization of the rabbits were low because there were not showed inflammatory effects.

Abundant viral particles were adsorbed to the prepared chitosan particles, as observed on transmission electron microscopy (Figure 1).

In Figure 2, phagocytosis of viral particles in the absence (a) or presence (b) of specific antibodies was observed. Adimod™ enhanced the nonspecific and specific phagocytotic response ($p < 0.05$). In addition to their role in neutralization, antibodies mediate additional functions including the recruitment of innate immune responses to eliminate antibody-opsonized material. Furthermore, antibodies promote phagocytosis, which may play an important role in the rapid containment and clearance of a pathogen following infection. However, robust assays that are able to capture differences in the quality of antibody-mediated phagocytosis are lacking [1] . Thus, in this study we developed a novel high- throughput assay by using monocytic cell blood to provide a platform to evaluate antigen-specific antibody-mediated phagocytosis. Briefly, the antibodies of interest are captured on the surface of chitosan beads (NANO) coated with highly fluorescent spiked viral particles, and the beads are then incubated overnight with monocytes prior to analysis by immunofluorescence. Because the beads used may be coated with an antigen of choice, this assay allows the evaluation of antigen-specific phagocytosis without requiring purification of the respective antibodies.

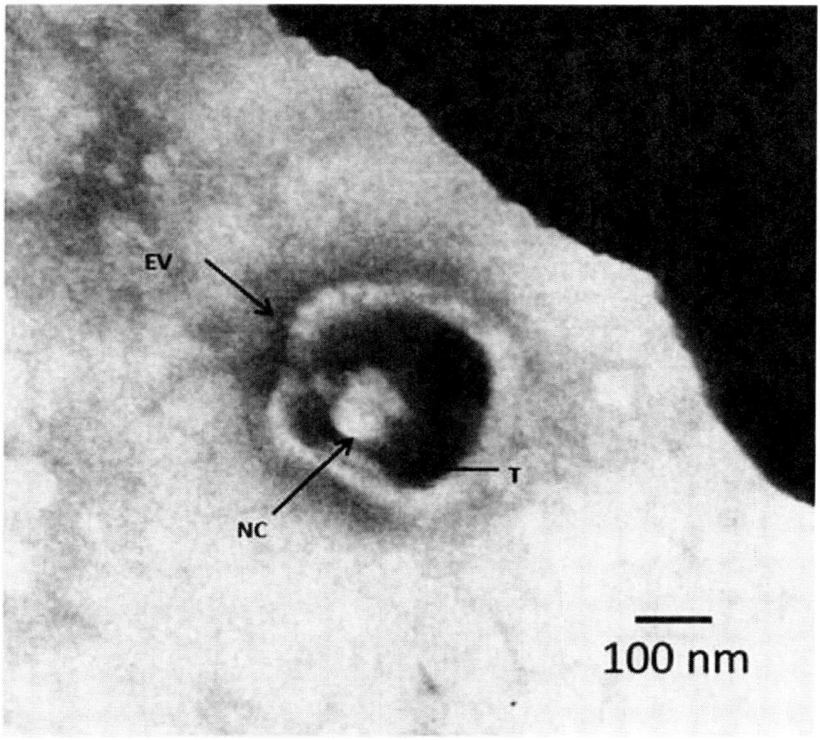

Figure 1. Micrograph taken by transmission electron microscopy showing adsorbed particles of herpesvirus (CHV) with an average diameter of 100 nm for the capsid and an average diameter of 240 nm for the sheath. Nucleocapsid (NC), the tegument (T) and the viral envelope (EV). Magnification: 20,000×.

We observed variation in the phagocytosis tests; however, comparison of data from microbiological in vitro studies on the phagocytosis and killing of C. albicans blastoconidia by phagocytes is hampered by divergent methodologies. The results of some studies are equivocal, and the reported differences may result from the various assay conditions and methods that have been used to determine phagocyte candidacidal activity [10] . However, to date, minor variations have been observed in other studies of the phagocytic ability of leukocytes.

We were not able to determine whether phagocytosis was performed by blood neutrophils or monocytes because both of these cell types have blood phagocytosis capability after opsonizing [11] ; however, all rabbits in this study showed an average cell content of 73% neutrophils and 4.5% monocytes; therefore, most of the fluorescent cells observed corresponded to neutrophils.

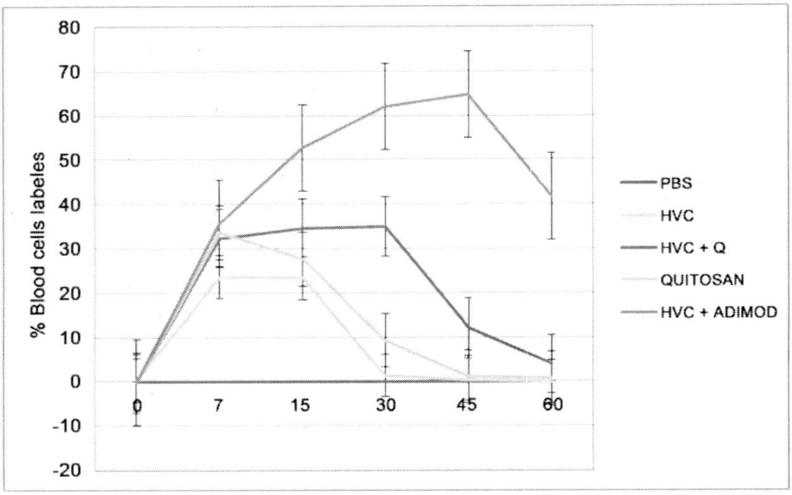

(a)

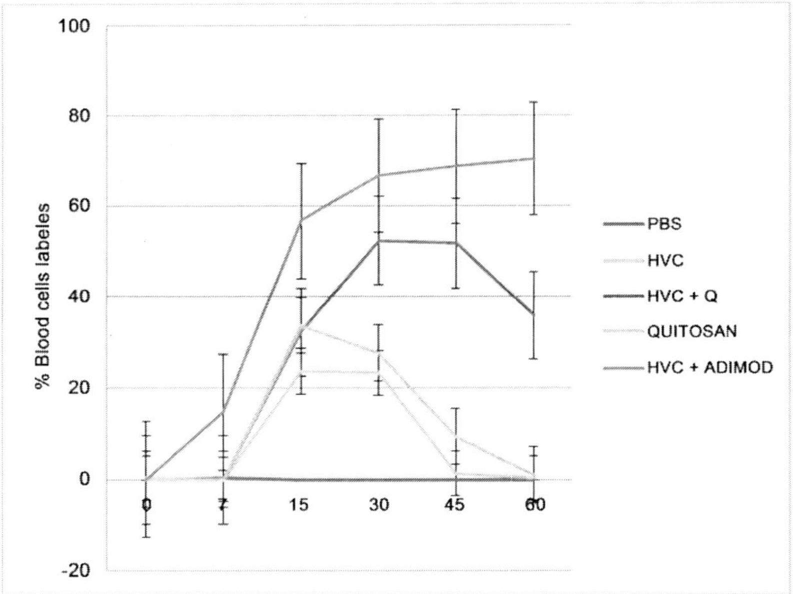

(b)

Figure 2. (a) Results indicate the average value for 12 rabbits per treatment. Chitosan particles adsorbed with inactivated canine herpesvirus (CHV) were placed in contact with ((a) blood leukocytes and evaluated by fluorescent antibodies against CHV. (b) the serum of rabbits and with leukocyte blood, and after incubation, they were evaluated using fluorescent antibodies against CHV).

In the present experiments, the association of the virus to the nanoparticles increased the phagocytic ability of blood cells; however, Adimod™ alone had a high effect on phagocytic activity. Pidotimod was also observed to generate a stronger immune response that corresponded to the increased phagocytosis ($p < 0.05$).

In Figure 3, the results of seroneutralization tests are shown for each group, and the mean of the inverse of the highest dilution that provided 50% protection is plotted. Our results confirm that Adimod™ enables a faster immune response ($p < 0.05$) than chitosan's response after 45 days, which has been observed to be as immunogenic as a vaccine adjuvant [12] [13] .

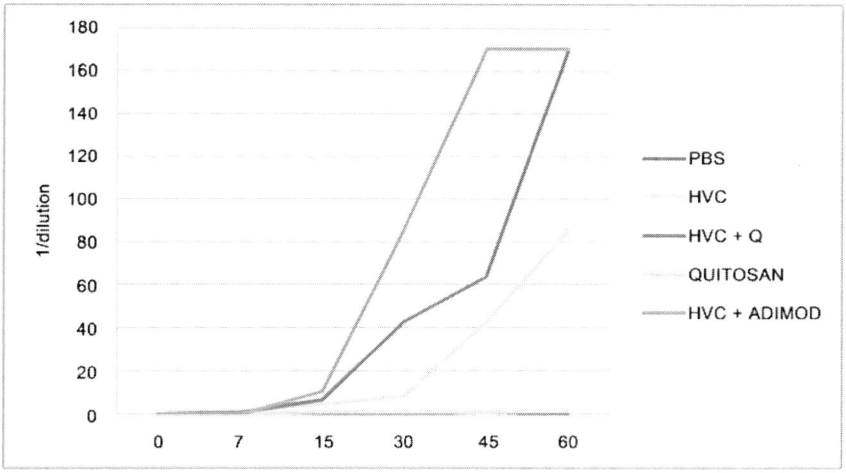

Figure 3. The results refer to 12 rabbits per treatment; the seroneutralization test was performed in triplicate using 103 viral particles per well, and the average of the inverse of the dilution that caused 50% lysis was plotted.

Pidotimod [(R)-3-(S)-(5-oxo-2-pyrrolidinyl)-carbonyl-thiazolidine-4-carboxylic acid] has been shown to significantly increase the survival time after challenge with low doses of herpes simplex virus and influenza virus [4] Pidotimod was also shown to induce dendritic cell (DC) maturation and up-regulate MHC class II cell surface receptor (HLA-DR) and the co-stimulatory molecules CD83 and CD86, which are essential for communication with adaptive immune cells, and it was demonstrated that Pidotimod promotes strong and specific humoral and cellular immune responses in vivo when co-administered intranasally with an antigen model; it was also shown that administration by oral route increases resistance to viral infections [5] [14] [15] [16] .

In this experimental design, we did not determine the combined effect of chitosan and Adimod™ because the purpose was to demonstrate Adimod's effect, but it is inferred that the immune response, using Adimod and nanoparticles combined, would possibly be higher and faster.

Studies developed to compare the intestinal absorption of various drugs in humans and experimental animals (e.g., rats and rabbits) have concluded that these animals are a good alternative model for preclinical studies for absorption in humans [17] .

Vaccination of pregnant bitches has been effectively used in other countries to lessen the impacts of canine herpes virus infection in puppies; however, to date, no data is available for the use of vaccination against canine herpesvirus to protect an open population. Although it is true that protection against viral diseases is more efficient when mediated by cellular immunity, the development of higher amounts of serum antibodies in animals can help to eliminate viral reactivation after chronic infections or latency observed in herpes viruses; thereby decreasing the possibility of childbirth transmittance [18] .

Evaluation of chemotaxis, phagocytosis and microbicidal killing may enable the differential diagnosis of immunodeficiency-related diseases as herpes virus infections observed in dogs and other species [11] . The methodology developed in our study may be useful to evaluate the effects of immunosuppressive diseases such as herpes virus and also to evaluate the effectiveness of immunomodulators and adjuvants.

The level of neutralizing antibodies was higher and increased more rapidly when Adimod™ was used (Figure 2).

ACKNOWLEDGEMENTS

The work was partially funded by the project PAPIIT IT202114 "Evaluation of canine herpesvirus disease in Mexico and development of measures for its control and prevention." of Universidad Nacional Autónoma de México, DGAPA.

CITE THIS PAPER

Lara, E.G.V., Aldana, A.Á., Solis, J.I.Á., Castro, P.M., García, J.C.D.R. and Anda, G.V. (2017) Sti- mulation of Phagocytosis and Production of Antibodies against Canine Herpesvirus Type 1 by Pidotimod (Adimod™). Open Journal of Veterinary Medicine, 7, 38-47. https://doi.org/ 10.4236 /ojvm.2017.74005

REFERENCES

1. Ackerman, M.E., Moldt, B., Wyatt, R.T., Dugast, A.S., McAndrew, E., Tsoukas, S., Jost, S., Berger, C.T., Sciaranghella, G., Liu, Q., Darrell, J., Irvine, D.J., Burton, D.R. and Alter, G. (2011) A Robust, High-Throughput Assay to Determine the Phagocytic Activity of Clinical Antibody Samples. Journal of Immunological Methods, 366, 8-19.
2. Van Oss, C.J. (1986) Phagocytosis: An Overview. Methods in Enzymology, 132, 3-15.
3. Cline, M.J. and Lehrer, R.I. (1968) Phagocytosis by Human Monocytes. Blood, 32, 423-435.
4. Masihi, K.N. (2000) Immunomodulatory Agents for Prophylaxis and Therapy of Infections. International Journal of Antimicrobial Agents, 14, 181-191.
5. Giagulli, C., Noerder, M., Avolio, M., Becker, P.D., Fiorentini, S., Guzmán, C.A. and Caruso, A. (2009) Pidotimod Promotes Functional Maturation of Dendritic Cells and Displays Adjuvant Properties at the Nasal Mucosa Level. International Immunopharmacology, 9, 1366-1373.
6. Valdivia Lara, E.G., Barrón Romero, B.L., Cobos Marín, L., ángeles Solis, J.I., González Gallardo, S., Cuenca Verde, C., Montaraz Crespo, J.A. and Valdivia Anda, G. (2016) Pathology Isolation and Identification of Canine Herpesvirus (CHV-1) in Mexico. Open Journal of Pathology, 6, Article ID: 67833.
7. Miranda, C.S.P. and Lizárraga, P.E.G. (2012) Is Chitosan a New Panacea? Areas of Application. In: Karunaratne, D.N., Ed., The Complex World of Polysaccharides, InTech, Croatia, 3-46.
8. Nothling, J.O., Hussy, D., Steckler, D. and Ackermann, M. (2008) Seroprevalence of Canine Herpesvirus in Breeding Kennels in the Gauteng Province of South Africa. Theriogenology, 69, 276-282.
9. Valdivia, L.E.G., Angeles, S.J.I., Cuenca, V.C., Montaraz, C.J.A., Cobos, M.L., Del Río, G.J.C. and Valdivia, A.G. (2016) Canine Herpesvirus Seroprevalence and Associated Factors in Dogs of Mexico. Open Journal of Veterinary Medicine, 6, 149-162.
10. Vonk, A.G., Wieland, C.W., Netea, M.G. and Kullberg, B.J. (2002) Phagocytosis and Intracellular Killing of Candida Albicans Blastoconidia by Neutrophils and Macrophages: A Comparison of Different Microbiological Test Systems. Journal of Microbiological Methods, 49, 55-62.
11. Chammas, P.P.C, Mitika, K. and Hagiwara, M.K. (1998) Evaluation of Neutrophilic Function (Chemotaxis, Phagocytosis and Microbicidal Activity) in Healthy Dogs and in Dogs Suffering from Recurrent Deep Pyoderma. Veterinary Immunology and Immunopathology, 64, 123-131.
12. Surya, K., Mallapragada, S.K. and Narasimhan, B. (2008) Immunomodulatory Biomaterials Review. International Journal of Pharmaceutics, 364, 265-271.
13. Gunbeyaza, M., Faraji, A., Ozkulb, A., Purali, N. and Senela, S. (2010) Chitosan Based Delivery Systems for Mucosal Immunization against Bovine Herpesvirus (BHV-1). European Journal of Pharmaceutical Sciences, 41, 531-545.

14. Artali, R., Bombieri, G., Meneghetti, F., Nava, D., Ragg, E. and Stradi, R. (2003) Structural Characterization of a Dipeptide Compound with Immunostimulant Activity: 3-(5-Thioxo-L-prolyl)-L-thiazolidine-4-carboxylic Acid Il. Farmaco, 58, 883-889.

15. Li, C., Meng, Chen, D., Wang, Z., Han, H., Chen, H., Zhang, J. and Li, X. (2013) The Immunostimulatory Effect of Novel Immunostimulator CH2b with a Thiazolidin-4-One Ring on the Functions of LPS-Activated RAW 264.7 Macrophages In Vitro LPS. International Immunopharmacology, 17, 698-703.

16. Meng, M., Li, C., Chen, D., Wang, Z., Hong, Y., Chen, H., Li, X., Zhang, J. and Wang, F.S. (2013) Novel Synthetic Immunostimulators with a Thiazolidin-4-One Ring Promote the Cytotoxicity of Human NK Cells via ERK1/2 Activation In Vitro. International Immunopharmacology, 15, 655-660.

17. Zhao, Y.H., Abraham, M.H., Le, J., Hersey, A., Luscombe, C.N., Beck, G., Sherborne, B. and Coope, I. (2003) Evaluation of Rat Intestinal Absorption Data and Correlation with Human Intestinal Absorption. European Journal of Medicinal Chemistry, 38, 233-243.

18. Virgin, H.W., Wherry, E.J. and Ahmed, R. (2009) Redefining Chronic Viral Infection. Cell, 138, 30-50.

CHAPTER 16

Comparative Evaluation of Different Treatment for Purulent Wounds in Dogs

Ahmad Massoud Niazi*, Mohammad Monir Tawfeeq, Amanullah Aziz, Jahid Zabuli, Shahpoor Rahmati, Abdul Razaq Irshad

Clinic Department, Faculty of Veterinary Science, Kabul University, Kabul, Afghanistan

ABSTRACT

To compare the healing of purulent wounds with Shilajit dressing vs. Vishnevsky dressing in dogs with purulent wounds of >2 weeks of duration, eight dogs were chosen from same ages in two groups i.e., Shilajit and Vishnevsky dressing group. Dressing was done on every day basis for more than two weeks of follow up period. Main outcome of healing measure was completed at three weeks. Wound healing status was assessed at three days intervals till end of three weeks. Shilajit treated achieved complete decrease in the wound surface area, effect of healing score in Shilajit dressing group in comparison to the Vishnevsky dressing group at $p > 0.05$ level of significance. Shilajit is highly effective in achieving a characteristic feature of regenerative and granulation healing of purulent wounds as compared to Vishnevsky dressing group in dogs.

Keywords: Shilajit Wound Dressing, Purulent Wound of Dog, Granulation Tissue

1. INTRODUCTION

Wounds often supervised openly in treatment procedure of veterinary medicine [1] . When a wound is treated in second stage from the wound, it is anticipating that the wound should have enough granulation tissue, contraction, and growth of smooth layer of epithelialization to get closed [2] [3] . For wound healing, a lot of medical agents are used; generally these agents have a great variation due to wound condition and host factors [4] [5] . Shilajits have many essential anti-inflammatory agents which are used as independently for treatment of open wounds in many centuries [6] . Generally, Shilajit is a blackish brown exudation found in the serene surroundings of Himalayas. It is also found in most of the sedimentary rocks especially in Afghanistan, Bhutan, China, Nepal, Pakistan, USSR (Tien-Shan, Ural and Caucasus), and Tibet as well in Norway [3] - [7] . Shilajit is a herbo-mineral drug, which oozes out from a special type of mountain rocks in the peak summer months. It is found at high altitudes ranging from 1000 to 5000 meters. The biologically important classes of compounds of Shilajit include dibenzo-alpha pyrenes phospholipids, triterpenes and phenolic acids of low molecular weight, fulvic acids: "carrier molecules", humins and humic acids and trace elements (Fe, Ca, Cu, Zn, Mg, Mn, Mo, and P) [8] [9] . These natural materials have been established as potent wound healing medicines, possessing anti-inflam- matory, antimicrobial, antioxidant, and proliferative properties [10] . The effectiveness of Shilajit pharmaceutical formulations on wound healing surely has been not proven in clinical studies in animals [11] . May confirmations of some findings have come from animal (mainly rats or mice) and in vitro studies [12] . Vishnevsky ointment contains essential agents such as birch tar, xeroformium and castor oil which are used for management of wounds, burns, skin ulcer and suppurations [13] . Vishnevsky ointment was first used in Russia by a Russian surgeon by the name of Alexander Alexandrovich Vishnevsky and it was broadly used in the Soviet army during the World War II [14] . Wound dressing with Vishnevsky ointment is commonly performed in veterinary medicine as part of open wound management [15] . Our hypothesis was that the use of a Shilajit based ointment may be more beneficial for second intention wound healing than dressing with Vishnevsky ointment. The aim of the present study was to evaluate the effectiveness of a Shilajit based ointment on second intention wound healing of surgically created wounds in the dog compared with Vishnevsky ointment.

2. MATERIAL AND METHODS

2.1. Preparation of Shilajit and Vishnevsky Ointments

Purified Shilajit tablets are bought from Pharmacy (Pure Mummia Asel, Bishkek City, Kargyzactan). Then, 30 gr of dried Shilajit tablets were collected and crushed thoroughly in a grinding stone. When they were converted into fine powder mixed with 70 gr of pure Vaseline, after that the mixed pasty preparation was kept overnight at room temperature and was allowed to cool. Now, the Shilajit ointment preparation was ready for application. Vishnevsky ointment which was prepared in 30 gr bought from Pharmacy (Mus Pharma, Russia). And, it is used in open wounds of dogs in research.

2.2. Animals

Research protocol was approved by Animal Care Committee of veterinary faculty. Eight mixed breed dogs (8 males and females) aged 3 to 4 months were used. In order to ascertain the good health of the dogs, a complete physical examination, complete blood (cell) count, and serum biochemical analyses were done on each dog before wound creation. Research animal adjust to a new environment 10 days prior to research. Animals were brought twice a day for exercise. Healthy foods were prepared and were given twice a day for each dog.

2.3. Presurgical Protocol

Preparation of animals for presurgical procedure (day 0), premedication such as Acepromazine in dose of 0.04 mg/kg bodyweight (BW) was given in every dog. In addition, Morphine was used in dose of 0.2 mg/kg Intramuscular (IM) in each dog. For Induction of anesthesia was used the Propofol in dose of 5mg/kg Intravenous and the maintenance stage of anesthesia was kept with isoflurane in oxygen. The dorsal side of each dog was clipped and cleaned with antiseptic solution for operation.

2.4. Surgical and Postoperative Procedure

First of all 5 cm full-thickness wound was created in the dorsal side of each dog, for clinical observation and treatment. The wounds were treated once daily until total healing with Shilajit based ointment in four dogs and other four dogs the wounds are dressed with Vishnevsky ointment until healing of wounds. Both groups' wounds are flushed with normal saline 0.9%.

During the dressing time each wound covered with clean bandage for prevention of flies and other infected materials. Complete wound healing was done in 21 days of research.

2.5. Evaluation of Physiological Parameters, Hematology and Biochemistry

During the wound healing physiological parameters, wound sizes, hematology and biochemistry were evaluated in under research dogs. Physiological parameters were collected and recorded everyday basis in all healing period. In mean time, the wound sizes were measured by ruler which had size of 10 cm in both groups in daily basis up to more than three weeks in dogs. Moreover, hematology tests were done in Blood analyzer (ERMA INC. Particle Counter, Model PCE-170) and the biochemistry element tests such as Calcium, Albumin and Glucose were completed by assistance of LiquickCor Laboratory Kit (PZ CORMAY S. A. Wiosenna 22, 05-092 Lomiank, Poland) and also other elements such as Total protein and Iron measured by Bio-TEST Kit (ErbaLachema s. r. o Karasek 1d, 621 33 Brno, CZ). Hence all biochemistry elements were weighted accurately in PD-303 Spectrophotometer (APEL Japan).

3. RESULTS

3.1. Statistical Finding

All parameters were evaluated one by one before moving to the next which allowed for more precise comparison of individual. Data were presented as the mean value ± standard deviation of the mean for wound healing size, hematology and biochemistry. Statistical analysis was performed on one-way ANOVA and T-test. Values of $P < 0.05$ were considered statistically significant.

3.2. Wound Size Observation

There was significant ($P < 0.05$) increase in Size of wounds in Shilajit group in day of 9 and 15 in wound healing compared to Vishnevsky group (Table 1).

3.3. Physiological Observation

There was significant ($P < 0.05$) decrease in Pulse rate (per min), in 15 day of research. However, the other parameters remained in normal range in period of wounds healing in Shilajit group compared to Vishnevsky group (Table 2 and Table 3).

3.4. Hematological Observation

There was significant ($P < 0.05$) increase of leukocytes in 3 days of research but it decrease in the on the sixth day after treatment almost doubled in subsequent observation days and downward trend is maintained Shilajit group compared to the Vishnevsky group. In mean time, There was significant ($P < 0.05$) decrease in the number of erythrocytes on the 15 day, significant ($P < 0.05$) reduction in concentration of hemoglobin in red blood cells observed on the 3 day and subsequently maintained in normal level other days. Therefore, it was significant ($P < 0.05$) increase in the number of platelets in the 9 days after the start of treatment by almost 2.5 times, apparently due to the healing process in Shilajit group compared Vishnevsky group (Table 4 and Table 5).

Table 1. Observation on wound size changes during the wound healings process (Shilajit and Vishnevsky group) (mean ± SD) (n = 4), *Significant at 5% level with Shilajit group value ($P < 0.05$).

Groups	Post-wound (3 day)	Post-wound (9 day)	Post-wound (15 day)
Shilajit	5.7 ± 0.2	2.2 ± 0..2*	0.5 ± 0.1*
Vishnevsky	6 ± 0.3	3 ± 0.3	1 ± 0.2

Table 2. Observations on physiological parameters (Shilajit group) (mean ± SD) (n = 4), *Significant at 5% level with pre- wound value ($P < 0.05$).

Parameters with units	Pre-wound	Post-wound (3 day)	Post-wound (9 day)	Post-wound (15 day)
Rectal temperature(°C)	39.43 ± 0.19	39.6 ± 0.04	39.3 ± 0.3	38.8 ± 0.2
Pulse (per min) ± 1.7	118	119.5 ± 1.6	104 ± 1.3	100 ± 5.5*
Respiration (per min)	19.5 ± 0.7	20 ± 0.7	20 ± 0.6	21.5 ± 0.7
Mucous membrane	pale roseate	Pale roseate	Pale roseate	Pale roseate
Capillary refill time (sec)	<2	<2	<2	<2

Table 3. Observations on physiological parameters (Vishnevsky group) (mean ± SD) (n = 4), *Significant at 5% level with pre-wound value ($P < 0.05$).

Parameters with units	Pre-wound	Post-wound (3 day)	Post-wound (9 day)	Post-wound (15 day)
Rectal temperature (°C)	39.2 ± 0.2	39.6 ± 0.05	39.6 ± 0.07	39.3 ± 0.08
Pulse (per min)	118 ± 2.3	114 ± 2.1	104 ± 2.2	100 ± 1.3*
Respiration (per min)	22 ± 0.6	20 ± 0.7	20 ± 0.09	20 ± 0.1
Mucous membrane	Pale roseate	Pale roseate	Pale roseate	Pale roseate
Capillary refill time (sec)	<2	<2	<2	<2

Table 4. Observations on hematological parameters (Shilajit group) (mean ± SD) (n = 4), *Significant at 5% level with preoperative value ($P < 0.05$).

Parameters with units	Pre-wound	Post-wound (3 day)	Post-wound (9 day)	Post-wound (15 day)
WBC ($\times 10^3/\mu L$)	129.9 ± 51.6	63.1 ± 4.5*	59 ± 7	86.7 ± 8.8
RBC ($\times 10^6/\mu L$)	8.4 ± 1.4	12.1 ± 0.1	13 ± 0.8	10.7 ± 0.3*
HGB (g/dl)	85.1 ± 12.4	106.5 ± 7.4	114.2 ± 10.3	111.3 ± 1.1
LYM (%)	23.5 ± 16.5	6.3 ± 1.7	3.9 ± 0.6	8.7 ± 1.5
MON (%)	13.2 ± 7.8	3.2 ± 0.6	3.3 ± 0.9	3.9 ± 0.5
MCV (f L)	71.5 ± 1.2	78.2 ± 2.7	77.8 ± 3.7	77.6 ± 1.8
MCH (p g)	29.5 ± 1.8	25.7 ± 0.5	33 ± 5.7	25.8 ± 1.2
MCHC (g/d L)	41.5 ± 3	33.3 ± 1.7	46 ± 6.1	33.5 ± 1.9
PLT ($\times 10^3/\mu L$)	830 ± 162.1	584 ± 74.2	680 ± 171	701 ± 194.8
PDW (f L)	16.73 ± 0.6	17.4 ± 0.1	17.4 ± 0.4	16.4 ± 0.3

WBC (White Blood Cell); MCV (Mean Corpuscular Volume); RBC (Red Blood Cell); MCH (Mean Corpuscular hemoglobin); HGB (Hemoglobin); MCHC (Mean Corpuscular Hemoglobin Concentrations); LYM (Lymphocytes); PLT (Platelets); MON (Monocytes); PDW (Platelet Distribution Width).

Table 5. Observations on hematological parameters (Vishnevsky group) (mean ± SD) (n = 4), * Significant at 5% level with preoperative value ($P < 0.05$).

Parameters with units	Pre-wound	Post-wound (3 day)	Post-wound (9 day)	Post-wound (15 day)
WBC ($\times 10^3/\mu L$)	81.8 ± 9.8	90.2 ± 3.5	92.3 ± 3.8	89.3 ± 4.3
RBC ($\times 10^6/\mu L$)	11.6 ± 1.04	12.2 ± 1.3	11.5 ± 1.06	13.6 ± 1.7
HGB (g/dl)	3.4 ± 0.3	3.6 ± 0.1	1.9 ± 0.3	4.8 ± 0.4
LYM (%)	6.5 ± 1.9	7.2 ± 2.9	6.9 ± 2.1	5.9 ± 1.3
MON (%)	3.85 ± 0.6	3.9 ± 1.2	4.5 ± 0.8	3.7 ± 0.6
MCV (f L)	80.8 ± 2.5	83.9 ± 2.5	85 ± 2.6	82.4 ± 1.9
MCH (p g)	25.45 ± 1.3	25.8 ± 1.8	27.2 ± 2.1	23.3 ± 1.2
MCHC (g/d L)	32.85 ± 1.5	34.2 ± 1.7	33.4 ± 1.6	30.7 ± 1.2
PLT ($\times 10^3/\mu L$)	809.25 ± 79	812.3 ± 81	808.9 ± 79.1	802 ± 75.3
PDW (f L)	16.5 ± 0.2	18.3 ± 0.5	19.6 ± 0.7	14.7 ± 0.1

WBC (White Blood Cell), MCV (Mean Corpuscular Volume); RBC (Red Blood Cell); MCH (Mean Corpuscular hemoglobin); HGB (Hemoglobin); MCHC (Mean Corpuscular Hemoglobin Concentrations); LYM (Lymphocytes); PLT (Platelets); MON (Monocytes); PDW(Platelet Distribution Width).

3.5. Biochemistry Observation

There was significant ($P < 0.05$) increase total protein content on Shilajit group in the 3 day compared to pre-wound. It should be noted that at 12 days there is increase in the amount of albumin, these changes indicate normalization general condition of the animal due to the wound healing. There was significant ($P < 0.05$) decline on the 9 day in the amount of calcium and iron, but in the following days of study these indicators have tended to increase. The amount of glucose was varied in the opposite direction it mean that, there was significant ($P < 0.05$) increase on the 3, day, and this trend has continued in the following days of observation compared to pre-wound. Changes in the biochemical composition of blood serum in Vishnevsky group during the observation period remained in the normal physiological level (Table 6 and Table 7).

3.6. Clinical Wound Observation (Day 4)

Clinical observation of wounds in Shilajit group was found on day 4 after the start of treatment mentioned cleansing wounds from purulent exudate and dead tissue monitored, and the wound was covered with granulation tissue (Figure 1). Subsequently, as a result of investigations at day 4 in Vishnevsky group the surface of the wound contains a certain amount of purulent exudate and the edges of the wound had edema (Figure 2). Our study demonstrated that growth of granulation tissues were more prominent in regions treated with particularly Shilajit dressing in comparison with Vishnesvsky dressing.

Table 6. Observations on biochemical parameters (Shilajit group) (mean ± SD) (n = 4), *Significant at 5% level with preoperative value ($P < 0.05$).

Parameters with units	Pre-wound	Post-wound (3 day)	Post-wound (9 day)	Post-wound (15 day)
Total protein (g/L)	43.5 ± 7.4	67.2 ± 8*	63.2 ± 5.2	55 ± 3.6
Albumin (%)	3.5 ± 0.1	3.5 ± 0.09	2.3 ± 0.3	4.5 ± 0.2
Calcium (mmol/L)	2.77 ± 0.4	2.1 ± 0.1	0.6 ± 0.08*	2.3 ± 0.2
Glucose (mmol/L)	3.3 ± 0.21	5.9 ± 0.6	4.9 ± 0.1	6.2 ± 0.3
Iron (mmol/L)	31.3 ± 7.4	24.4 ± 0.9	21.2 ± 2.1*	28.4 ± 3.2

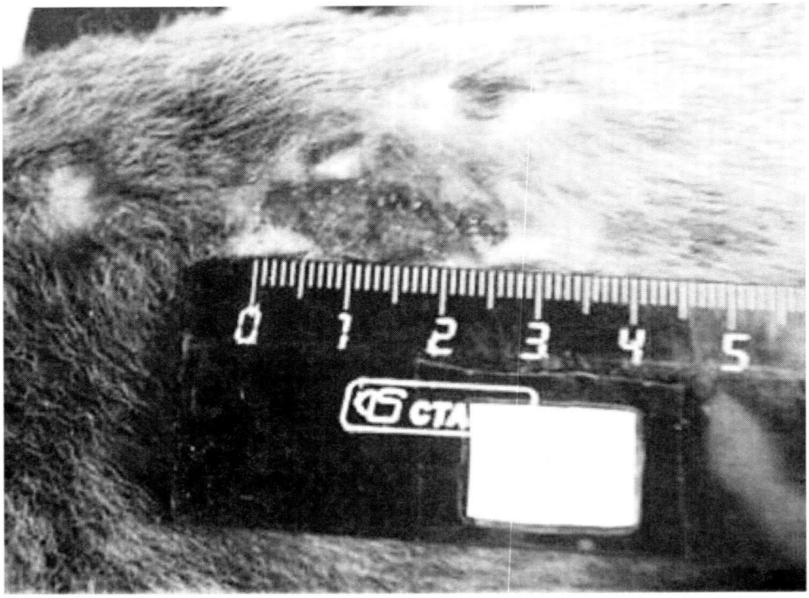

Figure 1. Clinical observation of Shilajit group at day 4 after operation which was showed growth of granulation tissues with contracted wound edges. Healing of wound passed normal stage of wound healing. In addition, there was no sign of exudate in wound bed and wound size is reduced to 2.5 cm (original image by Nikon Coolpix 16.0 Megapixals).

Table 7. Observations on biochemical parameters (Vishnevsky group) (mean ± SD) (n = 4), *Significant at 5% level with preoperative value ($P < 0.05$).

Parameters with units	Pre-wound	Post-wound (3 day)	Post-wound (9 day)	Post-wound (15 day)
Total protein (g/L)	34.9 ± 6.1	83.2 ± 5.2	79.6 ± 3.1	52.5 ± 4.9
Albumin (%)	3.4 ± 0.3	3.6 ± 0.1	1.9 ± 0.3	4.8 ± 0.4
Calcium (mmol/L)	2.8 ± 0.5	1.7 ± 0.6	0.6 ± 0.4	1.9 ± 0.2
Glucose (mmol/L)	3.7 ± 0.3	5.4 ± 0.4	5.2 ± 0.2	6.8 ± 0.5
Iron (mmol/L)	55.4 ± 8.1	21 ± 3.2	27.2 ± 4.6	34.2 ± 5.3

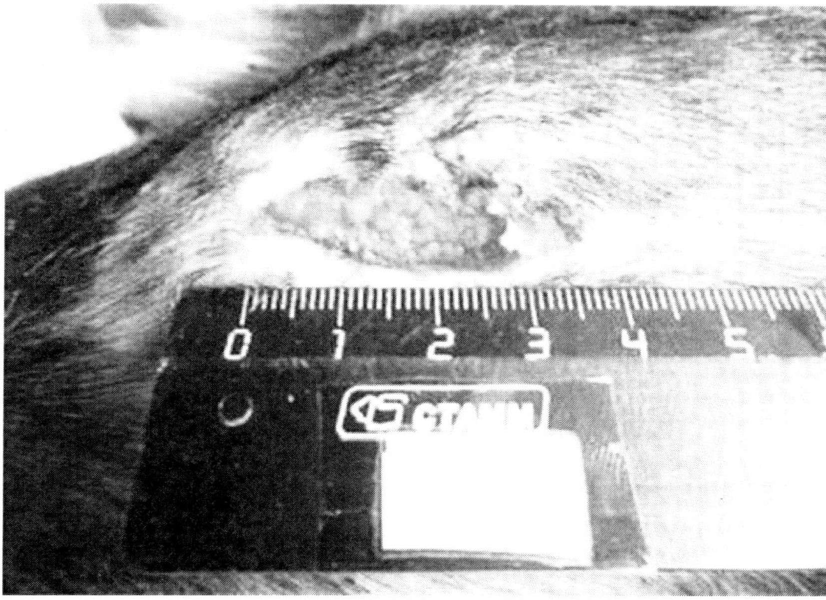

Figure 2. Clinical observation of Vishnevsky group at day 4 after operation the surface of the wound contains a certain amount of purulent exudate and the edges of the wound had edema the wound size is reduced to 3.5 cm (original image by Nikon Coolpix 16.0 Megapixals).

3.7. Clinical Wound Observation (One and Half Week Post-Operation)

In the clinical wound observation at day 10 in the Shilajit group, after the start of treatment the wound is significantly reduced in size, surface crust in the wound is epithelialization process which result on prevails during healing epithelization planar in the wound bed (Figure 3). However, in the Vishnevsky group the wound surface was covered with granulation, and they are fine-grained, contains a certain amount of wound secretion. Swelling, anemic granulation was observed (Figure 4).

4. DISCUSSION

Healing of wound is a complicated and running process, however, it has three deferent phases which each phase extending over; the inflammatory,

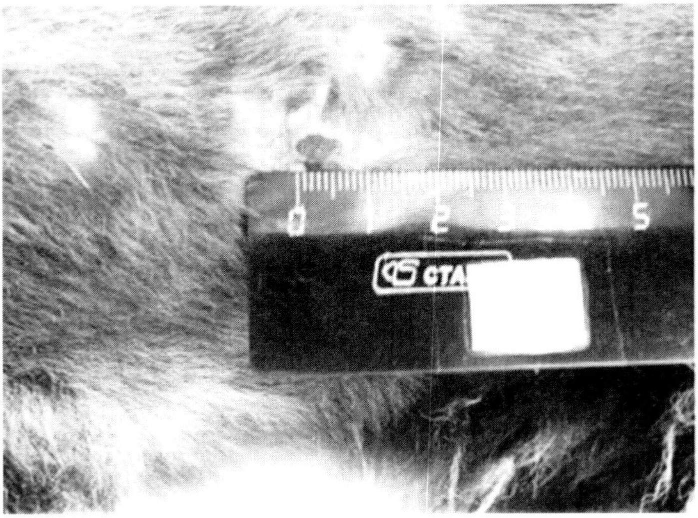

Figure 3. Clinical observation of Shilajit group at day 10 after operation, wound is significantly reduced in size up to 0.5 cm and the surface crust in the wound is the epitheliailzation process which in result introduced the planer epithelialiazation (Original image by Nikon Coolpix 16.0 Megapixals).

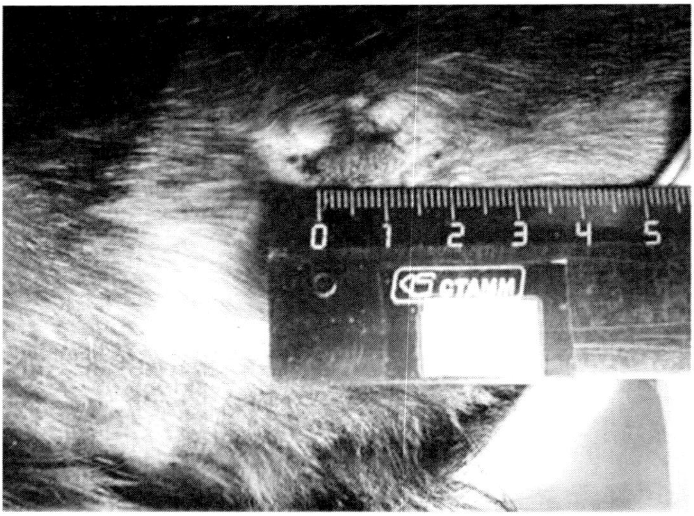

Figure 4. Clinical observation of Vishnevsky group at day 10 after operation, Although the wound surface is coverd by granulation tissue, the wound contains a certain amount of secration. In additon the wound size is reduced to 1.3 cm (Original image by Nikon Coolpix 16.0 Megapixals).

the repair or proliferative, and the remodeling phase [16] . Therapeutic agents has important role in some stage of healing of wound and mainly they have variation in some ways [17] . As mentioned earlier, Shilajit and their derivatives act as potent antimicrobial, anti-inflammatory, and wound healing agents. Their effectiveness in wound healing is attributed not only to their proliferative properties, but also to their anti-inflammatory activity. Therefore, Shilajit based wound healing pharmaceutical preparations modulate both the inflammatory and the proliferative stage of wound healing [18] . In this research, in full-thickness wound not seen any sign of contamination during the healing process. However, the Shilajit dressing had significant healing effect on Shilajit group of wounds. In addition, the swelling of wounds in Shilajit based group was noticeably decreased in days 9 to 12 compared to Vishnevsky dressing group. The most important role of granulation tissue is the shortening of wound, moreover, granulation tissue guard the wound, furnish the border to microbes, and a prepare the smooth layer to epithelialization in wound [6] - [19] . In this research, on day 3 the granulation tissue appeared in Shilajit based wounds and this process completed up to days of 15 of research compared to Vishnevsky based wounds. In the present study, the Shilajit based ointment did significantly improve the rate of wound healing compared to Vishnevsky; the mean wound size as evaluated by planimetry decreased significantly from day 0 to day 15 on Shilajit based wounds. Similar results were obtained from many other therapeutic agents or procedures on open wound healing in dogs; they allowed normal healing or showed minor advantages in some of the proliferative processes on certain days, but they did not actually accelerate the overall wound healing, which usually lasted 3 to 4 week [20] - [24] .

5. CONCLUSION

Based on the observations, the study concludes that the significant changes in the physiological, hematological and biochemical parameters indicate that Vishnevsky is a lavage solution capable of achieving closure of an open wound once it is clean. On the other hand, using of the Shilajit based ointments in clinical canine cases of chronic or contaminated wounds enhances the epithelization and granulation of wounds, which is better demonstrating the effectiveness Shilajit ointment on second intention wound healing in dogs.

ACKNOWLEDGEMENTS

The authors acknowledge Kabul University, Faculty of Veterinary Science and Kabul Animal Health Clinic, for providing scientific material. Dr. Amanullah Aziz and Associate Professor Jahed Zabuli (Department of Clinic, Faculty of Veterinary Science) are also gratefully acknowledged for their valuable scientific cooperation and advice.

CITE THIS PAPER

Ahmad Massoud Niazi, Mohammad Monir Tawfeeq, Amanullah Aziz, Jahid Zabuli, Shahpoor Rahmati, Abdul Razaq Irshad (2016) Comparative Evaluation of Different Treatment for Purulent Wounds in Dogs. *Open Journal of Veterinary Medicine*,**06**,119-126.

REFERENCES

1. Fahie, M.A. and Shettko, D. (2007) Evidence-Based Wound Management: A Systematic Review of Therapeutic Agents to Enhance Granulation and Epithelialization. Veterinary Clinics of North America: Small Animal Practice, 37, 559-577.
2. Swaim, S.F., Gillette, R.L., Sartin, E.A., Hinkle, S.H. and Coolman, S.L. (2000) Effects of a Hydrolyzed Collagen Dressing on the Healing of Open Wounds in Dogs. American Journal of Veterinary Research, 61, 1574-1578.
3. Waldron, D.R. and Zimmerman-Pope, N. (2003) Superficial Skin Wounds. In: Slatter, D, Ed., Textbook of Small Animal Surgery, 3rd Edition, WB Saunders, Philadelphia, 259-273.
4. Gul, N.Y., Topal, A., Cangul, T. and Yanik, K. (2008) The Effects of Topical Tripeptide Copper Complex and Helium-Neon Laser on Wound Healing in Rabbits. Veterinary Dermatology, 19, 7-14.
5. Krahwinkel, D.J. and Boothe, H.W. (2006) Topical and Systemic Medications for Wounds. Veterinary Clinics of North America: Small Animal Practice, 36, 739-757.
6. Goel, R.K., Banerjee, R.S. and Acharva, S.B. (1990) Antiulcerogenic and Anti-Inflammatory Studies with Shilajit. Journal of Ethnopharmacology, 29, 95-103.
7. Jaiswal, A.K. and Bhattacharya, S.K. (1992) Effects of Shilajit on Memory, Anxiety and Brain Monoamines in Rats. Indian Journal of Pharmacology, 24, 12-17.
8. Ghosal, S., Reddy, J.P. and Lal, V.K. (1976) Shilajit Part 1—Chemical Constituents. Journal of Pharmaceutical Sciences, 65, 772-773.
9. Ghosal, S. (1990) Shilajit Part 7—Chemistry of Shilajit, an Immunomodulatory Ayurvedic Rasayana. Pure and Applied Chemistry (IUPAC), 62, 1285-1288.
10. Schepetkin, I.A., Xie, G., Jutila, M.A. and Quinn, M.T. (2009) Complement-Fixing Activity of Fulvic Acid from Shilajit and Other Natural Sources. Phytotherapy Research, 23, 373-384.
11. Stohs, S.J. (2014) Safety and Efficacy of Shilajit (Mumie, Moomiyo). Phytotherapy Research, 28, 475-479.

12. Surapaneni, D.K., Adapa, S.R., Preeti, K., et al. (2012) Shilajit Attenuates Behavioral Symptoms of Chronic Fatigue Syndrome by Modulating the Hypothalamic-Pituitary-Adrenal Axis and Mitochondrial Bioenergetics in Rats. Journal of Ethnopharmacology, 143, 91-99.
13. Faivishevskii, V.A. and Shabad, A.L. (1956) Examination of Possible Blastomogenic Action of Vishnevskii's Ointment. Eksp Khirurgiia, 16, 53-57.
14. Milashkin, A.G. (1982) Use of Vishnevskii Liniment during the Second World War. Voenno-Meditsinskii Zhurnal, 2, 76.
15. Vardaanian, A. and Ohanjanian, S. (2015) Researching of Purulent Wounds during Treatment with Muhin's Extract. Animals of Agrarian Science, 7, 95-96.
16. Pavletic, M.M. (1999) Atlas of Small Animal Reconstructive Surgery. 2nd Edition, WB Saunders, Philadelphia, 12-19.
17. Fossum, T.W. (2007) Small Animal Surgery. 3rd Edition, Mosby, St. Louis, 159-176.
18. Swaim, S.F. and Henderson, R.A. (1997) Small Animal Wound Management. 2nd Edition, Williams &Wilkins, Baltimore, 1-12.
19. Hosgood, G. (2006) Stages of Wound Healing and Their Clinical Relevance. Veterinary Clinics of North America: Small Animal Practice, 36, 667-685.
20. Okamoto, Y., Shibazaki, K., Minami, S., Matsuhashi, A., Tanioka, S. and Shigemasa, Y. (1995) Evaluation of Chitin and Chitosan on Open Wound Healing in Dogs. The Journal of Veterinary Medical Science, 57, 851-854.
21. Swaim, S.F. and Gillette, R.L. (1998) An Update on wound Medications and Dressings. Compendium on Continuing Education for the Practising Veterinarian, 20, 1133-1144.
22. Chung, V.Q., Kelley, L., Marra, D. and Jiang, S.B. (2006) Onion Extract Gel versus Petrolatum Emollient on New Surgical Scars: Prospective Double Blinded Study. Dermatologic Surgery, 32, 193-197.
23. Rasik, A.M., Raghubir, R., Gupta, A., et al. (1999) Healing Potential of Calotropis procera on Dermal Wounds in Guinea Pigs. Journal of Ethnopharmacology, 68, 261-266.
24. Wilson, E., Rajamanickam, G.V., Dubey, G.P., et al. (2011) Review on Shilajit Used in Traditional Indian Medicine. Journal of Ethnopharmacology, 136, 1-9.

CHAPTER 17

Animal Models of Cancer-Associated Hypercalcemia

Nicole A. Kohart [1], Said M. Elshafae [1,2], Justin T. Breitbach [1] and Thomas J. Rosol [1,*]

[1]*Department of Veterinary Biosciences, College of Veterinary Medicine, The Ohio State University, Columbus, OH 43210, USA*

[2]*Department of Pathology, Faculty of Veterinary Medicine, Benha University, Banha 13511, Egypt*

ABSTRACT

Cancer-associated hypercalcemia (CAH) is a frequently-occurring paraneoplastic syndrome that contributes to substantial patient morbidity and occurs in both humans and animals. Patients with CAH are often characterized by markedly elevated serum calcium concentrations that result in a range of clinical symptoms involving the nervous, gastrointestinal and urinary systems. CAH is caused by two principle mechanisms; humorally-mediated and/or through local osteolytic bone metastasis resulting in excessive calcium release from resorbed bone. Humoral hypercalcemia of malignancy (HHM) is the most common mechanism and is due to the production and release of tumor-associated cytokines and humoral factors, such as parathyroid hormone-related protein (PTHrP), that act at distant sites to increase serum calcium concentrations. Local osteolytic hypercalcemia (LOH) occurs when primary or metastatic bone tumors act locally by releasing factors that stimulate osteoclast activity and bone resorption. LOH is a less frequent cause of CAH and in some cases can induce hypercalcemia in concert with HHM. Rarely, ectopic production of parathyroid hormone has been described. PTHrP-mediated hypercalcemia is the most common mechanism of CAH in human and canine malignancies and is recognized in other domestic species. Spontaneous and experimentally-induced animal models have been developed to study the mechanisms of CAH. These models have been essential for the evaluation of novel approaches and adjuvant therapies to manage CAH. This review will highlight the comparative aspects of CAH in humans and animals with a discussion of the available animal models used to study the pathogenesis of this important clinical syndrome.

Keywords: hypercalcemia; paraneoplastic syndrome; cat; comparative oncology; dog; HHM; PTHrP; bone metastasis

1. INTRODUCTION

Cancer-associated hypercalcemia (CAH) is associated with a wide variety of cancers in both humans and animals (Table 1). CAH is often associated with disseminated cancers, and 80% of affected human patients die within a year of diagnosis [1]. The incidence is reported to be between 20% and 30% in human patients with cancer. In humans, CAH occurs with hematopoietic malignancies, such as adult T-cell leukemia/lymphoma, diffuse B cell lymphoma and multiple myeloma, as well as carcinomas of the lung, kidney, breast, larynx, esophagus, cervix and vulva and occasionally in prostate cancer, colorectal cancers and neuroendocrine malignancies [2,3,4,5,6]. Greater than 50% of all CAH cases are associated with lung cancer, breast cancer and multiple myeloma [7]. Classic biochemical abnormalities include hypercalcemia, hypophosphatemia, hypercalciuria (often with decreased fractional calcium excretion), increased fractional excretion of phosphorous and increased nephrogenous cyclic adenosine monophosphate (cAMP) [8,9]. Serum 1,25-dihydroxyvitamin D (calcitriol) concentrations are usually low, but can be normal or high depending on the etiology of the hypercalcemia [10,11]. Patients with high serum ionized calcium concentrations present with symptoms affecting various body organs, including the gastrointestinal, nervous or urinary systems. Ionized serum calcium (Ca^{2+}) is the biologically-active form of calcium and should be measured when the test is available since it is more informative than total serum calcium concentration, which also includes the protein bound and complexed forms of serum calcium. The severity of disease often correlates with the magnitude of hypercalcemia, the rapidity of increase in calcium and the overall medical condition of the patient [12]. Early signs include anorexia, vomiting, nausea, lethargy, fatigue, polyuria and polydipsia. As serum calcium levels rise, patients can progress to renal failure, cardiac arrhythmias, coma and ultimately death [2,4].

Table 1. Cancers associated with hypercalcemia (CAH) in humans and animals.

Species	Tumors Associated with CAH	Reference
Humans	Lung cancer	[2,5,7,13]
	Breast cancer	[2,5,7,14]
	Squamous cell carcinoma	[15,16]
	Multiple myeloma	[2,17,18]
	Adult T-cell lymphoma	[2–4,19,20]
	Less common: colorectal cancer, multiple endocrine neoplasia, prostate cancer, urinary tract cancers, pancreatic cancer, non-Hodgkin's lymphoma	[2,7,21,22]
Dog	Lymphoma (usually T-cell type)	[19,23–25]
	Apocrine gland adenocarcinoma of the anal sac	[25–27]
	Multiple myeloma	[28–30]
	Less common: thyroid carcinoma, thymoma, various bone tumors, squamous cell carcinoma, mammary carcinoma/adenocarcinoma, melanoma, primary lung tumors, chronic lymphocytic leukemia, renal angiomyxoma, nasal carcinoma, renal cell carcinoma	[29,31,32]
Cat	Squamous cell carcinoma	[33–35]
	Lymphoma	[36–38]
	Multiple myeloma	[34,39]
	Less common: osteosarcoma, fibrosarcoma, bronchogenic carcinoma	[34,39,40]
Horse	Squamous cell carcinoma, especially gastric	[41,42]
	Lymphoma	[40]
	Less common: multiple myeloma, ameloblastoma	[43–45]

Cancer-associated hypercalcemia is mediated by two principal mechanisms that include: humoral hypercalcemia of malignancy (HHM) due to circulating tumor-produced factors and hypercalcemia resulting from primary bone tumors or metastases with local osteolytic bone resorption (LOH) (concepts are illustrated in Figure 1 below). Additional mechanisms have been described, although they are rare. These include ectopic parathyroid hormone (PTH) production or primary hyperparathyroidism that coexists with malignancy [4,8,46,47]. This review will focus on HHM; however, it is worth mentioning that metastatic bone tumors release factors that can function to induce both local osteolytic bone resorption and systemic HHM.

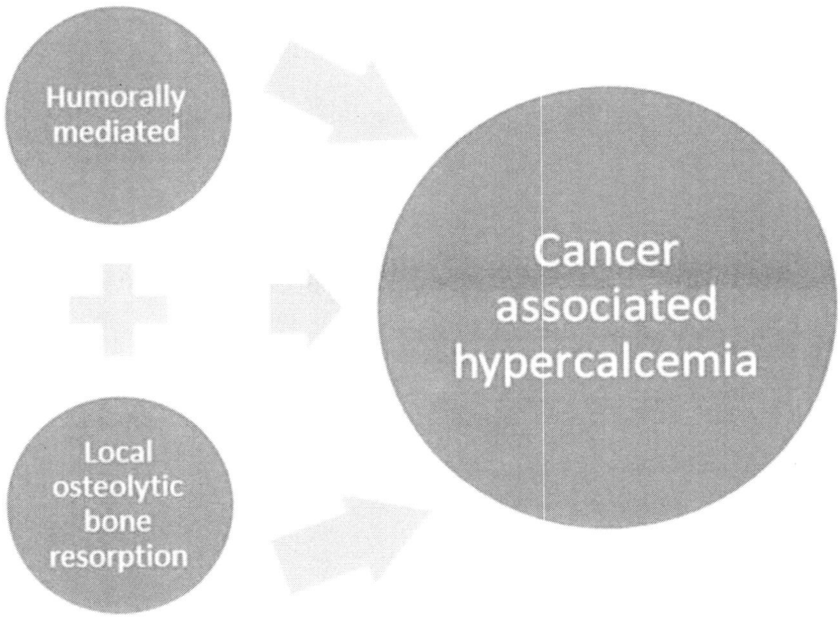

Figure 1. Cancer-associated hypercalcemia is mediated by two principle mechanisms: humoral hypercalcemia of malignancy with the production and release of tumor-associated hormones and/or cytokines that disrupt calcium homeostasis at the level of the bone, kidney and intestine and pathologic bone resorption from metastatic and primary bone tumors that produce factors that act locally to promote osteoclast differentiation and activity. Similar tumor-produced factors play a role in both mechanisms. Metastatic bone lesions result in focal pathologic bone resorption and may also increase serum calcium by humoral mechanisms.

1.1. Factors Associated with Humoral Hypercalcemia of Malignancy

There are several tumor-produced hormones, factors and cytokines that are capable of acting at distant sites to promote widespread pathologic bone resorption and renal reabsorption of calcium. These factors may work solely or synergistically to act on bone cells or renal tubular epithelium. The discussion in this report focuses on factors that are known to play a significant role in paraneoplastic hypercalcemia.

Parathyroid hormone-related protein (PTHrP): Humoral hypercalcemia of malignancy accounts for over 80% of CAH in human patients [48].

HHM is caused by secretion of factors secreted by cancer cells that disrupt calcium homeostasis in the bone, kidney and intestine. Circulating parathyroid hormone-related protein (PTHrP) has been shown to be a central player in the pathogenesis of HHM. PTHrP is a polyhormone with alternative mRNA splicing and post-translational modifications that functions in an endocrine, paracrine, autocrine and intracrine fashion. Since the discovery of PTHrP in 1987, it has been appreciated that PTHrP has diverse roles in normal physiologic functions, including the regulation of cartilage differentiation, bone formation, bone resorption, tooth eruption, mammary gland differentiation and function, the immune system, fetal calcium balance, epithelial-stromal interactions and multiple other actions [49,50,51]. PTHrP has a notable role in cancer to promote tumor cell proliferation, survival, invasion and mediate hypercalcemia [52,53]. Increased circulating plasma concentrations of PTHrP occur in patients with CAH with and without evidence of bone metastasis [12,46,54,55,56,57]. The endocrine and paracrine role of PTHrP in CAH has been well defined through the use of animal models of HHM and clinical investigations in veterinary and human medicine. Eight of the first thirteen amino acids of PTHrP are homologous to parathyroid hormone (PTH), a hormone that regulates serum calcium concentration. PTHrP induces hypercalcemia in a parathyroid hormone (PTH)-like manner by binding to and activating the common PTH/PTHrP (PTH1) receptor at the surface of renal tubular epithelial cells and osteoblasts. PTHrP enhances renal reabsorption of calcium in the ascending loop of Henle and distal convoluted tubule. In bone, PTHrP binds to the PTH1 receptor on osteoblasts, osteocytes and bone stromal cells and activates signal transduction through the cyclic adenosine monophosphate/protein kinase A (cAMP/PKA) cascade [58]. This pathway upregulates the expression of the receptor activator for NF-kB ligand (RANKL) in osteoblasts. When RANKL binds to the RANK receptor on osteoclasts, the NFAT, JNK and p38 pathways are induced and stimulate osteoclast differentiation, survival and bone resorbing activity [59]. Other factors secreted by cancer cells in patients with HHM may have a synergistic and/or additive effect with PTHrP. Such factors include osteoclast stimulatory factors, such as interleukin 1 (IL-1), IL-6, tumor necrosis factor-α (TNF-α), granulocyte-macrophage colony stimulating factor (GM-CSF), macrophage colony stimulating factor (M-CSF), transforming growth factor-β (TGF-β), calcitriol, RANKL and macrophage inflammatory protein-1α (MIP-1α) [12].

1,25-dihydroxyvitamin D_3 (1,25-$(OH)_2$ vitamin D_3, calcitriol): Vitamin D is a precursor that is first activated by hydroxylation in the liver to 25-$(OH)_2$ vitamin D_3 and then to 1,25-$(OH)_2$ vitamin D_3 (calcitriol) in the kidney by renal 1α-hydroxylase [60]. Calcitriol is the active form of vitamin D and increases plasma calcium by stimulating the absorption of dietary calcium in the intestines, increasing osteoclastic bone resorption when in high concentrations and augmenting the ability of parathyroid hormone to enhance renal calcium reabsorption in the kidney. Typically, in patients with HHM, calcitriol concentrations are low due to a compensatory response to the pathologic increases in serum ionized calcium. In contrast, increased serum calcitriol concentrations in patients with HHM have been recognized as an important humoral factor responsible for HHM in patients with non-Hodgkin's and Hodgkin's lymphoma [21,22]. Increased expression of the 1-α-hydroxylase enzyme that forms calcitriol has been identified in lymphoma cells, which has been speculated to contribute to the hypercalcemia in some patients with Hodgkin's disease and malignant histiocytosis [61].

Macrophage inflammatory protein-1α (MIP-1α): MIP-1α is a cytokine that binds to chemokine G-protein coupled receptors, CCR1, CCR5 and CCR9 that are expressed on several cell types, including bone marrow stromal cells. MIP-1α activity has also been described as a chemotactic factor for osteoclasts and osteoclast precursors [62]. Increased concentrations of MIP-1α have been identified as clinical biomarkers in patients with adult T-cell leukemia/lymphoma (ATL), an aggressive hematological malignancy, and HHM occurs in >80% of patients [20]. MIP-1α stimulates calcium release from fetal mouse long bones [20]. One mechanism by which MIP-1α induces hypercalcemia in ATL patients is by upregulating RANKL on ATL cells [20]. It is unknown how MIP-1α stimulates osteoclastogenesis and HHM in other cancers. RANKL-dependent and independent mechanisms have been proposed [63,64].

Tumor necrosis factor-α (TNF-α): TNF-α is a pro-inflammatory cytokine that is critical to the pathogenesis of several neoplastic and inflammatory diseases characterized by increased osteoclastic bone resorption [65]. Constitutive overexpression and secretion of TNF-α by Chinese hamster ovarian cells resulted in increased osteoclastic bone resorption and HHM in mice, whereas bone resorption or hypercalcemia did not occur with control cells [66]. TNF-α stimulates osteoclastogenesis at two distinct time points in vivo. Early effects are due to priming of osteoclast progenitors in the bone marrow through RANKL-independent mechanisms. This increases the number of bone marrow osteoclast precursors by stimulating a proliferative response to M-CSF. The osteoclast precursors enter circulation and deposit at sites of bone resorption and complete differentiation into mature osteoclasts in the presence of RANKL [67]. The expression of RANKL is upregulated in surrounding osteoblasts, bone stromal cells and lymphocytes by TNF-α and binds to RANK on osteoclasts to promote increased osteoclast differentiation and bone resorption. In addition to stimulating systemic osteoclastic activity and bone resorption, the role of TNF-α as a sole agent that induces HHM remains obscure. Elevated levels of TNF-α have been identified in patients with HHM in the absence of PTHrP, suggesting it should be further investigated as a single mediator of HHM. However, synergistic and/or cooperative actions with PTHrP and other cytokines are primarily thought to result in TNF-α-associated HHM [15,68].

Interleukin-6 (IL-6): Increased concentrations of IL-6 have been identified in hypercalcemic patients with multiple myeloma, squamous cell carcinoma and ATLL. The mechanism by which IL-6 stimulates osteoclastic bone resorption is independent of RANKL. IL-6 recruits osteoclasts by activating signal transducer and activator of transcription (STAT)-1/3 and mitogen-activated protein kinase (MAPK) signaling pathways in osteoblasts [69]. Increased circulating IL-6 has been reported in hypercalcemic and normocalcemic patients and does not correlate with the severity of hypercalcemia. Therefore, IL-6 likely serves a synergistic or additive effect in the development of HHM [70]. This is supported by the observation that Chinese hamster ovarian cells overexpressing both IL-6 and PTHrP, in contrast to those with IL-6 overexpression alone, had marked hypercalcemia [71].

Interleukin-1 (IL-1): The IL-1 family consists of several cytokines involved in regulating inflammation and immunity. Circulating IL-1 has been observed in patients with hypercalcemia and pathologic bone loss due to increase osteoclast number and activity [16,72,73]. IL-1α and IL-1β are potent stimulators of osteoclastic bone resorption in vitro and in vivo [74,75,76,77]. Subcutaneous continuous infusions of IL-1α and -β into normal Swiss female mice resulted in a marked, dose-dependent increase in plasma calcium concentrations that was associated with osteoclastic bone resorption [77]. IL-1 is involved in the regulation of osteoclast differentiation, multinucleation and survival through intracellular signaling mediated by the IL-1 receptor (IL-1R1) expressed on maturing osteoclasts. Permissive levels of RANKL are necessary for IL-1 to induce osteoclast differentiation. IL-1 also acts on the cells of the bone microenvironment to stimulate osteoclast activity by inducing the expression of RANKL, macrophage-colony stimulating factor (M-CSF) and prostaglandin E_2 by osteoblasts [78,79,80]. Increased reabsorption of calcium by the kidney has also been shown to contribute to hypercalcemia in mice and rats infused with continuous IL-1 [76,81].

Other cytokines: Several cytokines have been identified as factors involved in the pathogenesis of CAH. Cytokine profiles differ among patient with hypercalcemia associated with various malignancies. IL-1, IL-3, IL-11, M-CSF, GM-CSF, TNF-β, among others, have been proposed to serve as contributors to HHM. Serum from patients with HHM usually reveal increased concentrations of one or more of these factors, which result in a disproportionate rate of osteoclastic bone resorption to bone formation, breakdown of the mineralized matrix and release of calcium into systemic circulation.

1.2. Local Osteolytic Hypercalcemia

Local osteolytic bone resorption is caused by primary bone malignancies, such as multiple myeloma, or bone metastases that secrete paracrine factors, which act to stimulate osteoclast function and local bone resorption. The degree of hypercalcemia in these cases correlates with the degree of tumor burden and can be severe in cases of advanced disease [18]. The cytokines and osteoclastogenic factors listed above and in Table 2 also are secreted by tumor cells and have local effects on bone cells and stimulate osteoclastogenesis. It is important to recognize that the tumor factors that induce LOH can also contribute to hypercalcemia in a humoral manner. Therefore, LOH and HHM can coexist in the same patient. PTHrP was originally thought to only play a role in HHM; however, in normocalcemic patients, PTHrP was also found in local bone neoplasms and metastasis that caused local osteolysis in normocalcemic patients [82]. This confirmed the important paracrine role of PTHrP, since it induces local osteoclastic bone resorption when primary cancers or metastases are present in bone (Figure 2) [83,84]. This is particularly common in women with breast cancer, since 80% of breast cancer bone metastases express increased PTHrP in the bone microenvironment. At sites of bone metastasis, locally-acting PTHrP upregulates RANKL-RANK signaling, which results in pathologic bone resorption and the release of calcium and other growth factors, such as TGF-β, from the mineralized matrix of bone that upregulate PTHrP gene expression and promote cancer progression [85]. This has been called the “vicious cycle” of bone metastasis. In metastatic breast cancer, hypercalcemia is infrequent, despite a high incidence of local PTHrP expression and bone resorption. Patients with bone metastases of carcinomas are usually normocalcemic, because the degree of pathologic bone lysis is insufficient to overcome the efficient homeostatic regulation of serum calcium. Plasma PTHrP concentrations are greater in patients with breast cancer bone metastases and hypercalcemia when compared to early breast cancer patients and those with bone metastases and normal serum calcium concentrations, suggesting both a paracrine and endocrine role for PTHrP in breast cancer-associated hypercalcemia [14].

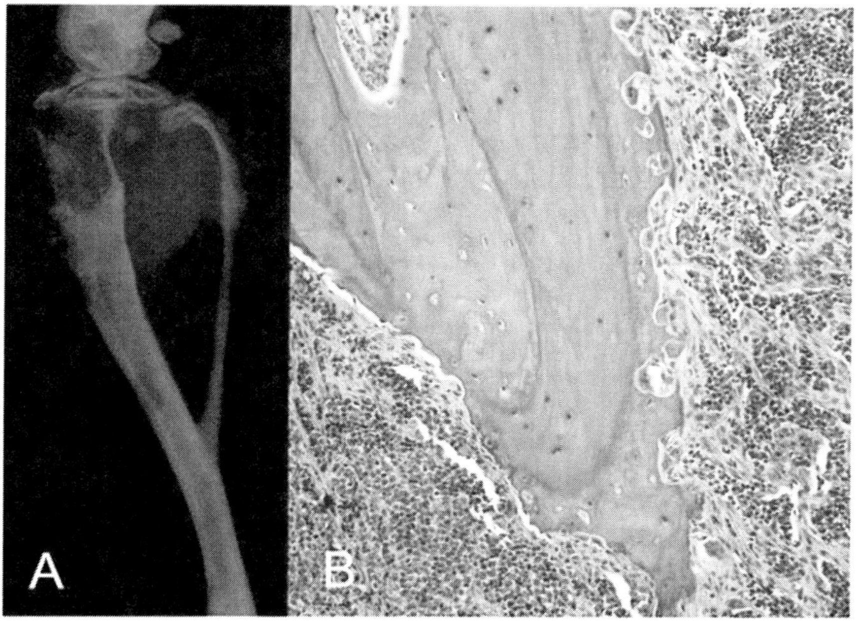

Figure 2. PTHrP-induced pathologic osteoclastic bone resorption. (**A**) The human adult T-cell leukemia cell line (RVATL) that secretes PTHrP was injected into the tibia of immunosuppressed mice and induced marked pathologic bone resorption observed radiographically; (**B**) histopathology revealed increased numbers of hypertrophic, active osteoclasts and extensive eroded bone surfaces. When mice were injected intraperitoneally with RVATL cells, there was increased circulating PTHrP and hypercalcemia due to HHM. This animal models support a dual endocrine and paracrine role of PTHrP in pathologic bone resorption and hypercalcemia.

In multiple myeloma, hypercalcemia is the most frequent metabolic complication and results from excessive local osteolysis and extensive tumor-induced bone resorption [17]. Myeloma cell secretion of RANKL, PTHrP, MIP-1α and TNFs (discussed above for HHM) act locally to enhance pathologic osteolysis. MIP-1α enhances osteoclast formation induced by IL-6, PTHrP and RANKL [63]. In addition to stimulating osteoclastogenesis, MIP-1α facilitates the survival of myeloma cells and increases their adhesion to bone [4,63].

Table 2. Factors and cytokines involved in humoral hypercalcemia of malignancy (HHM) and local osteolytic hypercalcemia (LOH).

Factor	Mechanism
Parathyroid hormone-related protein (PTHrP)	HHM; LOH
Macrophage inflammatory protein-1α (MIP-1α)	HHM; LOH
Calcitriol (1,25-$(OH)_2$ vitamin D_3)	HHM
Receptor activator of nuclear factor-κB ligand (RANKL)	LOH
Interleukin-1 (IL-1), IL-3, IL-6, IL-8, IL-11	LOH; HHM
Tumor necrosis factor-α (TNF-α), TNF-β	LOH; HHM
Macrophage colony stimulating factor (M-CSF)	LOH; HHM
Granulocyte macrophage-colony stimulating factor (GM-CSF)	LOH; HHM
Various, less common cytokines (prostaglandin E2, TGF-α)	LOH; HHM

1.3. Animal Models of HHM and LOH

HHM and LOH cause substantial patient morbidity and negatively impact patient prognosis and response to therapy. Similarly, veterinarians are faced with the challenge of implementing appropriate therapy and management of CAH in dogs and other animals. Because patients suffering from CAH are usually severely ill, efforts to investigate the underlying pathophysiologic mechanisms in affected patients are limited. Therefore, there has been much interest in using animal models and in vitro systems to identify the role of tumor-secreted factors involved in the pathogenesis of CAH. Identifying and characterizing the molecular and pathologic mechanisms of CAH are necessary for the development of novel adjuvant therapies and improved patient care. Spontaneous and experimental animal models have been invaluable to investigating these mechanisms. Examples of animal models include: (1) spontaneous cancers in domestic animals; (2) allogenic transplantation of rodent cancers; (3) xenogeneic transplantation of human and animal cancers to immunodeficient rodents; (4) drug-induced cancer in rodents; and (5) in vivo infusion of cancer-related humoral/paracrine factors responsible for CAH [84]. No single model can recapitulate all aspects of HHM or LOH; however, the information from the animal models has advanced the treatment of CAH and improved the understanding of the pathogenic mechanisms. Researchers should select the most appropriate animal model to investigate the specific question of interest. Additionally, while most models share in vivo features of the human syndrome, there are notable differences that must be considered. For example, rodent models with human or canine cancer xenografts and CAH consistently have high levels of serum calcitriol that contribute to CAH to varying degrees; however, humans with CAH usually have low serum calcitriol concentrations [45,86,87]. This review focuses on the comparative aspects of HHM and LOH in humans and animals, with a discussion of the available animal models for CAH research.

1.4. Spontaneous and Experimentally-Induced Animal Models of HHM and LOH

1.4.1. Spontaneous CAH in Dogs and Experimental Models

Cancer is the most common cause of hypercalcemia in dogs and accounts for 45%–65% of cases [30]. Lymphoma and adenocarcinoma of the apocrine gland of the anal sac are the most common causes of CAH; however, it also occurs with thyroid carcinoma, thymoma, multiple myeloma, various bone tumors, squamous cell carcinoma, mammary gland carcinoma, melanoma, primary lung cancers, chronic lymphocytic leukemia, renal angiomyxoma, nasal carcinoma, renal cell carcinoma, among other less common malignancies [28,29,31,32,88]. Dogs with hypercalcemia have increased total serum calcium, often greater than 12 mg/dL and an increase in ionized calcium (>1.3 mmol/L) [30]. The clinical signs of hypercalcemia can be nonspecific and are independent of the etiology. Clinical signs of hypercalcemia can be insidious if the increase is minimal or slow. Dogs with more severe elevations in serum calcium have lethargy, weakness, anorexia, vomiting, diarrhea, constipation, polyuria, polydipsia and can even progress to fatal renal failure [30].

The dog as a spontaneous model of human cancer has provided a valuable perspective that is distinct from studying human tissues or rodent models alone. The use of dogs as large animal, naturally-occurring models of disease has advanced our understanding of cancer pathogenesis, progression and therapy [89,90]. Several examples of CAH in dogs exist. These natural models have been used to investigate the pathogenesis of cancer and the development of novel therapeutics to advance both animal and human health [89]. Similarities exist between human and canine CAH, which demonstrates that the dog serves as an important large animal model for this syndrome [91]. Clinical trials in canine patients can advance treatment strategies for CAH and expedite development of novel therapies for both dogs and humans. Dogs with CAH may have advanced disease, which will preclude early studies to identify preventative strategies to mitigate hypercalcemia and its deleterious effects. Controlled clinical studies designed to investigate factors involved in the pathogenesis of CAH in dogs and novel treatment strategies can be expensive and challenging to recruit patients. Financial support and incentives for the owner are typical. However, the translational potential of using the dog as a large animal model of CAH is very promising.

Adenocarcinoma of the apocrine glands of the anal sac: Adenocarcinoma of the apocrine gland of the anal sac (AGAC) is a tumor of older dogs that occurs at a similar frequency in females and males, which has a high rate of metastasis, but infrequently to bone [26]. Up to 50–70% of dogs with AGAC develop HHM characterized by increased serum calcium, decreased serum phosphorous and increased circulating PTHrP that correlates with serum calcium levels [25,92]. There is no apparent relationship between HHM and median survival time [93]. The development and characterization of a mouse model of canine AGAC (CAC-8) has been described [25,27,92,94]. This tumor (CAC-8) originated from a dog with AGAC and hypercalcemia and results in a subcutaneous tumor when injected into nude mice. Mice with CAC-8 develop hypercalcemia, hypophosphatemia, increased serum calcitriol and increased circulating PTHrP. Urinary excretion of calcium, cAMP and hydroxyproline and the rate of bone resorption and bone formation were increased. Histomorphometric analysis of the lumbar vertebrae revealed increased osteoid, resorptive surface and number of osteoclasts [25,27,84]. Intestinal calcium absorption or decreased urinary calcium excretion could not completely explain the HHM in these mice. In this animal model, PTHrP was the principal hypercalcemic factor; however, the cosecretion of transforming growth factors may also play a role in the development of HHM [94].

Lymphoma: Canine lymphoma is a naturally-occurring cancer that closely resembles the biologic behavior of the human disease. Approximately 40% of dogs with T-cell lymphoma develop HHM or a combination of LOH and HHM due to lymphoma in bone [19,23,24]. Biochemistry changes include increased serum calcium concentrations and normal to low serum phosphorous [95]. Serum calcitriol concentrations are decreased, normal or increased [25]. In humans with Hodgkin's lymphoma and non-Hodgkin's lymphoma (NHL) with HHM, serum calcitriol levels are frequently increased and have been implicated to play a significant role in this syndrome [21,22,96]. Increased calcitriol concentrations may also play a role in CAH in dogs with lymphoma, because a subset of canine patients had increased serum calcitriol concentrations [25]. However, this has not been a consistent finding in dogs with lymphoma and HHM [19,97]. Several studies have examined the association between PTHrP and lymphoma-associated CAH in dogs [24,25,98]. Increased plasma PTHrP concentrations occur in dogs with lymphoma and HHM, but plasma PTHrP does not always correlate with the degree of hypercalcemia [25,99]. Increased plasma PTHrP measured by an immunoradiometric assay (6.1 pmol/L) was described in one dog with lymphoma and hypercalcemia and no detectable plasma PTHrP (<1.1 pmol/L) in dogs with lymphoma and normocalcemia and healthy dogs. In addition, PTHrP mRNA expression was reported in one dog with lymphoma and hypercalcemia and four dogs with lymphoma and normocalcemia [98]. These findings indicate that additional tumor-related cytokines, in addition to PTHrP, likely play a role in lymphoma-associated CAH [25]. In dogs with T-cell lymphoma and hypercalcemia, PTHrP mRNA was upregulated to varying degrees in all of the hypercalcemic dogs with lymphoma [24]. TNF-α, a proinflammatory cytokine that stimulates osteoclastogenesis through RANKL-dependent and independent pathways, was upregulated or downregulated. RANKL, another osteoclast stimulatory factor that is secreted by activated T lymphocytes, osteoblasts and bone marrow stromal cells, was increased in 3/6 hypercalcemic lymphoma [24]. Interestingly, IL-6, a proinflammatory cytokine implicated in the pathogenesis of HHM, was downregulated in all samples [100]. Unlike the primary role of PTHrP in HHM due to adenocarcinoma of the apocrine glands of the anal sac, PTHrP likely plays a cooperative or synergistic role with other cytokines in dogs with lymphoma and HHM.

A xenograft mouse model of canine T-cell lymphoma with HHM was developed to investigate the humoral factors and cytokines involved with the disease [24]. The cells were derived from a canine patient with lymphoma and hypercalcemia and resulted in multicentric lymphoma when xenografted intraperitoneally into NOD/SCID mice. The proposed mechanism of hypercalcemia was upregulation of tumor-produced PTHrP and TNF-α. Bone histomorphometry demonstrated that there was increased bone resorption, similar to what occurs in both humans and dogs with HHM. Future work using this model will provide valuable insight into the role of individual and synergistic cytokines in the development of HHM in dogs with T-cell lymphoma.

Multiple myeloma: Multiple myeloma (MM) is an uncommon plasma cell malignancy of older dogs that primarily involves bone and is associated with hypercalcemia in 15%–20% of cases [101]. Hypercalcemia is regarded as a poor prognostic factor in canine MM and is the result of several mechanisms similar to humans with MM-associated CAH [102]. These include extensive local tumor-induced bone resorption, secretion of PTHrP and/or an increase in M-protein-bound calcium. In humans with MM, more than 88% of patients develop extensive osteolytic bone disease and hypercalcemia [103]. Radiographic evidence of osteolysis was identified in 50% of canine patients; however, not all cases with bone resorption result in hypercalcemia [101,104]. Additional tumor-mediated mechanisms are responsible for stimulating osteoclasts in human MM, including IL-1, IL-6, tumor necrosis factor-α (TNF-α), TNF-β (lymphotoxin), RANKL, MIP-1α and PTHrP, among others [103]. In canine MM, increased PTHrP expression has been speculated to serve a role in CAH; however, some report undetectable circulating PTHrP [101,105,106]. Serum calcium exists in three fractions; ionized, complexed and protein bound. In human MM, negatively-charged myeloma proteins will bind to calcium resulting in an increase in total calcium concentrations, but no increase in serum ionized calcium [107]. This mechanism has been reported in a dog with MM and a monoclonal gammopathy [106].

1.4.2. Spontaneous CAH in Cats and Experimental Models

Cancer as a cause of hypercalcemia in cats accounts for up to 30% of cats with hypercalcemia [33,34]. In cats, hypercalcemia occurs with lymphoproliferative disorders, squamous cell carcinoma (SCC), multiple myeloma, osteosarcoma, fibrosarcoma, bronchogenic carcinoma and other malignancies [35,37,40,108]. Lymphoma and SCC are the most common causes. The clinical signs of hypercalcemia in cats are similar to those observed in dogs. The more common cancers associated with CAH in cats are described below. The advantage of using cats is they represent a large animal model of the disease. Spontaneous cancers in cats associated with hypercalcemia and the mechanisms by which they occur are similar to those in humans. However, CAH in cats is rare, and special attention to species-specific drug metabolism must be considered in therapeutic trials with cats, which can limit their translatability to humans.

Squamous cell carcinoma: Similarities exist between feline and human oral squamous cell carcinoma (OSCC), including local aggressive osteolytic bone resorption and PTHrP expression. In humans, squamous cell carcinoma of the head, neck and oropharyngeal region is a well-recognized cause of hypercalcemia, and tumors are often characterized by high PTHrP expression [108]. The majority of human cases with SCC and CAH are humorally mediated [108]. A similar mechanism may exist in cats. In a previous report of cats with SCC and CAH, osteolytic lesions were not detected [34,35]. Furthermore, increased circulating concentrations of PTHrP have been detected in cats with SCC-associated CAH [39]. Twenty five percent of humans with invasive head-and-neck SCC have hypercalcemia secondary to tumor-derived factors that increase local osteoclast-mediated bone resorption, rather than humoral mechanisms acting on bone distant from the primary tumor [109]. Calcium released from local bone resorption due to tumor invasion has been proposed as a mechanism to increase serum calcium in some cases of human SCC-associated CAH [110,111]. A similar mechanism may exist in feline oral SCC (OSCC) with CAH [112,113,114,115]. Both human and feline OSCC cell lines express high levels of osteoclast-stimulating factors, such as PTHrP, IL-6, IL-8 and IL-11 [114,115]. Martin et al. characterized the bone resorptive activity in a preclinical mouse model of bone-invasive feline OSCC [114]. The cell line derived from a feline SCC (SCCF2) resulted in maxillary bone invasion and osteoclastic bone resorption. Furthermore, the cells synthesized high levels of PTHrP mRNA, as was secreted PTHrP and stimulated osteoclastic bone resorption. Messenger RNA expression of RANKL, a potent inducer of osteoclast activity, was low in the SCCF2 cells. However, RANKL mRNA expression in murine pre-osteoblasts was upregulated when co-cultured with SCCF2 cells. These results suggest that SCCF2 expression of PTHrP plays an indirect role in stimulating pathologic osteolysis in OSCC-associated pathologic bone resorption. Therefore, PTHrP likely plays an important role in local tumor-mediated osteolysis and HHM in OSCC.

Lymphoma: Hypercalcemia has been associated with lymphoproliferative neoplasia in cats [34,36,37,38]. In a previous report, 82% of cats with neoplasia and hypercalcemia were due to PTH-independent mechanisms, and circulating PTHrP was increased in the majority of cases [39]. Increased serum PTHrP has been reported in cats with lymphoma and CAH. Cats with feline leukemia virus (FeLV)-associated leukemia-lymphoma can have hypercalcemia in the absence of skeletal metastasis and no renal or parathyroid pathology. The findings in this study suggest that humorally-mediated mechanisms were responsible for the development of hypercalcemia [39]. The mechanism of hypercalcemia in FeLV-induced lymphoma/leukemia in cats is unknown, but it may be similar to HHM in humans with human T-cell lymphotropic virus type 1 (HTLV-1)-induced lymphoma/leukemia, which is due to increased serum PTHrP and possibly synergistic cytokines.

1.4.3. Spontaneous CAH in Horses

Spontaneous CAH occurs in horses, but at a lower frequency than in humans and dogs [45]. HHM is a well-recognized equine paraneoplastic syndrome, which occurs in 25% of horses with gastric SCC [42]. The types of CAH-associated neoplasms are similar in horses to those of other species. HHM is likely the pathogenesis of hypercalcemia in most horses with CAH, since there are usually no osteolytic lesions or bone metastases, and there is a lack of parathyroid and renal pathology. Increased serum PTHrP has been reported and is likely an important mediator of equine HHM. Increased serum PTHrP concentrations have been documented in horses with CAH associated with MM and ameloblastoma [42,43,44,45,116]. Additional humoral mediators of CAH have been identified, including tumor-derived prostaglandins [41].

1.4.4. Spontaneous CAH in Exotic Species

Hypercalcemia of malignancy has been reported in Amazon parrots with malignant lymphoma [117].

1.4.5. Spontaneous and Experimentally-Induced CAH in Rodent Models

Rodent models have greatly advanced our understanding of CAH. Several strains of mice, rats and rabbits have been used to isolate and identify humoral and paracrine factors involved in the pathogenesis of CAH and investigate the pathophysiology of hypercalcemia in vivo. In addition, animal models have been instrumental in preclinical drug trials and the testing of novel therapeutics useful for patients with CAH. Investigation of CAH in rodent models involves four major methods: (1) allogenic transplantation of tumors; (2) xenogenic transplantation of tumors; (3) drug-induced tumor development; and (4) infusion of potential factor/factors responsible for CAH [84]. The use of each method varies between rodent species and the research hypothesis. It is also important to consider the advantages and disadvantages of selecting each animal model. Using rodent models has several general advantages, including they are less expensive, often readily available, have similar pathophysiologic mechanisms compared to humans and there are multiple strains to choose from that have been well characterized. Syngeneic transplantation of cancer cells or solid tumors resulting in CAH are advantageous because the host immune system will be competent, which more closely recapitulates the natural tumor, microenvironment and host response. The disadvantage is that there are few syngeneic rodent models of CAH, and they may not accurately represent the human condition. Xenogenic transplantation of tumors is relatively easy to perform, and multiple cell lines have been described from several species. Human cancer cells or tissue implanted into immunosuppressed mice often recapitulate the original human cancers, but the metastatic pattern may be absent or different. Disadvantages include the lack of host immunity against the cancer, which is an important component in spontaneous human cancer. This is particularly important in immunosuppressed rodents that have a knockout of one or more cytokines, which may play an integral role in the pathogenesis of CAH. It is interesting to note that athymic (nude) mice may have a robust B lymphocyte and innate macrophage response to cancer xenografts; whereas their T lymphocyte response is minimal. Modeling carcinogenesis using direct exposure to carcinogens enables investigations of premalignant lesions and the early pathogenesis of CAH. Chemical-induced tumor development also allows the use of an immunocompetent host. The disadvantages include the prolonged time for tumor development, laborious application procedures and potentially hazardous chemicals. Direct infusion of single or multiple CAH-inducing factors is an excellent method to model CAH and investigate the interactions of factors in the pathogenesis of the disease. Disadvantages include the challenges of administration using catheters or implantable infusion pumps and the limited duration of experiments.

2. MOUSE MODELS OF CAH

The majority of mouse models of CAH are derived from xenograft transplantation (Table 3). Either primary cancers or cancer cell lines have been used. The host species for the cancers include humans, dogs, cats, hamsters and other species. Few allogenic models have been developed and include the HSDM1 fibrosarcoma, the CE murine mammary carcinoma and the 5T multiple myeloma (MM) model. Additional techniques used to study the pathogenesis of HHM include the infusion of bioactive molecules (e.g., synthetic PTHrP) and the induction of squamous cell carcinoma by cutaneous application of dimethylbenzanthracene [50,118].

Table 3. Mouse models of cancer-associated hypercalcemia.

Type of model	Model Name and Tumor Origin	Strain	References
Allogenic	HSDM1 fibrosarcoma	Swiss albino mice	[119]
	CE mammary carcinoma	BALB/cfC3H	[120,121]
	5T2MM and 5T33MM multiple myeloma	C57BL/KalwRij	[122,123]
Transgenic	Tax+ lymphoma/leukemia	Tax+ C57B6/SJL	[124]
Infusion	PTHrP, IL-1, IL-11, TNF-α	-	[50,125,126]
Induction	Cutaneous administration of dimethylbenzanthracene induced squamous cell carcinoma	CD1 mice	[50,118]
Xenograft	Lung cancer	Nude mice	[127–129]
	Squamous cell carcinoma		[128,130]
	Renal cell carcinoma		[131,132]
	Urothelial cancer		[133]
	Esophageal carcinoma		[134,135]
	Melanoma		[136,137]
	Mammary cancer		[138]
	Pancreatic cancer		[139]
	Colon cancer		[140,141]
	Ovarian cancer		[142]
	Adult T-cell leukemia/lymphoma	SCID, NOD/scid	[143–145]
	Multiple myeloma		[146,147]
	Ovarian cancer		[148]

Allogenic models: The allogenic HSDM1 fibrosarcoma was derived from a mouse and results in progressive hypercalcemia 2–3 weeks following serial transfer into Swiss albino mice [119]. Evidence suggests that prostaglandin E_2(PGE_2) is responsible for mediating the hypercalcemia, which is a unique, but uncommon mechanism of HHM. Increased serum calcium concentrations were prevented by treatment with hydrocortisone and indomethacin, but not a low calcium diet [149,150]. Additional humoral factors and cytokines have not been described in this model, and it is unknown whether PGE_2 is solely capable of inducing the hypercalcemia.

The murine mammary carcinoma (CE maca) is a spontaneously derived cancer from the BALB/c mouse that can be transplanted to recipient mice and results in marked granulocytosis and hypercalcemia 3–4 weeks following subcutaneous or intramuscular injection [120,121]. Serum calcium concentrations return to the reference range following tumor removal. In this model, hypercalcemia is associated with a significant increase in urinary hydroxyproline, hypercalciuria and an increase in urinary cAMP excretion [151,152]. Increased intestinal calcium absorption as the cause of hypercalcemia was excluded based on the finding that the hypercalcemia was not corrected with a calcium-deficient diet [152]. Marrow granulocytic hyperplasia is associated with an increase in the number of osteoclasts lining endosteal surfaces and increased bone resorption [151,152]. Granulocyte colony stimulating factor (G-CSF) and macrophage colony stimulating factor (M-CSF) were detected by Northern blot analysis in CE maca tumors and treatment with anti-M-CSF reduced the neutrophilia, but did not affect the hypercalcemia [153]. Granulocytic hyperplasia has been associated with increased osteoclasts and osteoclast activity [154]. It could be speculated that the increased granulocytic hematopoiesis involves cytokines that could indirectly influence neighboring bone cells to promote osteoclastogenesis.

The 5TMM cell lines were derived from aged inbred C57BL/KalwRij mice with spontaneous multiple myeloma (MM). These tumor lines have been extensively characterized and are useful to investigate the pathogenesis of MM, including CAH and associated osteolytic lesions. When sub-clones of the 5T2MM and 5T33MM cell lines are injected intravenously into mice, hypercalcemia and osteolytic bone lesions are observed [17,122,123]. The tumor cells express IL-6, which is an important factor in the development of HHM. Inhibition of CCR1 and CCR5 receptors and their master ligand, MIP-1α, reduce osteolysis and bone resorption in vivo in the 5TMM model [155].

Transgenic model: A subset of HTLV-1 infected humans will develop adult T-cell leukemia/lymphoma, and HHM occurs in 50%–90% of patients. HTLV1 encodes for two viral oncogenes, Tax and HBZ, with Tax being more extensively studied. Tax is a transcriptional activator that can transactivate several cellular and viral genes and includes genes involved in CAH (e.g., IL-1, IL-6, TNF-α, and PTHrP). An HTLV-1 Tax+ transgenic mouse model has been developed to study the roles of these factors in vivo. Under the human granzyme-B promoter (expressed in cytotoxic T-cells and natural killer cells), Tax+ transgenic mice develop spontaneous ear and tail tumors associated with hypercalcemia and vertebral osteolytic bone lesions [124]. Tax+ tumors express several factors, including macrophage colony stimulating factor (M-CSF), IL-1, IL-6, TGF-β, TNF-α, MIP1α, RANKL and PTHrP [124]. These factors are known to have a role in the development of CAH through both humoral mechanisms and local stimulation of osteoclastogenesis and bone resorption.

Xenograft mouse models: Many human and animal cancer cell lines have been used in immunosuppressed mice to recapitulate and investigate the pathogenesis of HHM in vivo. Cancer cells are most often implanted in the subcutis, but other routes of administration include intraperitoneal, intratibial, intracardiac, orthotopic and intravenous.

Lung cancers: HARA, RWGT2, BEN, Lu-65A, Lu-61, OKa-C-1: In humans, hypercalcemia occurs in 8%–12% of patients with advanced lung cancer, with an incidence of up to 23% in patients with the pulmonary squamous cell carcinoma [13]. To recapitulate HHM in vivo, several human-derived cancer cell lines have been transplanted into immunodeficient mice. Mice bearing RWGT2, Lu-61, HARA and/or BEN xenografts develop HHM, and the tumor cells have high PTHrP expression. Hypercalcemia induced by RWGT2 is lowered in tumor-bearing mice following administration of IL-6 neutralizing antibodies. Inhibition of RANK decreases serum calcium in mice with RWGT2 tumors despite no change in circulating PTHrP concentrations [127,128,129,156]. The HARA cell line was derived from a human squamous cell lung carcinoma that did not have evidence of bone metastases, but had increased serum PTHrP and calcium concentrations. The HARA cells express high levels of PTHrP mRNA and following left ventricular intracardiac injection result in bone metastasis and high levels of serum PTHrP and calcium, similar to the syndrome observed in human patients. In this model, PTHrP plays a role in both HHM and local bone resorption at sites of bone metastasis [157]. The HARA-B cell line was established from a bone metastasis that developed following intracardiac injection of HARA cells. Nude mice bearing HARA-B tumors develop cachexia, hypercalcemia and increased serum PTHrP 3–4 weeks following inoculation. Combined therapy with OPG and anti-PTHrP antibodies reversed the hypercalcemia in these mice [158]. The role of the calcium sensing receptor (CaR) in lung SCC was explored using a xenograft model of BEN SCC. Rapid development of HHM in BEN mice was the result of PTHrP secretion by the cancer cells. In addition, extracellular calcium signaling through CaR was found to increase PTHrP production by stimulating the MAPK pathway. Therefore, hypercalcemia may be responsible for a positive feedback loop to induce increased PTHrP production by the cancer cells [159]. The Lu-65 cell line was derived from a patient with an anaplastic lung cancer and caused increased bone resorption and hypercalcemia in vivo. The Lu-65 cells produce and secrete large amounts of prostaglandin E_2, which may be responsible for the increased bone resorption and HHM [160].

Squamous cell cancers: T3M-1, MH-85, OKa-N-1: The incidence of HHM in squamous cell carcinoma (SCC) of the head and neck, oral cavity and skin has been reported in up to 50% of cancers, with the majority being humorally mediated [161]. Several cell lines have been derived from these cancers and implanted into nude mice to investigate the mechanisms of circulating tumor factors that result in hypercalcemia. Leukocytosis, cachexia and hypercalcemia have been observed in patients with SCC and are reproduced in several animal models engrafted with cell lines from patients. The T3M-1 cell line was established from a human SCC-bearing nude mouse that developed marked leukocytosis and hypercalcemia. The cells express mRNA for both PTHrP and granulocyte colony stimulating factor (G-CSF), which mediate the paraneoplastic changes. Additionally, IL-1α may be, in part, responsible for the HHM [130]. MH-85 cells cause hypercalcemia, leukocytosis and cachexia in tumor-bearing nude mice [128]. The OKa-N-1 cell line is a squamous cell carcinoma cell line derived from a patient with leukocytosis and hypercalcemia. Following subcutaneous injection, nude mice develop cachexia, leukocytosis and hypercalcemia, and tumors express G-CSF and PTHrP mRNA [162]. The relationship between G-CSF and PTHrP in these cancers is uncertain.

Renal carcinoma: 786-0: Nude mice with subcutaneous xenografts of the human renal carcinoma cell line, 786-0, develop hypercalcemia, hypophosphatemia and increased serum calcitriol. Urinary cAMP excretion was unchanged, which is usually increased in patients with HHM [132]. Following resection of the subcutaneous tumor, serum calcium concentrations returned to the normal range, consistent with a humoral mechanism for the hypercalcemia. A circulating tumor factor (likely PTHrP) resulted in the increased serum calcium and was associated with increased bone resorption and high bone turnover [131].

Pancreatic cancer: FA-6: The FA-6 cell line was derived from a human pancreatic cancer and resulted in HHM due to the production of PTH-like (likely PTHrP) and TGF-α activities after being injected subcutaneously in nude mice. These factors caused increased osteoclastic bone resorption [139].

Ovarian cancer: Small cell ovarian carcinoma of the hypercalcemic type (SCCOHT-1): The SCCOHT-1 cell line was derived from a tumor of a 31-year-old woman with a small cell ovarian carcinoma of the hypercalcemic type [142]. After subcutaneous injection, female NOD/scid mice developed subcutaneous tumors and hypercalcemia. Interestingly, when the combination of calcium and epothilone B, an inhibitor of mitosis was administered subcutaneously to the hypercalcemic mice, serum calcium concentration returned to normal; however, there was no effect with either agent alone [148]. Increased calcium levels can result in cell death in ovarian cancer cells in vitro, and hypercalcemia may represent a mechanism to antagonize tumor growth. The effects of hypercalcemia may synergize with epothilone B, however, remain unknown.

Urothelial cancer: Transitional Cell Carcinoma (TCC): Patients with urothelial cancers may develop HHM. The human TCC cell line causes HHM in nude mice when injected subcutaneously [133]. When conditioned medium was collected and added to neonatal mouse calvaria, marked bone resorption occurred in vitro. The mechanism of bone resorption by the TCC cells was a macromolecular secreted factor that stimulated prostaglandin synthesis by bone cells.

Esophageal carcinoma: KEsC-2, EC-GI: Subcutaneous injections of nude mice with human KEsC-2 cells derived from a patient with esophageal carcinoma, caused HHM, and the cancer cells expressed PTHrP mRNA [134]. In addition, EC-GI (a cell line derived from a human patient with esophageal carcinoma and hypercalcemia) cancer xenograftsin nude mice resulted in HHM that was mediated by the synergistic actions of PTHrP and IL-1α [135].

Melanoma: SEKI, A375: The human-derived SEKI cell line caused HHM following subcutaneous injection in nude mice. The SEKI cells produced PTHrP mRNA and secreted biologically-active PTHrP protein [134]. The human A375 cell line caused osteolytic bone metastases and hypercalcemia following intracardiac injection in nude mice. Bone resorption at the sites of metastasis was attributed to increased expression of TGF-α [137]. Hypercalcemia also developed when the A375 cells were injected into the mammary fat pad of SCID mice despite a lack of bone metastases. The degree of hypercalcemia correlated with serum concentrations of PTHrP [136].

Mammary tumor: S180-A: The ascites sarcoma 180 is a transplantable human malignancy that results in HHM following subcutaneous injection in nude mice. The mechanism of HHM and increased bone resorption was independent of PTHrP and PGE_2 and was mediated by TGF-α, IL-1α and IL-6 [138].

Colorectal cancer: C-26: The human C-26 cell line was derived from a colon adenocarcinoma, has high PTHrP expression and caused increased circulating PTHrP, hypercalcemia and increased osteoclastic bone resorption in normal mice (CDF1 mice) [140,141]. Blood ionized calcium concentrations returned to normal following treatment with the RANKL inhibitor, osteoprotegerin (OPG), and were reduced with the bisphosphonate, zoledronic acid. Histomorphometry of the tibia revealed a greater reduction in osteoclast surface and serum TRAP5C in mice treated with OPG compared to zoledronic acid [141].

Multiple myeloma: ARH-77, primary multiple myeloma patient-derived cells: Multiple myeloma is one of the most common causes of CAH in humans and is the result of extensive bone lysis, secretion of PTHrP and increased protein-bound serum calcium. ARH-77 cells are an IgG_{κ}-secreting human plasma cell leukemia cell line that mimics human myeloma when injected into SCID mice [146]. ARH-77 tumor-bearing mice develop hypercalcemia following intravenous injection that resulted from increased osteoclastic bone resorption. ARH-77 hypercalcemic mice did not have detectable TGF-α, IL-6, TNF-α or PTHrP mRNA expression or circulating plasma proteins. Furthermore, in vitro bone resorption was not inhibited following treatment with anti-IL-6 or TNF-β antibodies [3]. Dikkof-1 (DKK-1) was identified as a significant contributor to local bone loss in myeloma through the inhibition of osteoblasts. Following engraftment of primary multiple myeloma cells from patients with clinical disease into SCID-rab, anti-DKK1 therapy significantly increased bone mineral density, increased osteoblast number and reduced osteoclast TRAP staining [147].

Adult T-cell leukemia/lymphoma (ATL): RVATL, MET-1, primary ATL cells: Immunodeficient SCID mice injected intraperitoneally with primary human ATL cells develop hypercalcemia and increased serum PTHrP concentrations [145]. The RV-ATL cells were derived from an ATL patient [163] and, when injected intraperitoneally into SCID/beige or NOD/scid mice, developed multicentric lymphoma, ascites and HHM. The mice also had increased plasma PTHrP levels and increased osteoclastic bone resorption [143]. This model is an excellent tool to study the pathogenesis of ATL and HHM, as it closely recapitulates the disease observed in ATL patients. The RV-ATL cells were stably transduced with the luciferase gene for in vivo bioluminescent imaging. Tumor-bearing mice treated with zoledronic acid (a bisphosphonate) and PS-341 (a selective proteasome inhibitor) had decreased tumor burden and HHM [144]. NOD/scid mice injected intraperitoneally with the MET-1 ATL cells develop leukemia, multicentric lymphoma involving multiple organs and

HHM. Tumor-bearing mice had increased plasma PTHrP concentrations, and the MET-1 cells expressed PTHrP and RANKL mRNA [164,165].

Several additional ATL primary tumors have been inoculated into nude mice to understand the mechanisms that regulate HHM in ATL [134,166]. These experiments have been instrumental to advance the discovery of new therapeutic agents, including vitamin D analogues, bisphosphonates, OPG, soluble RANKL and proteasome inhibitors, among others [127,140,144,158,167].

Chemical-induced model: Carcinogenic compounds have been directly applied to mice to investigate the induction of cancer development following exposure. When dimethylbenzanthracene was applied to the skin of female CD1 mice, ninety percent of the mice develop SCC and HHM. The mice had hypercalcemia, hypophosphatemia, hypercalciuria, increased serum calcitriol, increased urinary cAMP and increased bone resorption [118]. Tumor extracts contained a biologically-active protein, slightly larger than PTH, yet had similar activity to PTH. Given these findings, the HHM was likely mediated by PTHrP.

Infusion of bioactive peptides and molecules: Infusion of synthetic peptides of PTHrP-(1–40, 1–34, 1–36) in both rat and mouse models has improved the understanding of the role PTHrP has in the pathogenesis of HHM [50,125,126,168]. Infusion of PTHrP in nude mice resulted in increased serum calcium and calcitriol, decreased serum phosphorous, increased urinary excretion of calcium, phosphorous and cAMP. Unlike the increased bone resorption observed in patients with HHM, there were increases in both bone resorption and bone formation in the mice [50]. Therefore, mouse models of HHM may have subtle differences compared to human patients.

2.1. Rat Models of CAH

Allogenic models: Two spontaneously-derived, transplantable rat tumors, Rice H-500 and the Walker rat carcinosarcoma 256, have been developed as models of CAH [169,170,171,172,173]. The testicular tumor of Fischer rats (Rice H-500) does not metastasize following subcutaneous injection, and rats develop progressive hypercalcemia within 8–10 days with increased circulating PTHrP concentrations [174]. Additional clinicopathologic changes include hypophosphatemia, hypercalciuria, hyperphosphaturia and increased urine cyclic adenosine monophosphate (cAMP) excretion [169,173]. Osteoclastic bone resorption was increased, and bone formation was decreased. There was increased renal tubular reabsorption of calcium despite hypercalcemia and hypercalciuria, as seen in human patients with HHM. Renal tubular calcium reabsorption is thought to be necessary for the pathogenesis of HHM in this model. This is supported by the observation that serum calcium levels were increased despite inhibition of osteoclastic bone resorption following treatment with dichlorodimethylene bisphosphonate [175]. Furthermore, renal phosphate uptake was specifically and persistently inhibited, which contributed to the hypophosphatemia [176]. While serum PGE has been shown to parallel the increase in serum calcium in the Rice model, inhibition of PGE did not affect the degree of hypercalcemia [169].

Immunoreactive serum PTH levels were decreased, and PTHrP and calcitriol concentrations were increased in rats with H-500 tumors [169]. The Rice H-500 model was useful in the identification and purification of PTHrP and in the development of an immunoradiometric assay [177,178,179]. It has been demonstrated that PTHrP plays a central role in the development of HHM in this model. Plasma calcium levels were normalized for several days following passive intravenous immunization with antiserum to PTHrP. The use of this model has been useful to investigate the role of PTHrP and the calcium-sensing receptor (CaR) relationship. Treatment of H-500-bearing mice with cinacalcet (an allosteric activator of CaR) decreased hypercalcemia in a dose-dependent manner, but did not affect serum PTHrP. Plasma calcitonin levels decreased with the treatment-mediated decrease in ionized serum calcium [167]. The Rice H-500 model has been used to identify the role of additional factors that mediate HHM (e.g., transforming growth factors (TGFs)). Conditioned medium from tumor cell cultures and tumor extracts revealed EGFR binding with TGF immunoreactivity and an absence of EGF activity. A potential for TGF stimulation of bone resorption through EGFR binding was proposed in this model [179,180]. TGF-α binds to the epidermal growth factor (EGF) receptor, and both TGF-α and EGFs are capable of stimulating bone resorption through this receptor.

The second allogenic rat model is the Walker 256 rat carcinosarcoma model. There are two variants of the Walker rat carcinosarcoma 256 and only one that results in hypercalcemia. The hypercalcemic variant is characterized by increased serum calcium, tartrate-resistant acid phosphatase, calcitriol and plasma PTHrP concentrations. Tumor-bearing rats had decreased calcium content in their bones secondary to PTHrP-mediated osteolysis [181]. Despite high serum calcium levels, the rats had a normal fractional rate of intestinal calcium absorption, suggesting an intestinal contribution to the hypercalcemia [182]. This model demonstrated a role of the TGF-β-PTHrP axis in HHM. Treatment of Walker carcinosarcoma 256 cells with TGF-β1 increased mRNA expression of PTHrP, but there was a decrease in secretion of PTHrP

[183]. This suggests that the effect of TGF-β is mediated at the transcriptional level. The Walker carcinosarcoma model has been widely employed to test antihypercalcemic agents [184,185,186,187,188].

Several other rat models of HHM have been described (Table 4). Allogenic transplantation of a lung carcinoma line (IP) into female F344 rats resulted in HHM [189]. This model was used to elucidate the potential impact of age on HHM. Following intraperitoneal injection of the IP cell line into three different age groups of female Fisher rats, they developed hypercalcemia, increased osteoclastic bone resorption and increased circulating levels of PTHrP. The degree of HHM varied between the groups with the greatest difference being in tissue calcification. The incidence and severity was decreased to absent as mice aged. The mechanism of tissue calcium deposition as a result of HHM in this model may be influenced by age [189].

Table 4. Rat models of cancer-associated hypercalcemia.

Type of Model	Model Name	Tumor	Strain	References
Allogenic	Rice-500	Leydig cell	Fisher rat	[169–171]
	Walker 256	Mammary gland carcinoma	Fisher rat	[181–183]
	IP	Pulmonary carcinoma	Fisher rat	[189]
Infusion	PTHrP	-	-	[126,168]
Induction	Nitrosourea compounds	Mammary gland carcinoma	Sprague-Dawley	[192]
Xenograft	Uterine carcinoma	Uterine carcinoma	Human	[190,192,195]
Xenograft	OMC-1	Fibrosarcoma	Human	[193]

Xenogeneic models: Xenogeneic transplantation of human cancers into nude rats has been used to investigate HHM [190,191,192,193,194,195]. Human uterine carcinoma xenografts caused hypercalcemia, hypophosphatemia, increased urinary phosphate and cAMP secretion and suppressed serum calcitriol concentrations. HHM in this model has been speculated to be predominantly attributed to secreted humoral factors, such as PTHrP [191].

The PTHrP-producing human infantile fibrosarcoma (OMC-1) resulted in HHM following subcutaneous injection into nude rats. The effect of calcium and PTHrP in regulating calcitriol was examined in this model. OMC-1-bearing rats had severe hypercalcemia and decreased serum calcitriol concentrations. When rats are treated with bisphosphonate therapy, the calcium concentrations were restored with an accompanying increase in serum calcitriol levels. However, when the mice were treated with a neutralizing antibody to PTHrP, serum calcitriol levels remained low despite a reduction in serum calcium. 1α-hydroxylase is an enzyme in the kidney and is responsible for converting 25-h

ydroxyvitamin D into active vitamin D (calcitriol). 1α-hydroxylase levels in the kidney were reduced in OMC-1-bearing rats with severe hypercalcemia, but were increased following bisphosphonate therapy, and there was no change in 1α-hydroxylase expression following PTHrP treatment. The authors suggest that PTHrP regulation of 1α-hydroxylase was mediated by calcium concentrations, and hypercalcemia reduced serum calcitriol by suppressing PTHrP-induced 1α-hydroxylase expression [193]. Chemical-induced models: Intraperitoneal injection of nitrosourea compounds into Sprague-Dawley rats resulted in a 90% incidence of mammary gland carcinoma and hypercalcemia. The rats did not have bone involvement or evidence of bone metastasis, indicating a humoral mechanism, similar to what is observed in the pathogenesis of HHM in some women with breast cancer [192].

2.2. Rabbit Model of CAH

The VX-2 carcinoma developed from malignant transformation of a viral-induced Shope papilloma that caused HHM following transplantation into the hind limb of New Zealand white rabbits [196]. Following transplantation, the tumor grows rapidly with local infiltration and metastasis to regional lymph nodes and lungs. Humoral mechanisms are responsible for pathologic hypercalcemia. This was supported by an absence of bone metastasis, yet increased bone resorption was observed. Progressive hypercalcemia and hypophosphatemia occurred 2–4 weeks following tumor transplantation [197,198,199,200]. Serum PTH and urinary cAMP excretion did not change. Circulating PTHrP has not been reported. Histomorphometric analysis of trabecular bone at distant sites revealed a reduction in osteoblast activity and an increase in the osteoclast number and resorptive surfaces [197,201].

Conflicting reports have attempted to elucidate the mechanism for the HHM in this model. The VX-2 cells contain large amounts of prostaglandin E. Indomethacin (a nonsteroidal anti-inflammatory agent) decreased PGE_2 and prevented the hypercalcemia. These results suggest that the HHM was due to tumor cell secretion of prostaglandin E_2, resulting in generalized osteopenia and increased bone resorption [198,199]. However, dietary calcium restriction has been found to be more effective at normalizing plasma calcium levels then inhibiting PGE_2. It was concluded that the hypercalcemia in this model was dependent on an increase in gastrointestinal calcium absorption rather than prostaglandin-mediated bone resorption [196].

3. CONCLUSIONS

Spontaneous and experimental animal models have significantly advanced cancer research and our knowledge of tumor biology and paraneoplastic syndromes. The use of animal models of CAH will improve our understanding of the pathogenesis of HHM and local tumor-induced bone resorption and facilitate novel drug discovery. Spontaneous HHM in dogs closely resembles the syndrome in humans and will serve as an excellent large animal model for future research and testing new therapeutics. While similarities exist between spontaneous and experimentally-induced models and the human syndrome, important differences exist. The investigator must be aware of the clinicopathologic differences observed between the models, as well as the strengths and limitations of each so that the appropriate model may be selected for experimental investigations. While several models are described in this report, future studies should focus on improving current models and the development of additional models with translatability to both humans and animals. Humanized mice, transgenic rodents and constitutive, conditional or temporal gene knock-in/knock-out of CAH factors in rodents could be used to elucidate the complexity of the pathogenesis of CAH. Clinical researchers should attempt to identify additional factors or cytokines that could potentially be involved in the pathogenesis of CAH. The use of clinical trials in dogs with spontaneous CAH is also needed to advance our understanding of the pathogenesis and treatment of this important disorder in animals and humans.

ACKNOWLEDGMENTS

This work and the primary author (Nicole A. Kohart) are supported by the National Cancer Institute T32 Training Grant (Mouse Models of Human Disease; Grant Number T32OD010429). There were no funds received for covering the costs to publish in open access.

AUTHOR CONTRIBUTIONS

Nicole A. Kohart and Thomas J. Rosol drafted, wrote and revised the manuscript. Said M. Elshafae contributed to the experiments and revisions of the manuscript. Justin T. Breitbach read and revised the manuscript.

CONFLICTS OF INTEREST

The authors declare no conflict of interest.

REFERENCES

1. Seccareccia, D. Cancer-related hypercalcemia. Can. Fam. Phys. 2010, 56, 244–246.
2. Mundy, G.R.; Martin, T.J. The hypercalcemia of malignancy: Pathogenesis and management. Metabolism 1982, 31, 1247–1277.
3. Roodman, G.D. Mechanisms of bone lesions in multiple myeloma and lymphoma. Cancer 1997, 80, 1557–1563.
4. Shu, S. Pathogenesis and Treatments of Humoral Hypercalcemia of Malignancy in Adult T-Cell. Leukemia/Lymphoma Induced by Human T Lymphotropic Virus Type. Ph.D. Thesis, Ohio State University, Columbus, OH, USA, 2009.
5. Muggia, F.M. Overview of cancer-related hypercalcemia: Epidemiology and etiology. Semin. Oncol. 1990, 17, 3–9.
6. Szymanski, J.J.; Otrock, Z.K.; Patel, K.K.; Scott, M.G. Incidence of humoral hypercalcemia of malignancy among hypercalcemic patients with cancer. Clin. Chim. Acta 2016, 453, 190–193.
7. Gastanaga, V.M.; Schwartzberg, L.S.; Jain, R.K.; Pirolli, M.; Quach, D.; Quigley, J.M.; Mu, G.; Scott, W.; Liede, A. Prevalence of hypercalcemia among cancer patients in the United States. Cancer Med. 2016, 8, 2091–2100.
8. Mirrakhimov, A.E. Hypercalcemia of malignancy: An update on pathogenesis and management. N. Am. J. Med. Sci.2015, 7, 483–493.
9. Stewart, A.F.; Horst, R.; Deftos, L.J.; Cadman, E.C.; Lang, R.; Broadus, A.E. Biochemical evaluation of patients with cancer-associated hypercalcemia: Evidence for humoral and nonhumoral groups. N. Engl. J. Med. 1980, 303, 1377–1383.
10. Nakayama, K.; Fukumoto, S.; Takeda, S.; Takeuchi, Y.; Ishikawa, T.; Miura, M.; Hata, K.; Hane, M.; Tamura, Y.; Tanaka, Y.; et al. Differences in bone and vitamin D metabolism between primary hyperparathyroidism and malignancy-associated hypercalcemia. J. Clin. Endocrinol. Metab. 1996, 81, 607–611.
11. Shivnani, S.B.; Shelton, J.M.; Richardson, J.A.; Maalouf, N.M. Hypercalcemia of malignancy with simultaneous elevation in serum parathyroid hormone—Related peptide and 1,25-dihydroxyvitamin D in a patient with metastatic renal cell carcinoma. Endocr. Pract. 2009, 15, 234–239.
12. Richard, V. Regulation of Parathyroid Hormone-Related Protein in Adult T-Cell Leukemia/Lymphoma in a Severe Combined Immuno-Deficient/Beige Mouse Model of Humoral Hypercalcemia of Malignancy. Ph.D. Thesis, Ohio State University, Columbus, OH, USA, 2003.
13. Kanaji, N.; Watanabe, N.; Kita, N.; Bandoh, S.; Tadokoro, A.; Ishii, T.; Dobashi, H.; Matsunaga, T. Paraneoplastic syndromes associated with lung cancer. World J. Clin. Oncol. 2014, 5, 197–223.
14. Bundred, N.J.; Ratcliffe, W.A.; Walker, R.A.; Coley, S.; Morrison, J.M.; Ratcliffe, J.G. Parathyroid hormone related protein and hypercalcaemia in breast cancer. BMJ 1991, 303, 1506–1509.
15. Mori, H.; Aoki, K.; Katayama, I.; Nishioka, K.; Umeda, T. Humoral hypercalcemia of malignancy with elevated plasma PTHrP, TNF alpha and IL-6 in cutaneous squamous cell carcinoma. J. Dermatol. 1996, 23, 460–462.
16. Sato, K.; Fujii, Y.; Ono, M.; Nomura, H.; Shizume, K. Production of interleukin 1 alpha-like factor and colony-stimulating factor by a squamous cell carcinoma of the

thyroid (T3M-5) derived from a patient with hypercalcemia and leukocytosis. Cancer Res. 1987, 47, 6474–6480.
17. Oyajobi, B.O. Multiple myeloma/hypercalcemia. Arthritis Res. Ther. 2007, 9, S4.
18. Sternlicht, H.; Glezerman, I.G. Hypercalcemia of malignancy and new treatment options. Ther. Clin. Risk Manag. 2015, 11, 1779–1788.
19. Meuten, D.J.; Kociba, G.J.; Capen, C.C.; Chew, D.J.; Segre, G.V.; Levine, L.; Tashjian, A.H., Jr.; Nagode, L.A. Hypercalcemia in dogs with lymphosarcoma. Biochemical, ultrastructural, and histomorphometric investigations. Lab. Investig. 1983, 49, 553–562.
20. Okada, Y.; Tsukadal, J.; Nakano, K.; Tonai, S.; Mine, S.; Tanaka, Y. Macrophage inflammatory protein-1alpha induces hypercalcemia in adult T-cell leukemia. J. Bone Miner. Res. 2004, 19, 1105–1111.
21. Seymour, J.F.; Gagel, R.F. Calcitriol: The major humoral mediator of hypercalcemia in Hodgkin's disease and non-Hodgkin's lymphomas. Blood 1993, 82, 1383–1394.
22. Seymour, J.F.; Gagel, R.F.; Hagemeister, F.B.; Dimopoulos, M.A.; Cabanillas, F. Calcitriol production in hypercalcemic and normocalcemic patients with non-Hodgkin lymphoma. Ann. Intern. Med. 1994, 121, 633–640.
23. Fournel-Fleury, C.; Ponce, F.; Felman, P.; Blavier, A.; Bonnefont, C.; Chabanne, L.; Marchal, T.; Cadore, J.L.; Goy-Thollot, I.; Ledieu, D.; et al. Canine T-cell lymphomas: A morphological, immunological, and clinical study of 46 new cases. Vet. Pathol. 2002, 39, 92–109.
24. Nadella, M.V.; Kisseberth, W.C.; Nadella, K.S.; Thudi, N.K.; Thamm, D.H.; McNiel, E.A.; Yilmaz, A.; Boris-Lawrie, K.; Rosol, T.J. NOD/SCID mouse model of canine T-cell lymphoma with humoral hypercalcaemia of malignancy: Cytokine gene expression profiling and in vivo bioluminescent imaging. Vet. Comp. Oncol. 2008, 6, 39–54.
25. Rosol, T.J.; Nagode, L.A.; Couto, C.G.; Hammer, A.S.; Chew, D.J.; Peterson, J.L.; Ayl, R.D.; Steinmeyer, C.L.; Capen, C.C. Parathyroid hormone (PTH)-related protein, PTH, and 1,25-dihydroxyvitamin D in dogs with cancer-associated hypercalcemia. Endocrinology 1992, 131, 1157–1164.
26. Bennett, P.F.; DeNicola, D.B.; Bonney, P.; Glickman, N.W.; Knapp, D.W. Canine anal sac adenocarcinomas: Clinical presentation and response to therapy. J. Vet. Intern. Med. 2002, 16, 100–104.
27. Grone, A.; Weckmann, M.T.; Blomme, E.A.; Capen, C.C.; Rosol, T.J. Dependence of humoral hypercalcemia of malignancy on parathyroid hormone-related protein expression in the canine anal sac apocrine gland adenocarcinoma (CAC-8) nude mouse model. Vet. Pathol. 1998, 35, 344–351.
28. Bergman, P.J. Paraneoplastic hypercalcemia. Top. Companion Anim. Med. 2012, 27, 156–158.
29. Daniels, E.; Sakakeeny, C. Hypercalcemia: Pathophysiology, clinical signs, and emergent treatment. J. Am. Anim. Hosp. Assoc. 2015, 51, 291–299.
30. Messinger, J.S.; Windham, W.R.; Ward, C.R. Ionized hypercalcemia in dogs: A retrospective study of 109 cases (1998–2003). J. Vet. Intern. Med. 2009, 23, 514–519.
31. Bae, B.K.; Kim, C.W.; Choi, U.S.; Choi, E.W.; Jee, H.; Kim, D.Y.; Lee, C.W. Hypercalcemia and high parathyroid hormone-related peptide concentration in a dog with a complex mammary carcinoma. Vet. Clin. Pathol. 2007, 36, 376–378.
32. Scruggs, J.L.; Nobrega-Lee, M.; Fry, M.M.; Applegate, R. Hypercalcemia and parathyroid hormone-related peptide expression in a dog with thyroid carcinoma and histiocytic sarcoma. Vet. Clin. Pathol. 2015, 44, 249–252.
33. Finch, N.C. Hypercalcaemia in cats: The complexities of calcium regulation and associated clinical challenges. J. Feline Med. Surg. 2016, 18, 387–399.
34. Savary, K.C.; Price, G.S.; Vaden, S.L. Hypercalcemia in cats: A retrospective study of 71 cases (1991–1997). J. Vet. Intern. Med. 2000, 14, 184–189.
35. Klausner, J.S.; Bell, F.W.; Hayden, D.W.; Hegstad, R.L.; Johnston, S.D. Hypercalcemia in two cats with squamous cell carcinomas. J. Am. Vet. Med. Assoc. 1990, 196, 103–105.

36. Dust, A.; Norris, A.M.; Valli, V.E. Cutaneous lymphosarcoma with IgG monoclonal gammopathy, serum hyperviscosity and hypercalcemia in a cat. Can. Vet. J. 1982, 23, 235–239.
37. Engelman, R.W.; Tyler, R.D.; Good, R.A.; Day, N.K. Hypercalcemia in cats with feline-leukemia-virus-associated leukemia-lymphoma. Cancer 1985, 56, 777–781.
38. Zenoble, R.D.; Rowland, G.N. Hypercalcemia and proliferative, myelosclerotic bone reaction associated with feline leukovirus infection in a cat. J. Am. Vet. Med. Assoc. 1979, 175, 591–595.
39. Bolliger, A.P.; Graham, P.A.; Richard, V.; Rosol, T.J.; Nachreiner, R.F.; Refsal, K.R. Detection of parathyroid hormone-related protein in cats with humoral hypercalcemia of malignancy. Vet. Clin. Pathol. 2002, 31, 3–8.
40. Yuki, M.; Nitta, M.; Omachi, T. Parathyroid hormone-related protein-induced hypercalcemia due to osteosarcoma in a cat. Vet. Clin. Pathol. 2015, 44, 141–144.
41. Axiak, S.; Johnson, P.J. Paraneoplastic manifestations of cancer in horses. Equine Vet. Educ. 2012, 24, 367–376.
42. Taylor, S.D.; Haldorson, G.J.; Vaughan, B.; Pusterla, N. Gastric neoplasia in horses. J. Vet. Intern. Med. 2009, 23, 1097–1102.
43. Barton, M.H.; Sharma, P.; LeRoy, B.E.; Howerth, E.W. Hypercalcemia and high serum parathyroid hormone-related protein concentration in a horse with multiple myeloma. J. Am. Vet. Med. Assoc. 2004, 225, 409–413.
44. Pusterla, N.; Stacy, B.A.; Vernau, W.; de Cock, H.E.; Magdesian, K.G. Immunoglobulin A monoclonal gammopathy in two horses with multiple myeloma. Vet. Rec. 2004, 155, 19–23.
45. Rosol, T.J.; Nagode, L.A.; Robertson, J.T.; Leeth, B.D.; Steinmeyer, C.L.; Allen, C.M. Humoral hypercalcemia of malignancy associated with ameloblastoma in a horse. J. Am. Vet. Med. Assoc. 1994, 204, 1930–1933.
46. Stewart, A.F. Clinical practice. Hypercalcemia associated with cancer. N. Engl. J. Med. 2005, 352, 373–379.
47. Rosner, M.H.; Dalkin, A.C. Onco-nephrology: The pathophysiology and treatment of malignancy-associated hypercalcemia. Clin. J. Am. Soc. Nephrol 2012, 7, 1722–1729.
48. Clines, G.A. Mechanisms and treatment of hypercalcemia of malignancy. Curr. Opin. Endocrinol. Diabetes Obes. 2011, 18, 339–346.
49. Kronenberg, H.M. PTHrP and skeletal development. Ann. N. Y. Acad. Sci. 2006, 1068, 1–13.
50. Rosol, T.J.; Capen, C.C.; Horst, R.L. Effects of infusion of human parathyroid hormone-related protein-(1-40) in nude mice: Histomorphometric and biochemical investigations. J. Bone Miner. Res. 1988, 3, 699–706.
51. Toribio, R.E.; Brown, H.A.; Novince, C.M.; Marlow, B.; Hernon, K.; Lanigan, L.G.; Hildreth, B.E., 3rd; Werbeck, J.L.; Shu, S.T.; Lorch, G.; et al. The midregion, nuclear localization sequence, and C terminus of PTHrP regulate skeletal development, hematopoiesis, and survival in mice. FASEB J. 2010, 24, 1947–1957.
52. McCauley, L.K.; Martin, T.J. Twenty-five years of PTHrP progress: From cancer hormone to multifunctional cytokine. J. Bone Miner. Res. 2012, 27, 1231–1239.
53. Soki, F.N.; Park, S.I.; McCauley, L.K. The multifaceted actions of PTHrP in skeletal metastasis. Future Oncol. 2012, 8, 803–817.
54. Fraser, W.D.; Robinson, J.; Lawton, R.; Durham, B.; Gallacher, S.J.; Boyle, I.T.; Beastall, G.H.; Logue, F.C. Clinical and laboratory studies of a new immunoradiometric assay of parathyroid hormone-related protein. Clin. Chem. 1993, 39, 414–419.
55. Henderson, J.E.; Shustik, C.; Kremer, R.; Rabbani, S.A.; Hendy, G.N.; Goltzman, D. Circulating concentrations of parathyroid hormone-like peptide in malignancy and in hyperparathyroidism. J. Bone Miner. Res. 1990, 5, 105–113.
56. Ratcliffe, W.A.; Norbury, S.; Stott, R.A.; Heath, D.A.; Ratcliffe, J.G. Immunoreactivity of plasma parathyrin-related peptide: Three region-specific radioimmunoassays and a two-site immunoradiometric assay compared. Clin. Chem.1991, 37, 1781–1787.
57. Segura Dominguez, A.; Olivie, M.A.A.; Sousa, T.R.; Alvarez, M.L.T.; Perez, D.R.; Novoa, R.A.; Garcia-Mayor, R.V. Plasma parathyroid hormone related-protein levels in

patients with cancer, normocalcemic and hypercalcemic. Clin. Chim. Acta 1996, 244, 163–172.

58. Greenfield, E.M.; Shaw, S.M.; Gornik, S.A.; Banks, M.A. Adenyl cyclase and interleukin 6 are downstream effectors of parathyroid hormone resulting in stimulation of bone resorption. J. Clin. Investig. 1995, 96, 1238–1244.
59. Leibbrandt, A.; Penninger, J.M. RANK/RANKL: Regulators of immune responses and bone physiology. Ann. N. Y. Acad. Sci. 2008, 1143, 123–150.
60. Guise, T.A.; Mohammad, K.S.; Clines, G.; Stebbins, E.G.; Wong, D.H.; Higgins, L.S.; Vessella, R.; Corey, E.; Padalecki, S.; Suva, L.; et al. Basic mechanisms responsible for osteolytic and osteoblastic bone metastases. Clin. Cancer Res.2006, 12, 6213s–6216s.
61. Schaefer, K.; Saupe, J.; Pauls, A.; von Herrath, D. Hypercalcemia and elevated serum 1,25-dihydroxyvitamin D3 in a patient with Hodgkin's lymphoma. Klin. Wochenschr. 1986, 64, 89–91.
62. Lentzsch, S.; Gries, M.; Janz, M.; Bargou, R.; Dorken, B.; Mapara, M.Y. Macrophage inflammatory protein 1-alpha (MIP-1 alpha) triggers migration and signaling cascades mediating survival and proliferation in multiple myeloma (MM) cells. Blood 2003, 101, 3568–3573.
63. Han, J.H.; Choi, S.J.; Kurihara, N.; Koide, M.; Oba, Y.; Roodman, G.D. Macrophage inflammatory protein-1alpha is an osteoclastogenic factor in myeloma that is independent of receptor activator of nuclear factor kappaB ligand. Blood2001, 97, 3349–3353.
64. Tsubaki, M.; Kato, C.; Manno, M.; Ogaki, M.; Satou, T.; Itoh, T.; Kusunoki, T.; Tanimori, Y.; Fujiwara, K.; Matsuoka, H.; et al. Macrophage inflammatory protein-1alpha (MIP-1alpha) enhances a receptor activator of nuclear factor kappaB ligand (RANKL) expression in mouse bone marrow stromal cells and osteoblasts through MAPK and PI3K/Akt pathways. Mol. Cell. Biochem. 2007, 304, 53–60.
65. Blind, E.; Knappe, V.; Raue, F.; Pfeilschifter, J.; Ziegler, R. Tumor necrosis factor alpha inhibits the stimulatory effect of the parathyroid hormone-related protein on cyclic AMP formation in osteoblast-like cells via protein kinase C+. Biochem. Biophys. Res. Commun. 1992, 182, 341–347.
66. Johnson, R.A.; Boyce, B.F.; Mundy, G.R.; Roodman, G.D. Tumors producing human tumor necrosis factor induced hypercalcemia and osteoclastic bone resorption in nude mice. Endocrinology 1989, 124, 1424–1427.
67. Boyce, B.F.; Li, P.; Yao, Z.; Zhang, Q.; Badell, I.R.; Schwarz, E.M.; O'Keefe, R.J.; Xing, L. TNF-alpha and pathologic bone resorption. Keio J. Med. 2005, 54, 127–131.
68. Pepper, K.; Jaowattana, U.; Starsiak, M.D.; Halkar, R.; Hornaman, K.; Wang, W.; Dayamani, P.; Tangpricha, V. Renal cell carcinoma presenting with paraneoplastic hypercalcemic coma: A case report and review of the literature. J. Gen. Intern. Med. 2007, 22, 1042–1046.
69. Sims, N.A.; Jenkins, B.J.; Quinn, J.M.; Nakamura, A.; Glatt, M.; Gillespie, M.T.; Ernst, M.; Martin, T.J. Glycoprotein 130 regulates bone turnover and bone size by distinct downstream signaling pathways. J. Clin. Investig. 2004, 113, 379–389.
70. Vanderschueren, B.; Dumon, J.C.; Oleffe, V.; Heymans, C.; Gerain, J.; Body, J.J. Circulating concentrations of interleukin-6 in cancer patients and their pathogenic role in tumor-induced hypercalcemia. Cancer Immunol. Immunother. 1994, 39, 286–290.
71. de la Mata, J.; Uy, H.L.; Guise, T.A.; Story, B.; Boyce, B.F.; Mundy, G.R.; Roodman, G.D. Interleukin-6 enhances hypercalcemia and bone resorption mediated by parathyroid hormone-related protein in vivo. J. Clin. Investig. 1995, 95, 2846–2852.
72. Wano, Y.; Hattori, T.; Matsuoka, M.; Takatsuki, K.; Chua, A.O.; Gubler, U.; Greene, W.C. Interleukin 1 gene expression in adult T cell leukemia. J. Clin. Investig. 1987, 80, 911–916.
73. Yamamoto, I.; Kawano, M.; Sone, T.; Iwato, K.; Tanaka, H.; Ishikawa, H.; Kitamura, N.; Lee, K.; Shigeno, C.; Konishi, J.; et al. Production of interleukin 1 beta, a potent bone

resorbing cytokine, by cultured human myeloma cells. Cancer Res. 1989, 49, 4242–4246.

74. Akatsu, T.; Takahashi, N.; Udagawa, N.; Imamura, K.; Yamaguchi, A.; Sato, K.; Nagata, N.; Suda, T. Role of prostaglandins in interleukin-1-induced bone resorption in mice in vitro. J. Bone Miner. Res. 1991, 6, 183–189.

75. Guise, T.A.; Garrett, I.R.; Bonewald, L.F.; Mundy, G.R. Interleukin-1 receptor antagonist inhibits the hypercalcemia mediated by interleukin-1. J. Bone Miner. Res. 1993, 8, 583–587.

76. McCauley, L.K.; Rosol, T.J.; Stromberg, P.C.; Capen, C.C. Effects of interleukin-1 alpha and cyclosporin A in vivo and in vitro on bone and lymphoid tissues in mice. Toxicol Pathol. 1991, 19, 1–10.

77. Sabatini, M.; Boyce, B.; Aufdemorte, T.; Bonewald, L.; Mundy, G.R. Infusions of recombinant human interleukins 1 alpha and 1 beta cause hypercalcemia in normal mice. Proc. Natl. Acad. Sci. USA 1988, 85, 5235–5239.

78. Kim, J.H.; Jin, H.M.; Kim, K.; Song, I.; Youn, B.U.; Matsuo, K.; Kim, N. The mechanism of osteoclast differentiation induced by IL-1. J. Immunol. 2009, 183, 1862–1870.

79. Tanabe, N.; Maeno, M.; Suzuki, N.; Fujisaki, K.; Tanaka, H.; Ogiso, B.; Ito, K. IL-1 alpha stimulates the formation of osteoclast-like cells by increasing M-CSF and PGE2 production and decreasing OPG production by osteoblasts. Life Sci. 2005, 77, 615–626.

80. Wei, S.; Kitaura, H.; Zhou, P.; Ross, F.P.; Teitelbaum, S.L. IL-1 mediates TNF-induced osteoclastogenesis. J. Clin. Investig. 2005, 115, 282–290.

81. Caverzasio, J.; Rizzoli, R.; Vallotton, M.B.; Dayer, J.M.; Bonjour, J.P. Stimulation by interleukin-1 of renal calcium reabsorption in thyroparathyroidectomized rats. J. Bone Miner. Res. 1993, 8, 1219–1225.

82. Powell, G.J.; Southby, J.; Danks, J.A.; Stillwell, R.G.; Hayman, J.A.; Henderson, M.A.; Bennett, R.C.; Martin, T.J. Localization of parathyroid hormone-related protein in breast cancer metastases: Increased incidence in bone compared with other sites. Cancer Res. 1991, 51, 3059–3061.

83. Kohno, N.; Kitazawa, S.; Fukase, M.; Sakoda, Y.; Kanbara, Y.; Furuya, Y.; Ohashi, O.; Ishikawa, Y.; Saitoh, Y. The expression of parathyroid hormone-related protein in human breast cancer with skeletal metastases. Surg. Today1994, 24, 215–220.

84. Schilling, T. In vivo models of hypercalcemia of malignancy. In Hypercalcemia of Malignancy; Springer: Berlin/Heidelberg, Germany, 1994; pp. 44–75.

85. Guise, T.A.; Chirgwin, J.M. Transforming growth factor-beta in osteolytic breast cancer bone metastases. Clin. Orthop. Relat. Res. 2003, 415, S32–S38.

86. Insogna, K.L.; Stewart, A.F.; Vignery, A.M.; Weir, E.C.; Namnum, P.A.; Baron, R.E.; Kirkwood, J.M.; Deftos, L.M.; Broadus, A.E. Biochemical and histomorphometric characterization of a rat model for humoral hypercalcemia of malignancy. Endocrinology 1984, 114, 888–896.

87. Komatsu, Y.; Imai, Y.; Itoh, F.; Kojima, M.; Isaji, M.; Shibata, N. Rat model of the hypercalcaemia induced by parathyroid hormone-related protein: Characteristics of three bisphosphonates. Eur. J. Pharmacol. 2005, 507, 317–324.

88. Merrick, C.H.; Schleis, S.E.; Smith, A.N.; Mallett, C.L.; Graff, E.C.; Johnson, C. Hypercalcemia of malignancy associated with renal cell carcinoma in a dog. J. Am. Anim. Hosp. Assoc. 2013, 49, 385–388.

89. Paoloni, M.; Khanna, C. Translation of new cancer treatments from pet dogs to humans. Nat. Rev. Cancer 2008, 8, 147–156.

90. Rowell, J.L.; McCarthy, D.O.; Alvarez, C.E. Dog models of naturally occurring cancer. Trends Mol. Med. 2011, 17, 380–388.

91. Heath, H., 3rd; Weller, R.E.; Mundy, G.R. Canine lymphosarcoma: A model for study of the hypercalcemia of cancer. Calcif. Tissue Int. 1980, 30, 127–133.

92. Grone, A.; Rosol, T.J.; Baumgartner, W.; Capen, C.C. Effects of humoral hypercalcemia of malignancy on the parathyroid gland in nude mice. Vet. Pathol. 1992, 29, 343–350.

93. Meier, V.; Polton, G.; Cancedda, S.; Roos, M.; Laganga, P.; Emmerson, T.; Bley, C.R. Outcome in dogs with advanced (stage 3b) anal sac gland carcinoma treated with surgery or hypofractionated radiation therapy. Vet. Comp. Oncol.2016.
94. Merryman, J.I.; Rosol, T.J.; Brooks, C.L.; Capen, C.C. Separation of parathyroid hormone-like activity from transforming growth factor-alpha and -beta in the canine adenocarcinoma (CAC-8) model of humoral hypercalcemia of malignancy. Endocrinology 1989, 124, 2456–2463.
95. Yarrington, J.T.; Hoffman, W.; Macy, D. Animal model of human disease. Pseudohyperparathyroidism. Animal model: Hypercalcemia associated with lymphosarcoma (pseudohyperparathyroidism) in dogs. Am. J. Pathol. 1977, 89, 531–534.
96. Hewison, M.; Kantorovich, V.; Liker, H.R.; van Herle, A.J.; Cohan, P.; Zehnder, D.; Adams, J.S. Vitamin D-mediated hypercalcemia in lymphoma: Evidence for hormone production by tumor-adjacent macrophages. J. Bone Miner. Res.2003, 18, 579–582.
97. Gerber, B.; Hauser, B.; Reusch, C.E. Serum levels of 25-hydroxycholecalciferol and 1,25-dihydroxycholecalciferol in dogs with hypercalcaemia. Vet. Res. Commun. 2004, 28, 669–680.
98. Kubota, A.; Kano, R.; Mizuno, T.; Hisasue, M.; Moore, P.F.; Watari, T.; Tsujimoto, H.; Hasegawa, A. Parathyroid hormone-related protein (PTHrP) produced by dog lymphoma cells. J. Vet. Med. Sci. 2002, 64, 835–837.
99. Mellanby, R.J.; Craig, R.; Evans, H.; Herrtage, M.E. Plasma concentrations of parathyroid hormone-related protein in dogs with potential disorders of calcium metabolism. Vet. Rec. 2006, 159, 833–838.
100. Grivennikov, S.I.; Karin, M. Inflammatory cytokines in cancer: Tumour necrosis factor and interleukin 6 take the stage. Ann. Rheum. Dis. 2011, 70, i104–i108.
101. Geigy, C.; Riond, B.; Bley, C.R.; Grest, P.; Kircher, P.; Lutz, H. Multiple myeloma in a dog with multiple concurrent infectious diseases and persistent polyclonal gammopathy. Vet. Clin. Pathol. 2013, 42, 47–54.
102. Matus, R.E.; Leifer, C.E.; MacEwen, E.G.; Hurvitz, A.I. Prognostic factors for multiple myeloma in the dog. J. Am. Vet. Med. Assoc. 1986, 188, 1288–1292.
103. Galson, D.L.; Silbermann, R.; Roodman, G.D. Mechanisms of multiple myeloma bone disease. Bonekey Rep. 2012, 1, 135.
104. Muñoz, A.; Riber, C.; Satué, K.; Trigo, P.; Gómez-Díez, M.; Castejón, F.M. Multiple myeloma in horses, dogs and cats: A comparative review focused on clinical signs and pathogenesis. In Multiple Myeloma—A Quick Reflection on the Fast Progress; InTech.: Rijeka, Croatia, 2013; pp. 289–326.
105. Seelig, D.M.; Perry, J.A.; Avery, A.C.; Avery, P.R. Monoclonal gammopathy without hyperglobulinemia in 2 dogs with IgA secretory neoplasms. Vet. Clin. Pathol. 2010, 39, 447–453.
106. Tripp, C.D.; Bryan, J.N.; Wills, T.B. Presumptive increase in protein-bound serum calcium in a dog with multiple myeloma. Vet. Clin. Pathol. 2009, 38, 87–90.
107. Attaelmannan, M.; Levinson, S.S. Understanding and identifying monoclonal gammopathies. Clin. Chem. 2000, 46, 1230–1238.
108. Rikimaru, K.; Matsumoto, F.; Hayashi, E.; Bando, H.; Nagayama, M. Evaluation of serum concentration of parathyroid hormone-related protein and its implication in hypercalcemia in squamous cell carcinoma of the head and neck. Int. J. Oral Maxillofac. Surg. 1995, 24, 365–368.
109. Lumachi, F.; Brunello, A.; Roma, A.; Basso, U. Cancer-induced hypercalcemia. Anticancer Res. 2009, 29, 1551–1555.
110. Shu, S.T.; Martin, C.K.; Thudi, N.K.; Dirksen, W.P.; Rosol, T.J. Osteolytic bone resorption in adult T-cell leukemia/lymphoma. Leuk. Lymphoma 2010, 51, 702–714.
111. Yoneda, T.; Hiraga, T. Crosstalk between cancer cells and bone microenvironment in bone metastasis. Biochem. Biophys. Res. Commun. 2005, 328, 679–687.
112. Supsavhad, W.; Dirksen, W.P.; Martin, C.K.; Rosol, T.J. Animal models of head and neck squamous cell carcinoma. Vet. J. 2016, 210, 7–16.

113. Tannehill-Gregg, S.; Kergosien, E.; Rosol, T.J. Feline head and neck squamous cell carcinoma cell line: Characterization, production of parathyroid hormone-related protein, and regulation by transforming growth factor-beta. In Vitro Cell Dev. Biol. Anim. 2001, 37, 676–683.
114. Martin, C.K.; Dirksen, W.P.; Shu, S.T.; Werbeck, J.L.; Thudi, N.K.; Yamaguchi, M.; Wolfe, T.D.; Heller, K.N.; Rosol, T.J. Characterization of bone resorption in novel in vitro and in vivo models of oral squamous cell carcinoma. Oral Oncol.2012, 48, 491–499.
115. Martin, C.K.; Tannehill-Gregg, S.H.; Wolfe, T.D.; Rosol, T.J. Bone-invasive oral squamous cell carcinoma in cats: Pathology and expression of parathyroid hormone-related protein. Vet. Pathol. 2011, 48, 302–312.
116. Cook, G.; Divers, T.J.; Rowland, P.H. Hypercalcaemia and erythrocytosis in a mare associated with a metastatic carcinoma. Equine Vet. J. 1995, 27, 316–318.
117. De Wit, M.; Schoemaker, N.J.; Kik, M.J.; Westerhof, I. Hypercalcemia in two Amazon parrots with malignant lymphoma. Avian Dis. 2003, 47, 223–228.
118. Gkonos, P.J.; Hayes, T.; Burtis, W.; Jacoby, R.; McGuire, J.; Baron, R.; Stewart, A.F. Squamous carcinoma model of humoral hypercalcemia of malignancy. Endocrinology 1984, 115, 2384–2390.
119. Manishen, W.J.; Sivananthan, K.; Orr, F.W. Resorbing bone stimulates tumor cell growth. A role for the host microenvironment in bone metastasis. Am. J. Pathol. 1986, 123, 39–45.
120. Barrett, C.P.; Cubitt, J.G.; Donati, E.J. Elevated serum calcium concentrations in mammary tumor bearing C3H/Fg mice without bony metastases. Lab. Anim. Sci. 1978, 28, 34–37.
121. Delmonte, L.; Liebelt, A.G.; Liebelt, R.A. Granulopoiesis and thrombopoiesis in mice bearing transplanted mammary cancer. Cancer Res. 1966, 26, 149–159.
122. Asosingh, K.; Radl, J.; van Riet, I.; van Camp, B.; Vanderkerken, K. The 5TMM series: A useful in vivo mouse model of human multiple myeloma. Hematol J. 2000, 1, 351–356.
123. Garrett, I.R.; Dallas, S.; Radl, J.; Mundy, G.R. A murine model of human myeloma bone disease. Bone 1997, 20, 515–520.
124. Gao, L.; Deng, H.; Zhao, H.; Hirbe, A.; Harding, J.; Ratner, L.; Weilbaecher, K. HTLV-1 Tax transgenic mice develop spontaneous osteolytic bone metastases prevented by osteoclast inhibition. Blood 2005, 106, 4294–4302.
125. Nagasaki, K.; Yamaguchi, K.; Watanabe, K.; Eto, S.; Abe, K. Interleukin-4 blocks parathyroid hormone-related protein-induced hypercalcemia in vivo. Biochem. Biophys. Res. Commun. 1991, 178, 694–698.
126. Yates, A.J.; Gutierrez, G.E.; Smolens, P.; Travis, P.S.; Katz, M.S.; Aufdemorte, T.B.; Boyce, B.F.; Hymer, T.K.; Poser, J.W.; Mundy, G.R. Effects of a synthetic peptide of a parathyroid hormone-related protein on calcium homeostasis, renal tubular calcium reabsorption, and bone metabolism in vivo and in vitro in rodents. J. Clin. Investig. 1988, 81, 932–938.
127. Oyajobi, B.O.; Anderson, D.M.; Traianedes, K.; Williams, P.J.; Yoneda, T.; Mundy, G.R. Therapeutic efficacy of a soluble receptor activator of nuclear factor kappaB-IgG Fc fusion protein in suppressing bone resorption and hypercalcemia in a model of humoral hypercalcemia of malignancy. Cancer Res. 2001, 61, 2572–2578.
128. Yoneda, T.; Alsina, M.A.; Chavez, J.B.; Bonewald, L.; Nishimura, R.; Mundy, G.R. Evidence that tumor necrosis factor plays a pathogenetic role in the paraneoplastic syndromes of cachexia, hypercalcemia, and leukocytosis in a human tumor in nude mice. J. Clin. Investig. 1991, 87, 977–985.
129. Yoneda, T.; Nakai, M.; Moriyama, K.; Scott, L.; Ida, N.; Kunitomo, T.; Mundy, G.R. Neutralizing antibodies to human interleukin 6 reverse hypercalcemia associated with a human squamous carcinoma. Cancer Res. 1993, 53, 737–740.
130. Sato, K.; Mimura, H.; Han, D.C.; Kakiuchi, T.; Ueyama, Y.; Ohkawa, H.; Okabe, T.; Kondo, Y.; Ohsawa, N.; Tsushima, T.; et al. Production of bone-resorbing activity and

colony-stimulating activity in vivo and in vitro by a human squamous cell carcinoma associated with hypercalcemia and leukocytosis. J. Clin. Investig. 1986, 78, 145–154.
131. Strewler, G.J.; Williams, R.D.; Nissenson, R.A. Human renal carcinoma cells produce hypercalcemia in the nude mouse and a novel protein recognized by parathyroid hormone receptors. J. Clin. Investig. 1983, 71, 769–774.
132. Strewler, G.J.; Wronski, T.J.; Halloran, B.P.; Miller, S.C.; Leung, S.C.; Williams, R.D.; Nissenson, R.A. Pathogenesis of hypercalcemia in nude mice bearing a human renal carcinoma. Endocrinology 1986, 119, 303–310.
133. Bringhurst, F.R.; Bierer, B.E.; Godeau, F.; Neyhard, N.; Varner, V.; Segre, G.V. Humoral hypercalcemia of malignancy. Release of a prostaglandin-stimulating bone-resorbing factor in vitro by human transitional-cell carcinoma cells. J. Clin. Investig. 1986, 77, 456–464.
134. Miyake, Y.; Yamaguchi, K.; Honda, S.; Nagasaki, K.; Tsuchihashi, T.; Mori, M.; Kimura, S.; Abe, K. Production of parathyroid hormone-related protein in tumour xenografts in nude mice presenting with hypercalcaemia. Br. J. Cancer 1991, 63, 252–256.
135. Sato, K.; Satoh, T.; Shizume, K.; Yamakawa, Y.; Ono, Y.; Demura, H.; Akatsu, T.; Takahashi, N.; Suda, T. Prolonged decrease of serum calcium concentration by murine gamma-interferon in hypercalcemic, human tumor (EC-GI)-bearing nude mice. Cancer Res. 1992, 52, 444–449.
136. El Abdaimi, K.; Papavasiliou, V.; Goltzman, D.; Kremer, R. Expression and regulation of parathyroid hormone-related peptide in normal and malignant melanocytes. Am. J. Physiol. Cell Physiol. 2000, 279, C1230–C1238.
137. Hiraga, T.; Nakajima, T.; Ozawa, H. Bone resorption induced by a metastatic human melanoma cell line. Bone 1995, 16, 349–356.
138. Suzuki, K.; Yamada, S. Ascites sarcoma 180, a tumor associated with hypercalcemia, secretes potent bone-resorbing factors including transforming growth factor alpha, interleukin-1 alpha and interleukin-6. Bone Miner. 1994, 27, 219–233.
139. Nagata, N.; Akatsu, T.; Kugai, N.; Yasutomo, Y.; Kinoshita, T.; Kosano, H.; Shimauchi, T.; Takatani, O.; Ueyama, Y. The tumor cells (FA-6) established from a pancreatic cancer associated with humoral hypercalcemia of malignancy: A simultaneous production of parathyroid hormone-like activity and transforming growth factor activity. Endocrinol. Jpn. 1989, 36, 75–85.
140. Capparelli, C.; Kostenuik, P.J.; Morony, S.; Starnes, C.; Weimann, B.; Van, G.; Scully, S.; Qi, M.; Lacey, D.L.; Dunstan, C.R. Osteoprotegerin prevents and reverses hypercalcemia in a murine model of humoral hypercalcemia of malignancy. Cancer Res. 2000, 60, 783–787.
141. Morony, S.; Warmington, K.; Adamu, S.; Asuncion, F.; Geng, Z.; Grisanti, M.; Tan, H.L.; Capparelli, C.; Starnes, C.; Weimann, B.; et al. The inhibition of RANKL causes greater suppression of bone resorption and hypercalcemia compared with bisphosphonates in two models of humoral hypercalcemia of malignancy. Endocrinology 2005, 146, 3235–3243.
142. Otte, A.; Gohring, G.; Steinemann, D.; Schlegelberger, B.; Groos, S.; Langer, F.; Kreipe, H.H.; Schambach, A.; Neumann, T.; Hillemanns, P.; et al. A tumor-derived population (SCCOHT-1) as cellular model for a small cell ovarian carcinoma of the hypercalcemic type. Int. J. Oncol. 2012, 41, 765–775.
143. Richard, V.; Lairmore, M.D.; Green, P.L.; Feuer, G.; Erbe, R.S.; Albrecht, B.; D'Souza, C.; Keller, E.T.; Dai, J.; Rosol, T.J. Humoral hypercalcemia of malignancy: Severe combined immunodeficient/beige mouse model of adult T-cell lymphoma independent of human T-cell lymphotropic virus type-1 tax expression. Am. J. Pathol. 2001, 158, 2219–2228.
144. Shu, S.T.; Nadella, M.V.; Dirksen, W.P.; Fernandez, S.A.; Thudi, N.K.; Werbeck, J.L.; Lairmore, M.D.; Rosol, T.J. A novel bioluminescent mouse model and effective therapy for adult T-cell leukemia/lymphoma. Cancer Res. 2007, 67, 11859–11866.
145. Takaori-Kondo, A.; Imada, K.; Yamamoto, I.; Kunitomi, A.; Numata, Y.; Sawada, H.; Uchiyama, T. Parathyroid hormone-related protein-induced hypercalcemia in SCID mice engrafted with adult T-cell leukemia cells. Blood 1998, 91, 4747–4751.

146. Alsina, M.; Boyce, B.; Devlin, R.D.; Anderson, J.L.; Craig, F.; Mundy, G.R.; Roodman, G.D. Development of an in vivo model of human multiple myeloma bone disease. Blood 1996, 87, 1495–1501.
147. Yaccoby, S.; Ling, W.; Zhan, F.; Walker, R.; Barlogie, B.; Shaughnessy, J.D., Jr. Antibody-based inhibition of DKK1 suppresses tumor-induced bone resorption and multiple myeloma growth in vivo. Blood 2007, 109, 2106–2111.
148. Otte, A.; Rauprich, F.; Hillemanns, P.; Park-Simon, T.W.; von der Ohe, J.; Hass, R. In vitro and in vivo therapeutic approach for a small cell carcinoma of the ovary hypercalcaemic type using a SCCOHT-1 cellular model. Orphanet J. Rare Dis. 2014, 9, 126.
149. Tashjian, A.H., Jr.; Voelkel, E.F.; Levine, L. Plasma concentrations of 13,14-dihydro-15-keto-prostaglandin E2 in rabbits bearing the VX2 carcinoma: Effects of hydrocortisone and indomethacin. Prostaglandins 1977, 14, 309–317.
150. Tashjian, A.H., Jr.; Voelkel, E.F., Jr.; Robinson, D.R.; Levine, L. Dietary menhaden oil lowers plasma prostaglandins and calcium in mice bearing the prostaglandin-producing HSDM1 fibrosarcoma. J. Clin. Investig. 1984, 74, 2042–2048.
151. Riddle, C.V.; Lee, M.Y.; Lee, M.J. Morphological characterization of a granulocytosis/hypercalcemia-inducing murine mammary carcinoma cell line. Anat. Rec. 1987, 219, 127–134.
152. Lee, M.Y.; Baylink, D.J. Hypercalcemia, excessive bone resorption, and neutrophilia in mice bearing a mammary carcinoma. Proc. Soc. Exp. Biol. Med. 1983, 172, 424–429.
153. Lee, M.Y.; Kaushansky, K.; Judkins, S.A.; Lottsfeldt, J.L.; Waheed, A.; Shadduck, R.K. Mechanisms of tumor-induced neutrophilia: Constitutive production of colony-stimulating factors and their synergistic actions. Blood 1989, 74, 115–122.
154. Lee, M.Y.; Rosse, C. Replacement of fatty marrow by active granulocytopoietic bone marrow following transplantation of mammary carcinoma into mice. Anat. Rec. 1979, 195, 31–46.
155. Menu, E.; de Leenheer, E.; de Raeve, H.; Coulton, L.; Imanishi, T.; Miyashita, K.; van Valckenborgh, E.; van Riet, I.; van Camp, B.; Horuk, R.; et al. . Role of CCR1 and CCR5 in homing and growth of multiple myeloma and in the development of osteolytic lesions: A study in the 5TMM model. Clin. Exp. Metastasis 2006, 23, 291–300.
156. Guise, T.A.; Yoneda, T.; Yates, A.J.; Mundy, G.R. The combined effect of tumor-produced parathyroid hormone-related protein and transforming growth factor-alpha enhance hypercalcemia in vivo and bone resorption in vitro. J. Clin. Endocrinol. Metab. 1993, 77, 40–45.
157. Iguchi, H.; Tanaka, S.; Ozawa, Y.; Kashiwakuma, T.; Kimura, T.; Hiraga, T.; Ozawa, H.; Kono, A. An experimental model of bone metastasis by human lung cancer cells: The role of parathyroid hormone-related protein in bone metastasis. Cancer Res. 1996, 56, 4040–4043.
158. Iguchi, H.; Aramaki, Y.; Maruta, S.; Takiguchi, S. Effects of anti-parathyroid hormone-related protein monoclonal antibody and osteoprotegerin on PTHrP-producing tumor-induced cachexia in nude mice. J. Bone Miner. Metab. 2006, 24, 16–19.
159. Lorch, G.; Viatchenko-Karpinski, S.; Ho, H.T.; Dirksen, W.P.; Toribio, R.E.; Foley, J.; Gyorke, S.; Rosol, T.J. The calcium-sensing receptor is necessary for the rapid development of hypercalcemia in human lung squamous cell carcinoma. Neoplasia 2011, 13, 428–438.
160. Bockman, R.S.; Bellin, A.; Repo, M.A.; Hickok, N.J.; Kameya, T. In vivo and in vitro biological activities of two human cell lines derived from anaplastic lung cancers. Cancer Res. 1983, 43, 4511–4516.
161. Wysolmerski, J.J.; Vasavada, R.; Foley, J.; Weir, E.C.; Burtis, W.J.; Kukreja, S.C.; Guise, T.A.; Broadus, A.E.; Philbrick, W.M. Transactivation of the PTHrP gene in squamous carcinomas predicts the occurrence of hypercalcemia in athymic mice. Cancer Res. 1996, 56, 1043–1049.
162. Asahi, Y.; Kubonishi, I.; Imamura, J.; Kamioka, M.; Matsushita, H.; Furihata, M.; Ohtsuki, Y.; Miyoshi, I. Establishment of a clonal cell line producing granulocyte

colony-stimulating factor and parathyroid hormone-related protein from a lung cancer patient with leukocytosis and hypercalcemia. Jpn. J. Cancer Res. 1996, 87, 451–458.

163. Feuer, G.; Zack, J.A.; Harrington, W.J., Jr.; Valderama, R.; Rosenblatt, J.D.; Wachsman, W.; Baird, S.M.; Chen, I.S. Establishment of human T-cell leukemia virus type I T-cell lymphomas in severe combined immunodeficient mice. Blood 1993, 82, 722–731.

164. Parrula, C.; Zimmerman, B.; Nadella, P.; Shu, S.; Rosol, T.; Fernandez, S.; Lairmore, M.; Niewiesk, S. Expression of tumor invasion factors determines systemic engraftment and induction of humoral hypercalcemia in a mouse model of adult T-cell leukemia. Vet. Pathol. 2009, 46, 1003–1014.

165. Phillips, K.E.; Herring, B.; Wilson, L.A.; Rickford, M.S.; Zhang, M.; Goldman, C.K.; Tso, J.Y.; Waldmann, T.A. IL-2Ralpha-Directed monoclonal antibodies provide effective therapy in a murine model of adult T-cell leukemia by a mechanism other than blockade of IL-2/IL-2Ralpha interaction. Cancer Res. 2000, 60, 6977–6984.

166. Linehan, W.M.; Kish, M.L.; Chen, S.L.; Andriole, G.L.; Santora, A.C. Human prostate carcinoma causes hypercalcemia in athymic nude mice and produces a factor with parathyroid hormone-like bioactivity. J. Urol. 1986, 135, 616–620.

167. Colloton, M.; Shatzen, E.; Wiemann, B.; Starnes, C.; Scully, S.; Henley, C.; Martin, D. Cinacalcet attenuates hypercalcemia observed in mice bearing either Rice H-500 Leydig cell or C26-DCT colon tumors. Eur. J. Pharmacol.2013, 712, 8–15.

168. Takahashi, K.; Shirahata, A.; Fukushima, S.; Kokubo, S.; Teramura, K.; Usuda, S. Effects of YM175, a new-generation bisphosphonate, on hypercalcemia induced by tumor-derived bone resorbing factors in rats. Jpn. J. Pharmacol. 1998, 76, 155–163.

169. Berger, M.E.; Golub, M.S.; Sowers, J.R.; Brickman, A.S.; Nyby, M.; Troyer, H.; Rude, R.K.; Singer, F.R.; Horst, R.; Deftos, L.J. Hypercalcemia in association with a Leydig cell tumor in the rat: A model for tumor-induced hypercalcemia in man. Life Sci. 1982, 30, 1509–1515.

170. Rice, B.F.; Roth, L.M.; Cole, F.E.; MacPhee, A.A.; Davis, K.; Ponthier, R.L.; Sternberg, W.H. Hypercalcemia and neoplasia. Biologic, biochemical, and ultrastructural studies of a hypercalcemia-producing Leydig cell tumor of the rat. Lab. Investig. 1975, 33, 428–439.

171. Sica, D.A.; Martodam, R.R.; Aronow, J.; Mundy, G.R. The hypercalcemic rat Leydig cell tumor—A model of the humoral hypercalcemia of malignancy. Calcif. Tissue Int. 1983, 35, 287–293.

172. Silve, C.; Santora, A.; Spiegel, A. A factor produced by cultured rat Leydig tumor (Rice 500) cells associated with humoral hypercalcemia stimulates adenosine 3′,5′-monophosphate production via the parathyroid hormone receptor in human skin fibroblasts. J. Clin. Endocrinol. Metab. 1985, 60, 1144–1147.

173. Troyer, H.; Sowers, J.R.; Babich, E. Leydig cell tumor induced hypercalcemia in the Fischer rat: Morphometric and histochemical evidence for a humoral factor that activates osteoclasts. Am. J. Pathol. 1982, 108, 284–290.

174. Gaich, G.; Burtis, W.J. Measurement of circulating parathyroid hormone-related protein in rats with humoral hypercalcemia of malignancy using a two-site immunoradiometric assay. Endocrinology 1990, 127, 1444–1449.

175. Martodam, R.R.; Thornton, K.S.; Sica, D.A.; D'Souza, S.M.; Flora, L.; Mundy, G.R. The effects of dichloromethylene diphosphonate on hypercalcemia and other parameters of the humoral hypercalcemia of malignancy in the rat Leydig cell tumor. Calcif. Tissue Int. 1983, 35, 512–519.

176. Sartori, L.; Insogna, K.L.; Barrett, P.Q. Renal phosphate transport in humoral hypercalcemia of malignancy. Am. J. Physiol. 1988, 255, F1078–F1084.

177. de Miguel, F.; Garcia-Canero, R.; Esbrit, P. Co-purification of calcium transport-stimulating and DNA synthesis-stimulating agents with parathormone-like activity isolated from the hypercalcaemic strain of the Walker 256 tumour. Eur. J. Cancer 1991, 27, 1022–1026.

178. D'Souza, S.M.; Ibbotson, K.J.; Smith, D.D.; Mundy, G.R. Production of a macromolecular bone-resorbing factor by the hypercalcemic variant of the walker rat carcinosarcoma. Endocrinology 1984, 115, 1746–1752.
179. Ibbotson, K.J.; D'Souza, S.M.; Ng, K.W.; Osborne, C.K.; Niall, M.; Martin, T.J.; Mundy, G.R. Tumor-derived growth factor increases bone resorption in a tumor associated with humoral hypercalcemia of malignancy. Science 1983, 221, 1292–1294.
180. Insogna, K.L.; Weir, E.C.; Wu, T.L.; Stewart, A.F.; Broadus, A.E.; Burtis, W.J.; Centrella, M. Co-purification of transforming growth factor beta-like activity with PTH-like and bone-resorbing activities from a tumor associated with humoral hypercalcemia of malignancy. Endocrinology 1987, 120, 2183–2185.
181. Juraschek, M.; Seibel, M.J.; Woitge, H.W.; Krempien, B.; Bauss, F. Association between histomorphometry and biochemical markers of bone turnover in a longitudinal rat model of parathyroid hormone-related peptide (PTHrP)-mediated tumor osteolysis. Bone 2000, 26, 475–483.
182. Torres, R.; de la Piedra, C.; Rapado, A. Osseous and intestinal compartments in the humoral hypercalcemia of malignancy associated to Walker 256 tumor in rats. Oncology 1993, 50, 148–152.
183. Benitez-Verguizas, J.; Loarte, D.; de Miguel, F.; Esbrit, P. Effects of transforming growth factor beta1 on cell growth and parathyroid hormone-related protein in Walker 256 tumor cells. Life Sci. 1999, 65, 1807–1816.
184. Guaitani, A.; Polentarutti, N.; Filippeschi, S.; Marmonti, L.; Corti, F.; Italia, C.; Coccioli, G.; Donelli, M.G.; Mantovani, A.; Garattini, S. Effects of disodium etidronate in murine tumor models. Eur. J. Cancer Clin. Oncol. 1984, 20, 685–693.
185. Guaitani, A.; Sabatini, M.; Coccioli, G.; Cristina, S.; Garattini, S.; Bartosek, I. An experimental rat model of local bone cancer invasion and its responsiveness to ethane-1-hydroxy-1,1-bis(phosphonate). Cancer Res. 1985, 45, 2206–2209.
186. Johannesson, A.J.; Onkelinx, C.; Rodan, G.A.; Raisz, L.G. Thionapthene-2-carboxylic acid: A new antihypercalcemic agent. Endocrinology 1985, 117, 1508–1511.
187. Krempien, B.; Manegold, C. Prophylactic treatment of skeletal metastases, tumor-induced osteolysis, and hypercalcemia in rats with the bisphosphonate Cl2MBP. Cancer 1993, 72, 91–98.
188. Weiss, J.; Walker, S.T.; Fallon, M.; Goldfarb, S. In vivo and in vitro effects of WR-2721 in experimental hypercalcemia in the rat. J. Pharmacol. Exp. Ther. 1986, 238, 969–973.
189. Nakanish, I.M.; Yamate, J.; Ide, M.; Sawamoto, O.; Kuwamur, M.; Kotani, T.; Nakatsuji, S. Humoral hypercalcemia of malignancy in female F344 rats implanted with a transplantable rat pulmonary carcinoma line (IP). J. Vet. Med. Sci.2001, 63, 1135–1138.
190. Gullino, P.M.; Pettigrew, H.M.; Grantham, F.H. N-nitrosomethylurea as mammary gland carcinogen in rats. J. Natl. Cancer Inst. 1975, 54, 401–414.
191. Ikeda, K.; Matsumoto, T.; Fukumoto, S.; Kurokawa, K.; Ueyama, Y.; Fujishige, K.; Tamaoki, N.; Saito, T.; Ohtake, K.; Ogata, E. A hypercalcemic nude rat model that completely mimics human syndrome of humoral hypercalcemia of malignancy. Calcif. Tissue Int. 1988, 43, 97–102.
192. Stoica, G.; Koestner, A.; Capen, C.C. Characterization of N-ethyl-N-nitrosourea-induced mammary tumors in the rat. Am. J. Pathol. 1983, 110, 161–169.
193. Michigami, T.; Yamato, H.; Suzuki, H.; Nagai-Itagaki, Y.; Sato, K.; Ozono, K. Conflicting actions of parathyroid hormone-related protein and serum calcium as regulators of 25-hydroxyvitamin D(3)-1 alpha-hydroxylase expression in a nude rat model of humoral hypercalcemia of malignancy. J. Endocrinol. 2001, 171, 249–257.
194. Onuma, E.; Azuma, Y.; Saito, H.; Tsunenari, T.; Watanabe, T.; Hirabayashi, M.; Sato, K.; Yamada-Okabe, H.; Ogata, E. Increased renal calcium reabsorption by parathyroid hormone-related protein is a causative factor in the development of humoral hypercalcemia of malignancy refractory to osteoclastic bone resorption inhibitors. Clin. Cancer Res. 2005, 11, 4198–4203.

195. Mehta, R.; Hawthorne, M.; Uselding, L.; Albinescu, D.; Moriarty, R.; Christov, K.; Mehta, R. Prevention of N-methyl-N-nitrosourea-induced mammary carcinogenesis in rats by 1alpha-hydroxyvitamin D(5). J. Natl. Cancer Inst. 2000, 92, 1836–1840.
196. Doppelt, S.H.; Slovik, D.M.; Neer, R.M.; Nolan, J.; Zusman, R.M.; Potts, J.T., Jr. Gut-mediated hypercalcemia in rabbits bearing VX2 carcinoma: New mechanism for tumor-induced hypercalcemia. Proc. Natl. Acad. Sci. USA 1982, 79, 640–644.
197. Hough, A., Jr.; Seyberth, H., Jr.; Oates, J.; Hartmann, W. Changes in bone and bone marrow of rabbits bearing the VX-2 carcinoma. A comparison of local and distant effects. Am. J. Pathol. 1977, 87, 537–552.
198. Seyberth, H.W.; Hubbard, W.C.; Oelz, O.; Sweetman, B.J.; Watson, J.T.; Oates, J.A. Prostaglandin-mediated hypercalcemia in the VX2 carcinoma-bearing rabbit. Prostaglandins 1977, 14, 319–331.
199. Voelkel, E.F.; Tashjian, A.H., Jr.; Franklin, R.; Wasserman, E.; Levine, L. Hypercalcemia and tumor-prostaglandins: The VX2 carcinoma model in the rabbit. Metabolism 1975, 24, 973–986.
200. Young, D.M.; Ward, J.M.; Prieur, D.J. Hypercalcemia of malignancy. Animal model: VX-2 carcinoma of rabbits. Am. J. Pathol. 1978, 93, 619–622.
201. Wolfe, H.J.; Bitman, W.R.; Voelkel, E.F.; Griffiths, H.J.; Tashjian, A.H. Systemic effects of the VX2 carcinoma on the osseous skeleton. A quantitative study of trabecular bone. Lab. Investig. 1978, 38, 208–215.

CHAPTER 18

Detection and Characterization of Histamine-Producing Strains of *Photobacterium damselae* subsp. *damselae* Isolated from Mullets

Marcello Trevisani [1,*], Rocco Mancusi [1], Matilde Cecchini [1], Claudia Costanza [1] and Marino Prearo [2]

[1]*Dipartimento di Scienze Mediche Veterinarie, Università degli Studi di Bologna, Alma Mater Studiorum, via Tolara di Sopra 50, Ozzano Emilia 40064, Italy*

[2]*S.S. Laboratorio Specialistico Ittiopatologia, Istituto Zooprofilattico Sperimentale del Piemonte, Liguria e Valle d'Aosta, Torino 10154, Italy*

ABSTRACT

Photobacterium damselae subsp. *damselae* (*Pdd*) is considered to be an emerging pathogen of marine fish and has also been implicated in cases of histamine food poisoning. In this study, eight strains isolated from mullets of the genera *Mugil* and *Liza* captured in the Ligurian Sea were characterized, and a method to detect histamine-producing *Pdd* from fish samples was developed. The histamine-producing potential of the strains was evaluated in culture media (TSB+) using a histamine biosensor. Subsequently, two strains were used to contaminate mackerel fillets (4 or 40 CFU/g), simulating a cross-contamination on the selling fish stalls. Sample homogenates were enriched in TSB+. The cultures were then inoculated on thiosulfate-citrate-bile salts-sucrose agar (TCBS) and the dark green colonies were cultured on Niven agar. The violet isolates were characterized using specific biochemical and PCR based tests. All *Pdd*strains were histamine producers, yielding concentration varying from 167 and 8977 μg/mL in TSB+ cultures incubated at 30 °C for 24 h. *Pdd* colonies were detected from the inoculated mackerel samples and their histidine decarboxylase gene was amplified using species-specific primer pairs designed for this study. The results indicate that mullets can be source of *Pdd* and the fish retailers needs to evaluate the risk posed by cross-contamination on the selling fish stalls.

Keywords: *Photobacterium damselae* subsp. *damselae*; histidine decarboxylase; cross contamination; mullets; histamine biosensor

1. INTRODUCTION

Histamine fish poisoning is among the most common food borne diseases related to fish consumption. Fifty-six of the 71 food borne disease outbreaks (78.9%) that have been notified in Europe in 2011 were due to histamine fish poisoning [1]. The risk is correlated with the number and the histidine decarboxylase activity of the contaminating bacteria that grow in the flesh of fishes that are rich of free histidine, such as tuna, mackerel, and bonito.

Bacteria of the genus *Photobacterium*, i.e., *P. damselae* subsp. *damselae* (*Pdd*) and *P. phosphoreum*, are strong histamine producers [2,3,4].

Photobacterium damselae subsp. *damselae* is considered to be an emerging pathogen of marine fish of importance in aquaculture, with a notable increase in its geographical distribution during the last several years [5].

Kanki et al. [2] demonstrated that *Pdd* inoculated on tuna can produce toxic levels of histamine even at 4 °C. These authors observed that *Pdd* displayed the highest performance in accumulating histamine in fish samples stored at refrigeration temperature in comparison with other psychrotolerant marine bacteria, namely *P. phosphoreum* and *Raoultella planticola*. They demonstrated that *Pdd* (strain JCM 8968) can produce more than 500 mg/kg histamine at 4 °C in 24 h and maintain 60% and 50% of the initial activity in tuna and dried saury for up to 12 weeks at −20 °C, respectively. The presence of *Pdd* in the fish that are stored in melting ice or at chilling temperature and even in the de-frozen and processed seafood can thus pose a significant hazard if the contamination is carried on fish species which are rich in free histidine.

The histidine decarboxylase activity of bacteria that are present in fish samples can be assessed in enrichment broth supplemented with histidine using different analytical methods, including immuno-enzymatic (ELISA) tests, chromatographic methods, or bio-sensing devices [3,6,7,8]. The detection of histamine-producing bacteria (HPB) is possible by plating the fish homogenates on differential grow media, such as the Niven agar plates [9] and the screening of the suspect colonies with PCR assays for the histidine decarboxylase encoding genes [10], but these analytical methods did not allow the isolation of HPB in a number of samples that developed high level of histamine in the enrichment broth supplemented with histidine [7]. Many studies have demonstrated that among HPB there are some species that are not able to grow on the Niven medium and consequently false negative results occur [11]. In order to isolate the halophilic *Photobacterium* spp., the use of specific culture medium is needed.

There are few studies concerning the histidine-decarboxylase encoding genes of *P. damselae* and *P. phosphoreum* [11], and the variability of phenotype expression (histidine-decarboxylase activity) [3,4]. Biosensor technology allows fast, cost effective, and specific detection of histamine in seafood spoilage [12] and histamine biosensors can be useful in assessing the microbiological quality of fish. The enzyme diaminoxidase is known to catalyse the conversion of histamine to imidazole acetaldehyde by means of a bi-enzyme system, used by various researchers [13,14,15]. The co-substrate, molecular oxygen, is reduced to hydrogen peroxide (Equation (1)). The reduction of hydrogen peroxide to water is catalysed by the enzyme peroxidase using potassium hexacyanoferrate(II) (or other redox systems) as the mediator (Equation (2)). The introduction of a mediator in the bi-enzymatic system (DAO/HRP) leads to an acceleration of the electron transfer, a decrease in the applied working potential (the reduction of the mediator is detected at the electrode), and an increase in sensitivity [15,16,17].

$$RCH_2NH_2 + O_2 + H_2O \rightarrow (DAO) \rightarrow RCHO + NH_3 + H_2O_2 \quad (1)$$

$$H_2O_2 + Fe(CN)_6{}^{-4} \rightarrow HRP \rightarrow H_2O + Fe(CN)_6{}^{-3} + e^- \quad (2)$$

The working electrode of a potentiostat acts as an oxidant and the oxidation current increases. The aims of this study were to develop a simple and rapid method to detect histamine-producing *P. damselae* subsp. *damselae* from fish samples and characterize their histidine decarboxylase activity.

2. MATERIALS AND METHODS

2.1. Characterization of Pdd Strains

Eight strains of *Pdd* that have been isolated from different samples of fish of the species *Mugil cephalus*, *Liza aurata*, *L. ramada,* and *L. saliens*, which were captured at the estuary of the river Magra in the Eastern Ligurian Sea (Bocche di Magra and Fiumaretta, Amelia municipality, La Spezia province, Italy) [18], were used in this study. The isolates were characterized with biochemical tests for their oxidase, catalase, urease activities, and the ability of fermenting glucose and galactose. In addition, the presence of specific target genes *ure*C and *16S* rRNA was assessed by PCR using the primers designed by Osorio et al. [19] (Table 1).

Table 1. Primers and PCR protocols used for the identification of *P. damselae* subsp. *damselae* and its histidine decarboxylase gene.

Gene/Primers	Nucleotide Sequences	Thermal Cycling Conditions	Reference
*ure*C			
Ure-5′	5′-TCCGGAATAGGTAAAGCGGG-3′	95 °C 4 min;	
Ure-3′	5′-CTTGAATATCCATCTCATCTGC-3′	30×	Osorio et al., 2000
16S rRNA		(95 °C 60 s; 60 °C 60 s; 72 °C 40 s);	
Car1	5′-GCTTGAAGAGATTCGAGT-3′	72 °C 5 min	
Car2	5′-CACCTCGCGGTCTTGCTG-3′		
hdc dp [a]		95 °C 3 min;	
HIS2-F	5′-AAYTSNTTYGAYTTYGARAARGARGT-3′	35×	De Las Rivas et al., 2005
HIS2-R	5′-TANGGNSANCCDATCATYTTRTGNCC-3′	(95 °C 15 s; 53 °C * 30 s; 72 °C 40 s);	
		72 °C 5 min	
*hdc*Pdd [b]		95 °C 3 min;	
*hdc*Pdd-F	5′-GGATTAGCGCATGGATTGGT-3′	30×	(Genebank accession number JCM 8968)
*hdc*Pdd-R	5′-AACGCCTAAGAAACCCCACA-3′	(95 °C 10 s; 60 °C 45 s)	

Notes: [a] dp, degenerate primers targeting multiple sequences of the Gram-histidine decarboxylase (*hdc*) genes; [b] primers targeting specific sequence of *Photobacterium damselae* subsp. *damselae hdc* gene. * gradient in the range from 49 to 53 °C.

In order to detect the histidine decarboxylase gene of *P. damselae* subsp. *damselae* two primer pairs were used. The primers *HIS*2-F and *HIS*2-R designed by De Las Rivas et al. [10] allow the amplification of a specific 531-bp DNA fragment from gram-negative histamine-producing bacteria including *P. damselae*, strain CECT 626T (ATCC 33539) [20].

In this study, target-specific primers for *Pdd* were designed to be specific for Genbank accession AB259289.1 (HDC, Histidine decarboxylase gene of *Photobacterium damselae* (strain JCM 8968)) [2]. The designed PCR primers (*hdc*Pdd, Table 1) were checked for specificity with Primer-Blast software (NCBI, Cambridge, UK).

2.2. DNA Extraction and PCR Procedures

Strains were grown on Tryptone Soy Agar (TSA, Thermofisher, Milano, Italy) supplemented with NaCl (2%) at 25 °C for 48 h. Two colonies were randomly picked for each strain using a sterile toothpick and the DNA was extracted by boiling suspensions of cells in a 5% suspension of Chelex 100 (Bio-Rad Laboratories, Hercules, CA, USA) following the producer protocol.

Real-Time PCR was performed on a MiniOpticon Real-Time PCR System (Bio-Rad Laboratories) using SsoFast EvaGreen Supermix (Bio-Rad). Total reaction volume of 20 μL included: 300 nM of each forward and reverse primers, 1 μL of template DNA and the SsoFast EvaGreen Supermix (according to the manufacturer's instructions). Each run included a non-template control (NTC). Two laboratory reference strains characterized as *Morganella morganii* and *Photobacterium damselae* subsp. *damselae* in accordance with their colony characteristics, biochemical tests (API20E and API20NE systems) and genetic characterization were used as positive controls [19,21]. Real-time amplification for each gene, including controls and samples, was performed in duplicate under identical PCR conditions. Thermal cycle was set as reported in Table 1. At the end of the PCR cycles, a melting curve was conducted between 65 °C and 95 °C with a 0.5 °C/5 s increment read and continuous fluorescence measurement. The effect of annealing temperature was analyzed using gradient PCR amplification in the range from 49 to 53 °C.

2.3. Assay for Histamine Production: A Biosensor Fabrication

To confirm histamine production by each HPB isolate, single colonies were suspended in Tryptone Soy Broth (Thermofisher, Milano, Italy) supplemented with NaCl (2%) and L-Histidine hydrochloride (1%) (Sigma-Aldrich, Milano, Italy) (TSB+) and incubated at 30 °C for 24 h. After the incubation, the cultures were sterilized at 121 °C for 15 min and the supernatant was used for histamine quantitative detection.

To comparatively assess histamine production ability of each strain, the microbial suspensions were diluted with TSB+ to have 40% optical density at 540 nm, corresponding to approximately 10^8 CFU/mL. The concentration of bacteria was subsequently assessed by plating the appropriate dilutions on Tryptone Soy Agar (TSA, Thermofisher, Milano, Italy) with 2% NaCl.

Quantitative detection of histamine in the enriched broth cultures and standardized microbial suspensions was made using an amperometric enzymatic biosensor. The histamine biosensor is based on the co-immobilization diamine-oxidase (DAO, from Porcine Kidney, 0.11 U/mg) and peroxidase (HRP, from horseradish, 25 KU/mg) (Sigma-Aldrich, Milano, Italy) on the surface of a glassy carbon electrode (diameter 3 mm, BAS, West Lafayette, IN, USA). The enzymes were immobilized by chemical cross-linking with glutaraldehyde (GA, 25%) and bovine serum albumin (BSA, fraction V, purity 96–99%). All reagents were purchased from Sigma-Aldrich (Milano, Italy). Briefly, 10 mg of DAO, 6 mg of HRP and 4 mg of BSA were dissolved in 100 μL of Phosphate Buffer (PB, 0.1 M Ph = 7.4). Five μL of GA diluted in water (2.5%) were mixed with 15 μL of the DAO-HRP-BSA solution. Ten μL of the final solution was put on the cleaned surface of the working electrode and allowed to dry for 30 min at 25 °C. After the BSA enzyme solution were cross-linked with GA and a consistent gel has developed on the electrode surface, the enzymatic sensor was washed in PB and stored in the same medium at refrigeration temperature.

The response of the biosensor to the histamine concentration in the bacterial cultures was calculated as the difference between the amperometric signal in inoculated and non-inoculated medium. The values were recorded in millivolts by a potentiostat (Metrhom Autolab PGSTAT10, EcoChemie, Utrecht, The Netherlands) connected to a personal computer using the Autolab GPES 4.9 software. The calibration data was obtained with a series of standard histamine solutions (1, 2, 4, 6, 8, 10, and 20 ppm) were fitted using the software Excel (Microsoft, Redmond, WA, USA). The concentration of histamine in the enriched cultures was calculated by interpolation of the amperometric responses for the samples into the calibration plot constructed with histamine standards.

2.4. Detection of *Pdd* from Fish Samples

A procedure to detect *Pdd* in fish was developed and tested. With this aim mackerel samples were inoculated in duplicate with two *Pdd* strains that produced more than 1000 μg/mL of histamine in TSB+ culture incubated at 30 °C for 24 h (high HPB). The test were carried out independently with each of the two strains. Fresh mackerels were purchased from the market and samples (25 g) were inoculated with either 100 μL of *Pdd* suspensions standardized to approximately 10^3 and 10^4 CFU/mL (4 and 40 CFU/g) and control samples (taken from the same mackerels) were analyzed in parallel (controls).

The samples were homogenized (1:10 w/v) in TSB+ and the contaminating bacteria were enriched at 25 °C for 48 h and subsequently plated on thiosulfate-citrate-bile salts-sucrose agar (TCBS, Thermofisher, Milano, Italy). Dark green colonies with smooth edges were picked and inoculated on plates of modified Niven Agar [9]. Isolated purple colonies developed within 48 h (suspect HPB) were picked and seeded on TSA with 2% NaCl.

Isolates were analyzed for the oxidase activity (oxidase detection strips Oxoid, Thermofisher, Milano, Italy) and by Gram stain, then the Gram-oxidase positive strains were characterized for the ability to ferment glucose and galactose and the urease activity with the API 20E kit (bioMérieux Italia, Bagno di Ripoli, FI, Italy). DNA of the strains that gave positive reactions was extracted by boiling suspensions of cells in a 5% suspension of Chelex 100 (Bio-Rad Laboratories) following the producer protocol. DNA extracts were analyzed to detect the presence or absence of specific nucleotide sequence of the genes *ure*C e *16S* rRNA of *Pdd* by using the PCR protocols described (Table 1).

3. RESULTS

All the strains isolated from the fish of the genera *Mugil* and *Liza* captured in the Ligurian Sea grew on TCBS producing dark green colonies with smooth edges, tested positive at the glucose, galactose, and urease tests, and produced typical amplicons after PCR, with melting temperatures (*Tm*) of 84 °C and 86 °C for the genes *ure*C and *16S* rRNA, respectively. The primers *his*1 and *his*2 produced positive PCR tests only with three of the eight strains analyzed and the level of amplification (cycle threshold, Ct values was higher than 33). The *Tm* of the amplicons of *Pdd* was 80 °C and was different from the amplicon produced by *M. morganii* (Tm 84.5 °C) (Figure 1). The level of amplification improved when the annealing temperature of PCR was reduced from 53 to 49 °C and amplicon melting temperature remained at 80 °C for *Pdd*strains.

All the strains produced good amplification levels by using the primers couple HDC*Pdd* designed for this study and the PCR amplicons produced had a melting temperature of 80.5 °C (Figure 2).

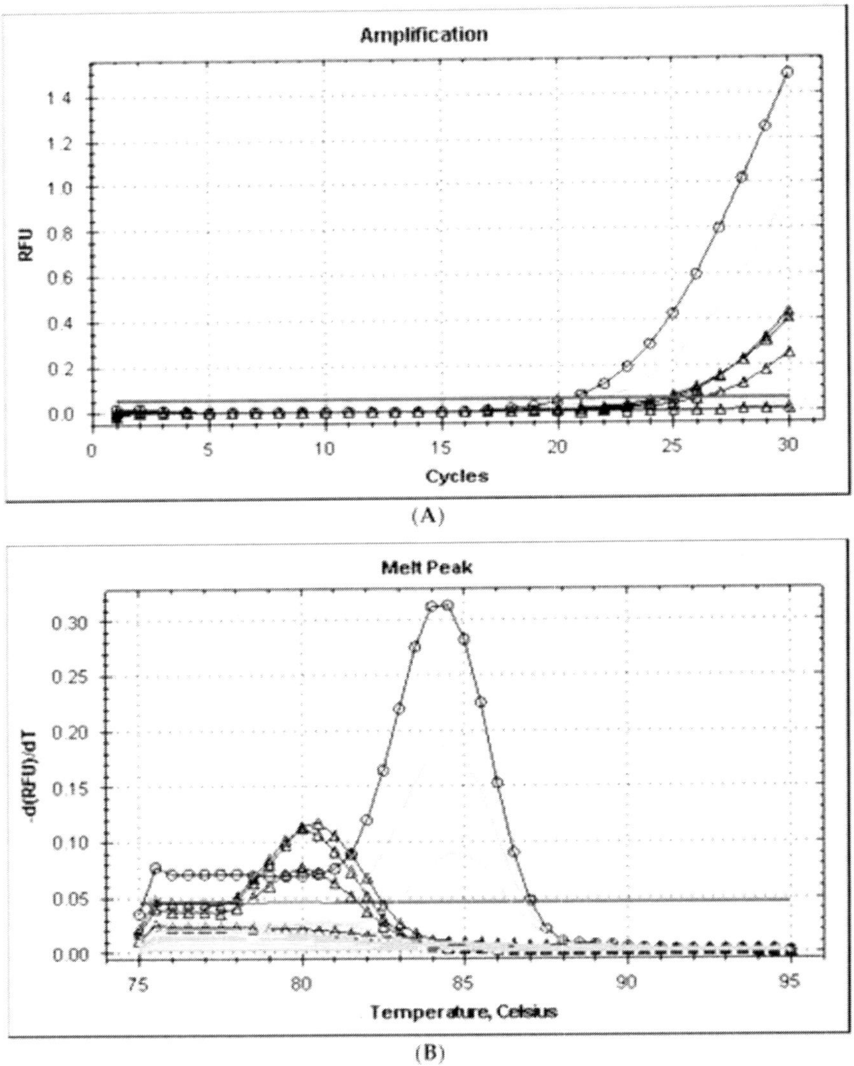

Figure 1. Amplification plot (**A**) and melt curves (**B**) from rt-PCR of histidine decarboxylase genes of Gram-negative bacteria (primers hdc-dp). Legend: annealing temperature for gradient PCR amplification ranged from 49 (dark lines) to 53 °C (grey lines); lines with symbols Δ and ○ correspond to *Pdd* (3 strains)and *M. morgani*, respectively.

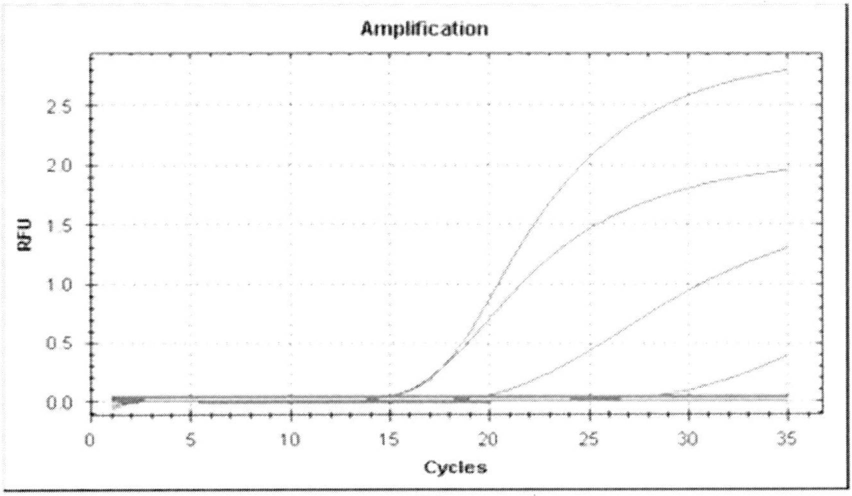

(A)

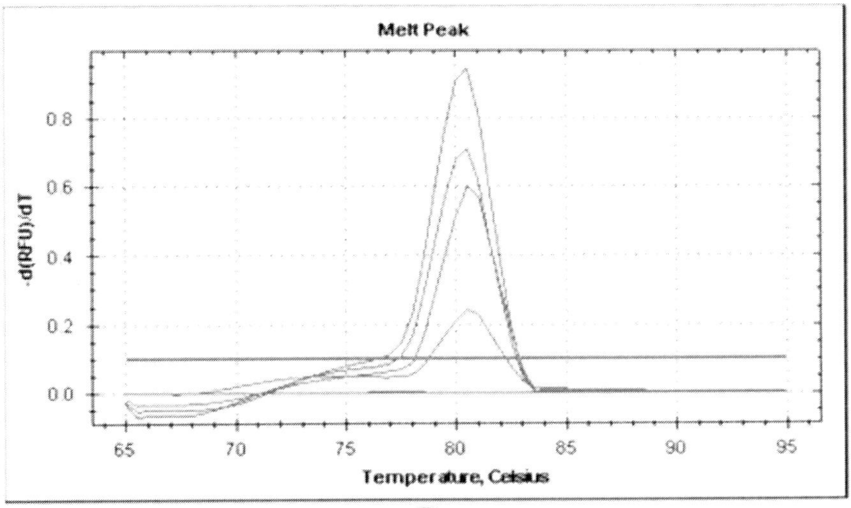

(B)

Figure 2. Amplification plot (**A**) and melt curves (**B**) from rt-PCR of histidine decarboxylase gene of *Photobacterium damselae* subsp. *damselae*.

Molecular sequence analysis with BLAST software were used to compare the HDC*Pdd* primers sequences with the genomes of Gram-bacteria. The primers showed 100% similarity with a known nucleotide sequence of *Pdd* histidine decarboxylase gene and no similarity with other known sequences of Gram-bacteria.

All these strains gave positive results with the assay for histamine production in TSB+ at 30 °C. The concentration of histamine produced in 24 h was between 167 and 8977 µg/mL (median = 1053).

Histamine producing *Pdd* strains were isolated from all the inoculated samples and typical colonies can be easily detected and characterized using the procedure described even at the lower concentration (4 CFU/g) used in this study. Other Gram-bacteria, subsequently identified as *Proteus* spp., grew on the TCBS and also on the Niven plates, but they did not show swarming behavior. The colonies on the TCBS plates were green, but not dark as those of *Pdd*. The PCR and biochemical assays can easily discriminate the *Pdd* strains.

4. DISCUSSION

Contamination of fish meat with histamine-producing bacteria that can grow at refrigeration temperature pose a risk for the marine species that have high concentration of free histidine.

The presence of psychrototrophic bacteria that are strong producers of histamine of the genus *Morganella*, *Photobacterium*e *Raoultella* was observed in fish during surveys or in the many outbreaks of intoxication [22,23,24]. Their presence has been frequently detected also in surveys carried out in the Italian fish market [7,25].

Photobacterium damselae subsp. *damselae* do not grow at the temperature of melting ice, but their role in accumulation of histamine in tuna and dried saury was clearly documented and was correlated to the high stability of its histidine decarboxylase also in defrozen and dried fish [2] and can persist even if its viability is reduced [21]. Recent studies have reported a notable increase in the *Pdd* geographical distribution during the last several years, especially in marine aquaculture [5,26].

Contamination can occur after the harvest, due to manipulations, and during processing. Development of histamine is favored by the spread of bacteria and HDC resulting from the loss of integrity and the post mortem decay of natural barriers (skin, gut, and gills). The risk of histamine (scombroid) intoxication is correlated with the number of HPB and their enzymatic activity of the microbial species, other than to the temperature and time [27,28].

The screening of fish samples to detect HPB and characterize their HDC activity is of utmost importance for the food business operators (FBO) and the food safety authorities that verify the effectiveness of the FBO's controls [29,30].

Differential culture method [9], conductance method [31], and also culture independent Real-Time PCR Method [32] have been used to detect and quantify HPB. Immunological and chromatographic methods, and also biosensors, can be used to measure the histamine produced in the culture media and distinguish the phenotype of HPB strains (i.e., strong or weak histamine producers). Biosensors offer a cheaper alternative to the immunoenzymatic methods [2] or HPLC [3]. The specific detection of HPB strains is needed to characterize the risk, but many studies have reported that the detection of HPB from enriched samples that have tested positive by the assays for histamine can be hampered by the use of the Niven differential culture media, which do not allow the growth of strict halophilic species or other species which are sensitive to its low pH, producing a number of false negative results [4,7,33]. In addition, Niven medium is not selective, producing false positive isolates that must be subsequently discriminated by PCR assays or labelled DNA probes for the *hdc* gene [4,10,34].

Multiple primers sets have been reported for amplification of the histidine decarboxylase gene from Gram-negative bacteria [10,33], but this study has revealed that the degenerate primers designed by De Las Rivas et al. [10] to amplify several similar genetic sequences of the Gram- *hdc* gene do not give a good performance with the *Pdd* strains analyzed in this study. A PCR amplification was obtained only by lowering annealing temperature, thus allowing primers with many mismatches with templates to provide amplification [35].

In order to detect *Pdd* from fish samples that have tested strongly positive by the assays for histamine, the use of a selective and differential medium like TCBS in combination with the Niven medium for the selective detection of HPB allowed improvement of the performance of the detection method and the presence of *Pdd* can be confirmed by a PCR assay targeted toward the species-specific sequence of *hdc* gene.

An excellent correlation between the histamine amount values obtained with biosensors and the histamine amount values obtained by ELISA (enzyme-linked immunosorbent assay) and HPLC has been established [14,36,37]. Therefore, its use for screening purpose is suitable to analyze the sample enrichments and detect those contaminated with strong HPB active at low or ambient temperatures. Nevertheless, comparative analysis of HPB are usually carried out at 30 °C for 24 h and the *Pdd* isolates can be categorized as high histamine (>1000 ppm) producers (three strains), intermediate (501 to 999 ppm) histamine producers (four strains), and low histamine (126 to 500 ppm) producer (one strain) [3].

The widespread distribution of *P. damselae* subsp. *damselae* was reported in farmed fish in Spain (*Pagrus auriga*) [38] and South-eastern Black Sea (*Dicentrarchus labrax*) [5] where it has been associated with high mortalities. Sea bream and sea bass or mullet are not fish species implicated in histamine poisoning cases, but these farmed fishes are commonly exposed on the selling fish stalls close to tuna and mackerel fillets and manipulated, and even the ice used in the stalls has been reported to be contaminated with HPB [39].

5. CONCLUSIONS

Fish retailers and other operators should review the risk management practices employed throughout the fish processing chain in relation to controlling histamine formation in at-risk fish species and consider that other fish species, such as mullets of the genera Mugil and Liza, can be considered potential carriers of Pdd. Therefore, they must ensure good hygienic practices to prevent cross contamination.

ACKNOWLEDGMENTS

We wish to thank the Ictiopathology Laboratory of the Istituto Zooprofilattico Sperimentale del Piemonte, Liguria e Valle d'Aosta for supplying *Photobacterium damselae* strains.

AUTHOR CONTRIBUTIONS

Marcello Trevisani conceived, designed the experiments, wrote the paper, contributed reagents/materials/analysis tools and analyzed the data; Rocco Mancusi performed the microbiological experiments and designed the set of primers for PCR detection of P. damselae; Matilde Cecchini performed the analysis for quantification of histamine; Claudia Costanza helped in the microbiological experiments; Marino Prearo provided the Pdd strains and helped reviewing the paper.

CONFLICTS OF INTEREST

The authors declare no conflict of interest.

REFERENCES

1. European Food Safety Authority (EFSA); European Centre for Disease Prevention and Control (ECDC). The European Union Summary Report on Trends and Sources of Zoonoses, Zoonotic Agents and Food-Borne Outbreaks in 2011. EFSA J. 2013, 11, 3129.
2. Kanki, M.; Yoda, T.; Tsukamoto, T.; Baba, E. Histidine Decarboxylases and Their Role in Accumulation of Histamine in Tuna and Dried Saury. Appl. Environ. Microbiol. 2007, 73, 1467–1473.
3. Bjornsdottir, K.; Bolton, G.E.; McClellan-Green, P.D.; Jaykus, L.-A.; Green, D.P. Detection of Gram-Negative Histamine-Producing Bacteria in Fish: A Comparative Study. J. Food Prot. 2009, 72, 1987–1991.
4. Björnsdóttir-Butler, K.; Bolton, G.E.; Jaykus, L.A.; McClellan-Green, P.D.; Green, D.P. Development of Molecular-Based Methods for Determination of High Histamine Producing Bacteria in Fish. Int. J. Food Microbiol. 2010, 139, 161–167.
5. Terceti, M.S.; Ogut, H.; Osorio, R. In the Black Sea: Evidence of a Multiclonal Origin. Appl. Environ. Microbiol. 2016, 82, 3736–3745.
6. Chen, H.C.; Kung, H.F.; Chen, W.C.; Lin, W.F.; Hwang, D.F.; Lee, Y.C.; Tsai, Y.H. Determination of Histamine and Histamine-Forming Bacteria in Tuna Dumpling Implicated in a Food-Borne Poisoning. Food Chem. 2008, 106, 612–618.
7. Costanza, C.; Cecchini, M.; Mancusi, R.; Mosso, A.; Giani, G.; Rosmini, R.; Trevisani, M. Evaluation and Identification of Histamine-Forming Bacteria on Fish Products of Middle Adriatic Sea. Ital. J. Food Saf. 2013, 1, 17–20.
8. Marcobal, A.; De Las Rivas, B.; Muñoz, R. Methods for the Detection of Bacteria Producing Biogenic Amines on Foods: A Survey. J. Verbraucherschutz Lebensmittelsicherheit 2006, 1, 187–196.
9. Mavromatis, P.; Quantick, P.C. Modication of Niven's Medium for the Enumeration of Histamine-Forming Bacteria and Discussion of the Parameters Associated with Its Use. J. Food Prot. 2002, 65, 546–551.
10. De Las Rivas, B.; Marcobal, Á.; Muñoz, R. Improved Multiplex-PCR Method for the Simultaneous Detection of Food Bacteria Producing Biogenic Amines. FEMS Microbiol. Lett. 2005, 244, 367–372.
11. Landete, J.M.; De las Rivas, B.; Marcobal, A.; Muñoz, R. Updated Molecular Knowledge about Histamine Biosynthesis by Bacteria. Crit. Rev. Food Sci. Nutr. 2008, 48, 697–714.
12. Male, K.B.; Bouvrette, P.; Luong, J.H.T.; Gibbs, B.F. Amperometric Biosensor for Total Histamine, Putrescine and Cadaverine Using Diamine Oxidase. J. Food Sci. 1996, 61, 1012–1016.
13. Niraj, M.M.G.; Pandey, S. Histamine Biosensor: A Review. Int. J. Pharm. Sci. Res. 2012, 3, 4158–4168.
14. Pérez, S.; Bartrolí, J.; Fàbregas, E. Amperometric Biosensor for the Determination of Histamine in Fish Samples. Food Chem. 2013, 141, 4066–4072.
15. Tombelli, S.; Mascini, M. Electrochemical Biosensors for Biogenic Amines—A Comparison between Different Approaches. Anal. Chim. Acta 1998, 358, 277–285.
16. Grieshaber, D.; Mackenzie, R.; Vörös, J.; Reimhult, E. Electrochemical Biosensors—Sensor Principles and Architectures. Sensors 2008, 8, 1400–1458.
17. Rondeau, A.; Larsson, N.; Boujtita, M.; Gorton, L.; El Murr, N. The Synergetic Effect of Redox Mediators and Peroxidase in a Bienzymatic Biosensor for Glucose Assays in FIA. Analusis 1999, 27, 649–656.
18. Serracca, L.; Ercolini, C.; Rossini, I.; Battistini, R.; Giorgi, I.; Prearo, M. E Photobacterium Damselae Subsp. Piscicida in Cefali Del Fiume Magra (Italia) Detection of Photobacterium Damselae Subsp. Damselae and Photobacterium Damselae Subsp. Piscicida in Mullets Caught in the Magra River (Italy). Ittiopatologia 2009, 6, 221–228.

19. Osorio, C.R.; Toranzo, A.E.; Romalde, J.L.; Barja, J.L. Multiplex PCR Assay for ureC and 16S rRNA Genes Clearly Discriminates between Both Subspecies of Photobacterium Damselae. Dis. Aquat. Org. 2000, 40, 177–183.
20. Landete, J.M.; de las Rivas, B.; Marcobal, A.; Muñoz, R. Molecular Methods for the Detection of Biogenic Amine-Producing Bacteria on Foods. Int. J. Food Microbiol. 2007, 117, 258–269.
21. Podeur, G.; Dalgaard, P.; Leroi, F.; Prévost, H.; Emborg, J.; Martinussen, J.; Hansen, L.H.; Pilet, M.F. Development of a Real-Time PCR Method Coupled with a Selective Pre-Enrichment Step for Quantification of Morganella Morganii and Morganella Psychrotolerans in Fish Products. Int. J. Food Microbiol. 2015, 203.
22. Lin, C.S.; Kung, H.F.; Lin, C.M.; Tsai, H.C.; Tsai, Y.H. Histamine Production by Raoultella Ornithinolytica in Mahi-Mahi Meat at Various Storage Temperatures. J. Food Drug Anal. 2016, 24, 305–310.
23. Emborg, J.; Dalgaard, P. Formation of Histamine and Biogenic Amines in Cold-Smoked Tuna: An Investigation of Psychrotolerant Bacteria from Samples Implicated in Cases of Histamine Fish Poisoning. J. Food Prot. 2006, 69, 897–906.
24. Kanki, M.; Yoda, T.; Ishibashi, M.; Tsukamoto, T. Photobacterium Phosphoreum Caused a Histamine Fish Poisoning Incident. Int. J. Food Microbiol. 2004, 92, 79–87.
25. Mancusi, R.; Bini, R.M.; Cecchini, M.; Delle Donne, G.; Rosmini, R.; Trevisani, M. Presenza D'istamina Nei Prodotti Ittici in Commercio. Ital. J. Food Saf. 2012, 1, 35–39.
26. Rivas, A.J.; Lemos, M.L.; Osorio, C.R. Photobacterium Damselae Subsp. Damselae, a Bacterium Pathogenic for Marine Animals and Humans. Front. Microbiol. 2013, 4, 1–6.
27. Agence Nationale de Sécurité Sanitaire de L'Alimentation, de L'Environnement et du Travail. Caractéristiques et Sources de L'histamine. In Fiches de Dangers Biologiques Transmissibles par les Aliments "Histamine"; ANSES: Maisons-Alfort Cedex, France, 2012; Available online: https://www.anses.fr/fr/ system/ files/MIC2010sa0261.pdf (accessed on 8 June 2017).
28. Canadian Food Inspection Agency. Lookup Results of CFIA Fish List; Canadian Food Inspection Agency: Ottawa, ON, Canada, 2016; Available online: http://www.inspection.gc.ca/active/scripts/fssa/fispoi/fplist/fpresults.asp?lang=e&q=&cmbIn=e&cbShowAll=on (accessed on 8 June 2017).
29. Hwang, C.C.; Kung, H.F.; Lin, C.S.; Hwang, D.F.; Tsai, Y.H. Bacteriological Quality and Histamine-Forming Bacteria Associated with Fish Meats and Environments in HACCP and Non-HACCP Fish Processing Factories. Food Control2011, 22, 1657–1662.
30. Hungerford, J.M. Scombroid Poisoning: A Review. Toxicon 2010, 56, 231–243.
31. Dalgaard, P.; Mejlhom, O.; Huss, H.H. Conductance Method for Quantitative Determination of Photobacterium Phosphoreum in Fish Products. J. Appl. Microbiol. 1996, 81, 57–64.
32. Macé, S.; Mamlouk, K.; Chipchakova, S.; Prévost, H.; Joffraud, J.J.; Dalgaard, P.; Pilet, M.F.; Dousset, X. Development of a Rapid Real-Time PCR Method as a Tool to Quantify Viable Photobacterium Phosphoreum Bacteria in Salmon (Salmo Salar) Steaks. Appl. Environ. Microbiol. 2013, 79, 2612–2619.
33. Takahashi, H.; Kimura, B.; Yoshikawa, M.; Fujii, T. Cloning and Sequencing of the Histidine Decarboxylase Genes of Gram-Negative, Histamine-Producing Bacteria and Their Application in Detection and Identification of These Organisms in Fish. Appl. Environ. Microbiol. 2003, 69, 2568–2579.
34. Wongsariya, K.; Bunyapraphatsara, N.; Yasawong, M. Development of Molecular Approach Based on PCR Assay for Detection of Histamine Producing Bacteria. J. Food Sci. Technol. 2016, 53, 640–648.
35. Green, S.J.; Venkatramanan, R.; Naqib, A. Deconstructing the Polymerase Chain Reaction: Understanding and Correcting Bias Associated with Primer Degeneracies and Primer-Template Mismatches. PLoS ONE 2015, 10, e0128122.

36. Apetrei, I.M.; Apetrei, C. Amperometric Biosensor Based on Diamine Oxidase/platinum Nanoparticles/graphene/chitosan Modified Screen-Printed Carbon Electrode for Histamine Detection. Sensors 2016, 16, 422.
37. Keow, C.M.; Bakar, F.A.; Salleh, A.B.; Heng, L.Y.; Wagiran, R.; Siddiquee, S. Screen-Printed Histamine Biosensors Fabricated from the Entrapment of Diamine Oxidase in a Photocured poly(HEMA) Film. Int. J. Electrochem. Sci. 2012, 7, 4702–4715.
38. Labella, A.; Berbel, C.; Manchado, M.; Castro, D.; Borrego, J.J. Photobacterium Damselae Subsp. Damselae, an Emerging Pathogen Affecting New Cultured Marine Fish Species in Southern Spain. Recent Adv. Fish Farms 2011, 136–152.
39. Economou, V.; Gousia, P.; Kemenetzi, D.; Sakkas, H.; Papadopoulou, C. Microbial Quality and Histamine Producing Microflora Analysis of the Ice Used for Fish Preservation. J. Food Saf. 2016, 37, e12285.

CHAPTER 19

Adenoviral Vectors in Veterinary Vaccine Development: Potential for Further Development

Olasumbo L. Ndi[1*], Mary D. Barton[1], Thiru Vanniasinkam[2]

[1]*Sansom Institute, School of Pharmacy and Medical Sciences, University of South Australia, Adelaide, Australia;*

[2]*School of Biomedical Sciences, Charles Sturt University, Wagga Wagga, Australia.*

ABSTRACT

Vaccines are an integral part of veterinary disease prevention. However there are still a significant number of veterinary diseases for which vaccines do not currently exist or where currently available vaccines do not provide adequate immunity. Adenoviruses have transitioned from tools for gene replacement therapy to bona fide vaccine delivery vehicles because of their ability to elicit potent cell-mediated and humoral responses making them ideal for use against viruses and other intracellular pathogens. Adenoviral vector based vaccines are likely to play a significant role in overcoming these problems in the future. However, this vector is under utilized in veterinary vaccine development at this time. This review focuses on adenoviral vector based vaccines developed to date and explores the potential for veterinary vaccine development based upon this platform: advantages and potential disadvantages of this technology are discussed as well as the potential for developing efficacious commercial veterinary adenoviral vector based vaccines.

Keywords: Veterinary Vaccines; Adenoviral Vector; Adenoviral Vector Vaccine; Vaccine Vector; Immune Response

1. VIRAL VECTORS IN VACCINE DEVELOPMENT

The term "vaccine" which is from the Latin term "vacca," means cow was first coined by Edward Jenner, an English doctor in 1796 to describe the inoculation of humans with the cowpox virus to confer protection against the related human smallpox virus. The history of veterinary vaccine development starts with the well-known story of Louis Pasteur and his rabbit spinal cord vaccine and continues to this day with the demonstration of protection in animals by rabies virus reverse transcriptase DNA plasmid vaccination [1].

Depending on the animal groups under consideration, the criteria for successful animal or veterinary vaccines can be very different from those for human vaccines.

In most cases, the criteria for companion animal vaccines are similar to those for human vaccines because the health and welfare of the individual animal is of primary concern. The emphasis of livestock vaccines, on the other hand, is to improve overall animal production for the primary producers, since this will produce some cost benefits for the industry.

Vaccination against zoonotic or food-borne infections is aimed at reducing or eliminating the risk for the consumer and in some cases to improve the productivity of the individual animal. Vaccination of wildlife is generally considered only with respect to infections that are transmittable to humans (zoonotic diseases), although welfare concerns are of increasing importance [2,3].

Live attenuated vaccines were the first form of vaccine and these have been successfully used for many years in economically important animal diseases as well as in companion animals [3,4]. A live attenuated virus vaccine is a live virus that has lost its virulence that is the ability to cause disease but still has the ability to induce protective immunity against the virulent virus. They are prepared either from a naturally occurring virus from another species (e.g. Jenner's cowpox) or are artificially attenuated. Although live attenuated vaccines are the most successful form of vaccine, they have been found to have some limitations or shortcoming. The reversion to virulence, differences in virulence in different hosts, preparation instability, heat lability are some of the risks associated with their use [3,5].

The recent advances in molecular biology and recombinant DNA technology have revolutionized the field of vaccine research with different cutting edge approaches being developed. This has helped in addressing some of the problems and concerns encountered in developing and using live attenuated vaccines. Today, both DNA and RNA viruses are extensively utilized in vaccine development. The use of viruses as vaccine vectors has a lot of advantages as well as disadvantages. Most viruses used in vector development such as adenoviruses and pox viruses have medium to fairly large genomes that are relatively easy to manipulate and are capable of infecting mammalian cells, making them ideal for use in veterinary and human vaccine development [6-9]. Some relatively small viruses such as Adeno-associated viruses have also been used in vaccine development [10].

Adenoviruses as Vaccine Vectors

Adenoviruses (AdV) are ubiquitous pathogens associated with most vertebrates such as reptiles, birds, mammals and amphibians [11-19]. They are non-enveloped, icosahedral viruses that replicate in the nucleus. They have a linear double stranded DNA of about 26 - 45 kbp. The viral capsid comprises 252 structural units (capsomers), including 240 Hexons (II) and 12 pentons (III). This capsid protects the viral genome ands plays a substantial role during virus entry into the host cell. They belong to the family Adenoviridae which is divided into 5 genera based on the host species and the DNA composition; Atadenovirus, Siadenovirus, Mastadenovirus, Aviadenovirus, Ichtadenovirus [12]. The human adenoviruses (HAdV) are classified within the genome Mastadenoviridae with 57 serotypes classified to date which are divided into seven sub groups (A-G).

Adenovirus vectors have become a very popular tool for gene transfer into mammalian cells [20-24]. The AdV genome is divided into early (E) and late (L) genes, expressed, respectively, before and after replication of the viral chromosome. E1 gene products are involved in the control of viral gene transcription, shutoff of cellular proteins and cellular transformation. The E2 gene codes for proteins involved in viral replication, including a DNA-binding protein involved in DNA elongation (E2A) and a DNA polymerase (E2B); E3 gene, dispensable for Ad replication, codes for proteins that interfere with the host immune response against virus infection; finally, the E4 genes are involved in the transition from early to late gene expression, the shutoff of host-cell gene expression, the viral replication and the assembly of the virion [25,26].

Human AdV are by far the best characterised, and the vast majority of gene transfer studies involving Ad, whether for therapeutic or vaccine purposes, have been carried out with vectors derived from serotype 5 (HAdV-5) [7, 27-29]. Recently, other serotypes such as adenovirus serotype 26 have been developed as vaccine vectors [30].

Different strategies have been employed in the construction of AdV vectors. The vectors can be constructed to be replication competent or replication defective with most viral vectors available in both systems. In order to render these viruses replication defective one or more of the early genes that are required for virus multiplication (e.g. the E1 gene in adenovirus) are normally excised [31,32]. Generally, these replication defective adenoviral vectors are preferred for many reasons. They are usually considered to be safe, more so than replication competent viruses [33,34]. They are often more effective in prime boost protocols than replication competent vectors which have been shown to induce potent immune responses against both the viral vector antigens and the vaccine (transgene encoded) antigen with the resultant neutralizing antibodies against the vaccine vector severely reducing the efficacy of the vaccine when used as a boost [35]. Replication competent vectors may also cause lytic infections in the host cells, leading to apoptosis of infected cells and a reduction in the duration of antigen presentation, potentially compromising the efficacy of the vaccine [36]. Importantly, the use of replication defective adenoviral vectors, does not generally lead to the potentially serious consequences that have been associated with replication competent vectors, especially in immunocompromised hosts and those with relatively immature immune systems [37]. Conversely, it should be noted that some studies have shown that replication competent vectors can elicit stronger immune responses in the host than their replication defective counterparts [38]. Nevertheless, replication competent vectors present a major drawback; in that their use in permissive hosts gives rise to the production of infectious particles that can be released into the environment. This is not favourable in view of the legislation governing the use of genetically modified organisms.

An important advantage of adenoviral vector vaccines is that they lend themselves to delivery by various routes including in-vivo, intramuscular and oral. In addition, AdV vector based vaccines have been utilized successfully in prime boost protocols with other vaccine modalities [39,40]. The type of immune response generated by these vaccines can also be modulated using different doses and delivery methods, with mucosal delivery for example leading to more effective protection against certain pathogens by providing protection at the site of pathogen entry [41,42]. In addition, cell-mediated responses to a vaccine in particular may be markedly affected by the route of viral vector based vaccine delivery [41]. Advantageously, adenoviral vector vaccine technology lends itself to the development of multi-epitope vaccines including those that express multiple genes or fusion genes thereby leading to the development of vaccines that can elicit broad immune responses against multiple antigens from one or more pathogens [43].

One potential shortcoming of adenoviral vector vaccines, however, is that these vaccines may be ineffective and perhaps potentially harmful if used in the presence of pre-existing immunity to the vaccine vector as demonstrated in clinical studies on an HIV vaccine candidate; in this study vaccines with pre-existing immunity to the vaccine carrier were shown to be at greater risk of acquiring a fulminant HIV infection [44]. A number of other studies have also indicated that pre-existing antibodies to AdV affected the induction of humoral immune response following immunization resulting in lower immune response or reduced protection after challenge [45-47].

This problem may be circumvented however by the use adenoviral vector with lower sero prevalence in the target animal species. In the study by Geisbert et al. [8] HAd26 and HAd35 which segregate genetically from HAd5 and exhibit lower sero prevalence in humans, was used as vaccine vector. Their study showed that HAd26 and HAd35 vectors generate robust antigen-specific cellmediated and humoral immune responses against Ebola Virus glycoprotein and that HAd5 immune status does not affect the generation of GP-specific immune responses by the vaccines. Another way to circumvent this will be to design vaccine vectors using adenoviruses isolated from a different host species.

2. IMMUNE RESPONSES GENERATED BY ADENOVIRAL VECTORS

Adenoviral vectors especially those first generation vectors that possess intact adenoviral genes have been shown to produce strong innate and adaptive responses in the host, these responses are made in response to viral gene products and the viral capsid. Not surprisingly the level of immune response is dose dependent. Depending on the number of viral genes deleted and especially with the gutless viral vectors that do not contain any adenoviral genes immune responses elicited by these vectors can be relatively weak. The potential for adenoviruses to elicit powerful B cell and T cell responses in the mammalian host are the main reason for the use of these vectors in vaccine development. The strong immune responses elicited by these vectors have been linked to their ability to infect immature dendritic cells thereby activating them to become mature antigen presenting dendritic cells thus promoting T cells responses [48,49].

Adenoviruses currently used in vaccine development or gene therapy are constructed as E1 or E3 gene deleted through to last generation vectors devoid of the entire adenoviral genome. Depending upon the genes deleted the immunogenicity of the vector can vary [48]. The effects of the gene deletions are taken into consideration in the creation of adenoviral vectors depending upon their use. In gene therapy for example, multiple genes are often deleted as gene therapy vectors unlike vaccine vectors should ideally not induce an immune response in the host. In vaccine development on the other hand potent immune responses that are specific for the vaccine antigen are required which requires long term expression of the vaccine antigen within the vaccinated host, this can be achieved by deleting the E1 gene and such vectors cannot cause apoptosis of the infected cells, therefore there is sustained presentation of the vaccine antigen.

Both innate and adaptive immune responses are induced by adenoviral vectors and responses are generated against the transgene but also the vector itself. Following systemic administration, adenoviral vectors will attach to the Coxsackie Adenovirus receptor (CAR) on host cells and are internalized. Initially entering the cell within an endosome the virus subsequently escapes from the endosome and from within the cytoplasm the viral DNA will enter the nucleus of the cell and is eventually transcribed [50]. Immune responses can be triggered by internalised viruses via various Toll-like receptors (TLR) [51]. The binding of the adenovirus fibre to the CAR receptor leads to the production of proteins such as P13K kinase and junctional adhesion molecule—like protein causing the production of various chemokines [52]. Significant levels of cytokines including IL-6 and TNF-α are usually detected within 24 hours following adenoviral vector administration. Importantly even when the vectors have few or no adenoviral genes they still elicit significant innate responses characterized by upregulated neutrophil and macrophage activity at the site of vector administration which are induced by the viral capsid [53]. These innate responses can also result in the elimination of significant numbers of viral vectors in a host [54]. Apart from their innate responses, significant humoral responses against the viral vector can result in the production of neutralizing antibodies against the vector. Overall, despite these issues adenoviral vectors are still considered to be amongst the best vectors for use in vaccine development.

3. ADENOVIRAL VECTOR BASED VACCINES IN PRIME BOOST PROTOCOLS

Prime-boost vaccination using plasmid DNA and replication-defective adenovirus vectors has emerged as a highly effective strategy for vaccinating against various pathogens. Often this has involved constructing a DNA vaccine and a viral vector based vaccines both containing the same transgene. DNA vaccines have been aggressively researched since the early 90s and following extensive testing in clinical trials it is now known that these vaccines, whilst being largely ineffective as a sole vaccine, can be successfully used in prime boost immunizations. This approach has been investigated with superior results obtained using a DNA vaccine to prime the immune system followed by an adenoviral vector vaccine boost [40,55,56]. Generally, studies have shown that better antibody responses and strong cell-mediated immune responses, characterized by high levels of Th1 cytokines such as IFN gamma, antigen-specific CD4þ and CD8þ cells, are generated using the DNA prime/viral vector boost approach compared with using a single vaccine modality [35,36]. In some cases the efficacy of a prime boost vaccination protocol can be further enhanced with the administration of adjuvants such as glycosyl chitosan or CpG oligodeoxynucleotides [57,58] and chicken CD154 [23]. Th1 cytokines such as IL-12 and IFN-gamma have also been used to enhance immune responses generated by prime boost vaccination protocols [59]. Successful use of an adenoviral vector vaccine in a prime boost regimen is exemplified in vaccination of cattle against bovine tuberculosis (TB). The current strategy on vaccine development for bovine TB is to improve on the existing vaccine, Bacille Calmette-Guerin (BCG). Studies conducted with BCG and a recombinant MVA or AdV vaccine expressing the 85A antigen in a prime boost approach demonstrate that this approach generates higher levels of cell-mediated immune responses than by using BCG alone [60,61]. Boosting with an adenovirus-vectored (Ad) tuberculosis vaccine AdAg 85A through the intranasal mucosal route was able to enhance the long-term survival of BCG-primed guinea pigs following pulmonary Mycobacterium tuberculosis challenge [62]. Hammond et al. [55] demonstrated that a prime-boost vaccination strategy which comprised a naked DNA (pCI-gp55) prime followed by a boost with a live recombinant porcine adenovirus (rPAV-gp55) completely protected weaned pigs from disease following experimental challenge with classical swine fever virus. This strategy also protected against the transient febrile response observed in pigs vaccinated with rPAV-gp55 alone, and prevented any pathology in the animal. The results obtained by Sun et al. [63] demonstrate that the heterologous DNA prime and recombinant adenovirus boost strategy can indeed induce solid protective immunity against the classical swine fever virus (CSFV) in pigs.

The results demonstrated that the pigs receiving pSFV1CS-E2/rAdV-E2 heterologous prime boost immunization developed significantly higher titres of CSFV-specific neutralizing antibodies and comparable $CD4^+$ and $CD8^+$ T-cell proliferation, compared to the pigs receiving double immunizations with rAdV-E2 alone. When challenged with virulent CSFV Shimen strain, the pigs of the heterologous prime boost group did not show clinical symptoms or viremia, which were observed in one of the 5 pigs immunized with rAdV-E2 alone and all the 5 control pigs immunized with an empty adenovirus.

Although the mechanism of action of prime boost vaccination protocols is not fully understood, it has been suggested that the process involves the ability of DNA vaccines to prime memory cells which are then able to respond fairly rapidly to the recombinant viral vector boost, resulting in heightened immune responses to the target antigen. If a recombinant virus was used to prime as well as boost, the effect may not be quite the same, as the immune responses generated by the prime would be primarily an inflammatory response that, when boosted with the same vector, would result in a rapid decline of the vaccine vector [64]. Some of these effects are brought about by the presence of vector specific antibodies formed during the priming response but there is evidence to suggest that T cell-mediated mechanisms may also play a significant role [65]. Some researchers, however, have reported success using multiple immunizations with viral vector based vaccines [66,67] but this approach has not been adopted in veterinary vaccine development. Based upon published studies it is clear that the prime boost strategy is important to keep in mind when developing adenoviral vector based vaccines for veterinary pathogens.

4. ADENOVIRAL VECTORS CURRENTLY USED IN VETERINARY VACCINES

The appeal of AdV-based vaccine vectors stems from their use as efficacious vaccines [68]. Most studies to date have focused on human serotype AdVs, particularly human AdV type 5 (HAd5). Human serotype AdV vaccine vectors are particularly useful for development of veterinary vaccines as neutralizing antibodies to the vector will not usually be present in the vaccinates and to date a range of veterinary vaccine candidates have been developed using this platform [23,63,69]. Adenoviruses from various species have been developed as vectors and some examples are listed in **Table 1**.

AdV vectors are acknowledged to be highly efficacious as vaccine carriers and are capable of inducing strong antibody, CD4+ and CD8+ T cell responses [70]. While some uses of viral vectors in cancer therapy rely on the production of a cytopathic effect within the host tissue [71] in the development of vaccines designed to protect the host against a range of pathogens such replication competent vectors are not required. Research to date shows that replication defective AdV vectors can induce higher B and T cell responses than other viral vectors (e.g. pox viral vectors), however, they may not induce significant CD4+ T cell responses as they do not establish a lytic infection in the host [72].

One of the major problems associated with vaccinating young animals is the presence of low levels of maternally derived antibodies; often insufficient to provide complete protection but enough to interfere with conventional vac cination. For instance, puppies are vulnerable to canine distemper virus (CDV) because conventional vaccines that are highly effective in older animals are not protective in these young animals due to the presence of low levels of antibodies acquired from the dams which have been immunized against CDV [73]. An important advantage of adenoviral vector based vaccines is that they can be used in neonates or young animals as they are not adversely affected by the presence of maternal antibodies [74]. This was clearly demonstrated in a study utilizing an E1-deleted replication defective AdH5 vector based swine influenza virus (SIV) vaccine. In this study, piglets born to dams previously exposed to SIV were immunized with a mixture of two AdH5 based vaccines expressing the hemagglutinin antigen or the nucleoprotein of SIV. Half of the vaccinated piglets were boosted 5 weeks later with commercial SIV H1N1 and H3N2 vaccine. A control group received a sham adenovirus prime at 1 week of age followed by the H1N1 and H3N2 boost at 5 weeks. Two weeks later all three groups

Table 1. Examples of adenoviruses that have been developed as veterinary vaccine vectors.

Vector/Transgene	Animal Model	Method of delivery; Dose	Results of animal studies	Reference
Canine adenovirus serotype 2 rabies virus glycoprotein	Dogs	Subcutaneous, $10^{7.8}$ $TCID_{50}$ (boost at 14 days)	100% protection following challenge with rabies virus	[78]
Human adenovirus serotype 5 avian influenza virus haemagglutinin (HA)	Chickens	Subcutaneous: 1×10^7, 1×10^8, 5×10^8 PFU (boost at 5 weeks)	High levels of HA specific antibodies; High levels of IFN-gamma	[23]
Bovine adenovirus serotype 3 herpesvirus glycoprotein D	Cattle	Intratracheal: 10^8 PFU (boost at 4 and 8 weeks)	Partial protection following challenge with bovine herpesvirus	[79]
Porcine adenovirus serotype 3 transmissible gastroenteritis virus spike gene	Pigs	Oral: 5×10^6 PFU	High titres of virus neutralizing antibody	[80]
Ovine adenovirus taenia ovis 45W antigen	Sheep			[81]
Fowl adenovirus serotype 1 (CELO) VP2 protein of infectious bursal disease virus (IBDV)	Chickens	Oro-nasal: 3×10^8 to 8×10^8 PFU per animal (boost 14 days after prime)	Partial protection following lethal IBDV challenge (complete protection was achieved when vaccine was delivered subcutaneously or intradermally)	[82]

of piglets were challenged with virulent H3N2; only piglets primed with the adenoviral vector based vaccines showed a significant increase in memory responses in the presence of maternally derived antibodies present against SIV H3N2 [75]. Adenoviral vectors have also been used in the construction of vaccines that can inhibit virus replication in the host. The principle of this technology is that the viral vector is engineered to express short hairpin RNAs (shRNA) which, when introduced into the host, will suppress the function of an important target protein expressed by the pathogenic virus. This approach has been used in the development of a replication defective AdH5 virus based vaccine [76] containing either the foot-andmouth disease virus (FMDV) structural protein 1D or polymerase 3D specific shRNA. When challenged with 100 ID_{50} of homologous FMDV partial protection was observed in guinea pigs injected with a mixture of both vaccines (2×10^9 pfu) or the vaccine containing the FMDV polymerase 3D gene (10^6 pfu) [76]. This method has not been widely utilised in vaccine development to date, however, the results of the Chen and co-workers' study suggest that this approach should be investigated further.

The route of vaccination has, in some cases, proved to be crucial to overcome interference from pre-existing neutralizing antibodies. In one study, the efficacy of a vaccination regimen using two replication competent canine adenovirus serotype 2 based CDV vaccines was investigated in pups that had maternally derived CDV and canine adenovirus specific antibodies. Pups vaccinated subcutaneously were protected while those vaccinated via the mucosa were not [73]. This suggests that in any study looking at the efficacy of viral vector vaccines multiple routes of administration should be evaluated.

One of the most important reasons to develop adenoviral vector based vaccines is that this platform would lend itself to the development of vaccines that can differentiate between infected and vaccinated animals (DIVA). For example, this approach has been used by Li et al. (2011) who developed a turkey herpesvirus vector based vaccine for avian influenza and Marek's disease [77]. This vaccine was designed so as not to elicit nucleoprotein specific antibody responses in the vaccinated host and could therefore be used in a DIVA protocol.

The Potential for Commercial Adenoviral Vector Based Veterinary Vaccines

Conventional vaccines have played an important role in the control and possible eradication of certain diseases e.g. rinderpest. However, new approaches to vaccine development are urgently required for those diseases where efficacious vaccines do not exist [83]. Currently however, veterinary vaccines comprise a fraction of the market size of human vaccines; consequently there are much lower levels of investment into veterinary vaccine research and development. These issues can have a negative impact on the utilization of new technologies in veterinary vaccine development.

The commercialization of vaccines whether for human or veterinary use usually begins following extensive testing of the vaccine candidate in animal models and field trials. The final stage of this arduous process would be market authorization which for a veterinary vaccine would involve obtaining permission from veterinary authorities to allow the vaccine to be made available to the relevant consumers [84]. Despite years of research into the technologically advanced DNA and viral/bacterial vector technologies, conventional vaccines based upon either killed or attenuated pathogens or recombinant antigens still form a major part of the veterinary vaccine market today. One of the main reasons for this is the relatively long process leading to commercialization of a new vaccine which involves establishing vaccine efficacy and safety prior to obtaining registration, which may be complicated if the vaccines were to be commercialized globally [85]. Furthermore, the need for global cooperation between governments and industry in veterinary vaccine development remains a pivotal issue in the future development of these vaccination agents; unlike human vaccine development for infections such as HIV and malaria, this arena has not yet been fully explored [30,86]. Over the years a number of adenoviral vector based veterinary vaccines have been developed but none are currently licensed. One such Adenoviral vector based veterinary vaccine is currently under consideration for commercialization. This is a replication-defective human adenovirus serotype 5 (Ad5) containing the capsid and 3C protease-coding regions of foot-and-mouth disease (FMD) virus as a vaccine candidate has been developed and gone through field trials and other testing in preparation for commercial production. Even though many of the trials have been successful; several US Department of Agriculture Centre for Veterinary Biologics regulatory requirements still need to be met in order to obtain product licensure [87,88].

Most of the currently available viral vector based veterinary vaccines are based on other platforms such as Canary Pox virus, Fowl Pox virus and Baculovirus [89-92]. Adenoviral vector based systems are clearly underutilized in veterinary vaccine development despite being well characterized and shown to be potent inducers of protective immune responses [72]. If these vectors are to be used widely to construct veterinary vaccines that can be commercialized, more research needs to be undertaken on this platform, particularly studies focusing on the comparison of this platform with other viral vector platforms and the use of adenoviral vaccines in prime boost immunization protocols with non-adenoviral vector based vaccines.

5. CONCLUSION AND CHALLENGES THAT REMAIN

Adenoviral vector vaccines have been shown to induce potent immune responses in the host. It has been hypothesized that if pre-existing immunity to the vaccine vector is not present, these vaccines are likely to be highly effective against viral pathogens in particular that require robust cell-mediated immune responses in the host for protection. There is evidence to suggest however that even in the presence of neutralizing antibodies adenoviral vector based vaccine could work [93].

Despite the many advantages of this technology there are some potential drawbacks to consider, such as the potential toxicity of adenoviral vaccine vectors. For example, viral based vectors elicit an inflammatory cytokine response (by stimulating both innate and adaptive immune responses), thus promoting harmful side effects in the host [94]. In light of this, these viral vaccine vectors must be thoroughly evaluated for potentially harmful immunological reactions on the host immune system, especially in the context of pre-existing immunity to the vaccine vector prior to conducting field trials.

Other important issues to be considered when developing adenoviral vector based on vaccines are the choice of transgene and the stability of the vector genome. In order for a vaccine to be protective, the pathogen specific transgene used in the construction of the vaccine should contain highly immunogenic epitopes otherwise the immune responses generated by the vaccine will be suboptimal. Ideally extensive research to identify the most immunogenic antigens of a pathogen should therefore be undertaken before vaccine development begins. Studies conducted to ensure the stability of the viral vector genome must also be conducted soon after the vaccine is constructed and well before large scale production of the vaccine as it has been reported that mutants which do not express the vaccine transgene can randomly arise when producing large quantities of viral vector based vaccines [95]. Vaccines therefore need to be constructed using strategies that avoid cloning the transgene into sections of the viral vector that are relatively unstable and more prone to mutation or deletions.

There are still many important veterinary pathogens for which no vaccines currently exist. In addition, issues such as maintaining herd immunity and the potential waning of vaccine induced protection remain a serious challenge for eradication of important diseases such as foot-and-mouth disease using existing vaccines [83]. As human vaccine studies have shown the adenoviral vector platform has the potential to be used in the development of efficacious vaccines for veterinary pathogens where other approaches have failed, however, many challenges remain in regard to the development of these vaccines.

6. ACKNOWLEDGEMENTS

Olasumbo L. Ndi, Mary D. Barton, Thiru Vanniasinkam We would like to thank the Australian Rural Industries Research and Development Corporation for providing research funding.

REFERENCES

1. P. Carnieli, A. M. Ventura and E. L. Durigon, "Construction of a Recombinant Plasmid pSH-G Containing the Rabies-Virus Glycoprotein G Gene," Journal of Venomous Animals and Toxins including Tropical Diseases, Vol. 13, No. 4, 2007, pp. 874-880.
2. M. Lombard, P. P. Pastoret and A. M. Moulin, "A Brief History of Vaccines and Vaccination," Revue et Scientifique Technique (International Office of Epizootics), Vol. 26, No. 1, 2007, pp. 29-48.
3. E. N. T. Meeusen, J. Walker, A. Peters, P. Pastoret and G. Jungersen, "Current Status of Veterinary Vaccines," Clinical Microbiology Reviews, Vol. 20, No. 3, 2007, pp. 489-510.
4. D. J. Alexander, "Newcastle Disease," In: H. G. Purchase, L. H. Arp, C. H. Domermuth and J. E. Pearson, Eds., A Laboratory Manual for the Isolation and Identification of Avian Pathogens, American Association for Avian Pathologists, Inc., Philadelphia, 1989, pp. 114-120.
5. A. Brun, E. Albina, T. Barret, D. A. G. Chapman, M. Czub, L. K. Dixon, G. M. Keil, B. Klonjkowski, M.-F. Le Potier, G. V. Libeau, J. Ortego, J. Richardson and H. H. Takamatsu, "Antigen Delivery Systems for Veterinary Vaccine Development: Viral-Vector Based Delivery Systems," Vaccine, Vol. 26, No. 51, 2008, pp. 6508-6528.
6. M. M. Erdmana, K. I. Kamrud, D. L. Harris and J. Smith, "Alphavirus Replicon Particle Vaccines Developed for Use in Humans Induce High Levels of Antibodies to Influenza Virus Hemagglutinin in Swine: Proof of Concept," Vaccine, Vol. 28, No. 3, 2010, pp. 594-596.
7. E. S. Gabitzsch, Y. Xu, L. H. Yoshida, J. Balint, A. Amalfitano and F. R. Jones, "Novel Adenovirus Type 5 Vaccine Platform Induces Cellular Immunity against HIV-1 Gag, Pol, Nef Despite the Presence of Ad5 Immunity," Vaccine, Vol. 27, No. 46, 2009, pp. 6394-6398.
8. T. W. Geisbert, M. Bailey, L. Hensley, C. Asiedu, J. Geisbert, D. Stanley, A. Honko, J. Johnson, S. Mulangu, M. G. Pau, J. Custers, J. Vellinga, J. Hendriks, P. Jahrling, M. Roederer, J. Goudsmit, R. Koup and N. J. Sullivan, "Recombinant Adenovirus Serotype 26 (Ad26) and Ad35 Vaccine Vectors Bypass Immunity to Ad5 and Protect Nonhuman Primates against Ebolavirus Challenge," Journal of Virology, Vol. 85, No. 9, 2011, pp. 4222-4233. doi:10.1128/JVI.02407-10
9. M. C. Pardo, J. E. Bauman and M. Mackowiak, "Protection of Dogs against Canine Distemper by Vaccination with a Canarypox Virus Recombinant Expressing Canine Distemper Virus Fusion and Hemoglutinin Glycoproteins," American Journal of Veterinary Research, Vol. 58, No. 8, 1997, pp. 833-836.

10. D. Kuck, T. Lau, B. Leuchs, A. Kern, M. Müller, L. Gissmann and J. A. Kleinschmidt, "Intranasal Vaccination with Recombinant Adeno-Associated Virus Type 5 against Human Papillomavirus Type 16 L1," Journal of Virology, Vol. 80, No. 6, 2006, pp. 2621-2630.
11. M. Benko, P. Élo, K. Ursu, W. Ahne, S. E. LaPatra, D. Thomson and B. Harrach, "First Molecular Evidence for the Existence of Distinct Fish and Snake Adenoviruses," Journal of Virology, Vol. 76, No. 19, 2002, pp. 10056- 10059.
12. A. J. Davison, M. Benko and B. Harrach, "Genetic Content and Evolution of Adenoviruses," Journal of General Virology, Vol. 84, No. 11, 2003, pp. 2895-2908.
13. J. Dean, K. S. Latimer, J. L. Oaks, M. Schrenzel, P. T. Redig and A. Wunschman, "Falcon Adenovirus Infection in Breeding Taita Falcons (Falco fasciinucha)," Journal of Veterinary Diagnostic Investigation, Vol. 18, No. 3, 2006, pp. 282-286.
14. K. Erles, E. J. Dubovi, H. W. Brooks and J. Brownlie, "Longitudinal Study of Viruses Associated with Canine Infectious Respiratory Disease," Journal of Clinical Microbiology, Vol. 42, No. 10, 2004, pp. 4524-4529.
15. R. P. Hedrick, J. Speas, M. L. Kent and T. McDowell, "Adenovirus Like Particles Associated with a Disease of Cultured White Sturgeon Acipenser transmontanus," Canadian Journal of Fish and Aquatic Science, Vol. 42, No. 7, 1985, pp. 1231-1235.
16. E. R. Kovacs, J. Mate, D. Adam, H. Balazs and M. Benko, "Recognition and Partial Genome Characterization by Non-Specific DNA Amplification and PCR of a New Siadenovirus Species in a Sample Originating from Parus major, a Great Tit," Journal of Virological Methods, Vol. 163, No. 2, 2010, pp. 262-268.
17. T. H. Lim, B. Y. Kim, M. S. Kim, J. H. Jang, D. H. Lee, Y. K. Kwon, J. B Lee, S. Y. Park, I. S. Choi and C. S. Song, "Outbreak of Gizzard Erosion Associated with Fowl Adenovirus Infection in Korea," Poultry Science, Vol. 91, No. 5, 2012, pp. 1113-1117.
18. H. D. Lehmkuhl and R. C. Cutlip, "A New Goat Adenovirus Isolate Proposed as the Prototype Strain for Goat Adenovirus Serotype 1," Archives of Virology, Vol. 144, No. 8, 1999, pp. 1611-1618.
19. J. Smyth and M. McNulty, "Adenoviridae," In: M. Pattison, P. McMullin, J. Bradbury and D. Alexander, Eds., Poultry Diseases, Elsevier, New York, 2008, pp. 367-381.
20. P. Ketzer, S. F. Haas, S. Engelhardt, J. S. Hartig and D. M. Nettelbeck, "Synthetic Riboswitches for External Regulation of Genes Transferred by Replication-Deficient and Oncolytic Adenoviruses," Nucleic Acids Research, Vol. 40, No. 21, 2012, p. e167.
21. B. Klonjkowski, D. Klein, S. Galea, F. Gavard, M. Monteil, L. Duarte, A. Fournier, S. Sayon, K. Gorrna, R. Ertl, N. Cordonnier, P. Sonigo, M. Eloit and J. Richardson, "Gag-Specific Immune Enhancement of Lentiviral Infection after Vaccination with an Adenoviral Vector in an Animal Model of AIDS," Vaccine, Vol. 27, No. 6, 2009, pp. 928-939.
22. H. Murakami, R. Q. Miao, L. Chao and J. Chao, "Adenovirus-Mediated Kallikrein Gene Transfer Inhibits Neointima Formation via Increased Production of Nitric Oxide in Rat Artery," Immunopharmacology, Vol. 44, No. 1-2, 1999, pp. 137-143.
23. O. S. Ramos, A. G. Pose, S. Gomez-Puerta, J. N. Gomez, A. V. Redondo, J. C. A. Benites, L. S. Amaran, N. C. Parra and J. R. T.Alonso, "Avian CD154 Enhances Humoral and Cellular Immune Responses Induced by an Adenovirus Vector-Based Vaccine in Chickens," Comparative Immunology, Microbiology and Infectious Diseases, Vol. 34, No. 3, 2011, pp. 259-265.
24. B. Trapnell and M. Gorziglia, "Gene Therapy Using Adenoviral Vectors," Current Opinion in Biotechnology, Vol. 5, No. 6, 1994, pp. 617-625.
25. J. L. Imler, "Adenovirus Vectors as Recombinant Viral Vaccines," Vaccine, Vol. 13, No. 13, 1995, pp. 1143- 1151.
26. W. Russell, "Update on Adenovirus and Its Vectors," Journal of General Virology, Vol. 81, No. 11, 2000, pp. 2573-2604.
27. R. R. Bradley, D. M. Lynch, M. J. Iampietro, E. N. Borducchi and D. H. Barouch, "Adenovirus Serotype 5 Neutralizing Antibodies Target Both Hexon and Fiber Following Vaccination and Natural Infection," Journal of Virology, Vol. 86, No. 1, 2012, pp. 625-629.

28. M. K. Knowles, D. Roberts, S. Craig, M. Sheen, S. A. Nadin-Davis and A. I. Wandeler, "In Vitro and In Vivo Genetic Stability Studies of a Human Adenovirus Type 5 Recombinant Rabies Glycoprotein Vaccine (ONRAB)," Vaccine, Vol. 27, No. 20, 2009, pp. 2662-2668.
29. M. Suleman, S. Galea, F. Gavard, N. Merillon, B. Klonjkowski, E. Tartour and J. Richardson, "Antigen Encoded by Vaccine Vectors Derived from Human Adenovirus Serotype 5 Is Preferentially Presented to CD8+ T Lymphocytes by the CD8+ Dendritic Cell Subset," Vaccine, Vol. 29, No. 35, 2011, pp. 5892-5903.
30. J. Esparza, "The Global HIV Vaccine Enterprise," International Microbiology, Vol. 8, No. 2, 2005, pp. 93-101.
31. A. Lieber, C. Y. He, I. IKirillova and M. A. Kay, "Recombinant Adenoviruses with Large Deletions Generated by Cre-Mediated Excision Exhibit Different Biological Properties Compared with First-Generation Vectors in Vitro and in Vivo," Journal of Virology, Vol. 70, No. 12, 1996, pp. 8944-8960.
32. J. Schaack, "Adenovirus Vectors Deleted for Genes Essential for Viral DNA Replication," Frontiers in Bioscience, Vol. 10, 2005, pp. 1146-1155.
33. W. Oualikene, P. Gonin and M. Eloit, "Short and Long Term Dissemination of Deletion Mutants of Adenovirus in Permissive (Cotton Rat) and Non-Permissive (Mouse) Species," Journal of General Virology, Vol. 75, No. 10, 1994, pp. 2765-2768.
34. L. Peiperl, C. Morgan, Z. Moodie, H. Li, N. Russell, B. S. Graham, G. D. Tomaras, S. C. De Rosa, M. J. McElrath and NIAID Vaccines Trials Network, "Safety and Immunogenicity of a Replication-Defective Adenovirus Type 5 HIV Vaccine in Ad5-Seronegative Persons: A Randomized Clinical Trial (HVTN 054)," PLoS ONE, Vol. 5, No. 10, 2010, p. e13579.
35. L. Wu, W. Kong and G. Nabel, "Enhanced Breadth of CD4 T-Cell Immunity by DNA Prime and Adenovirus Boost Immunization to Human Immunodeficiency Virus Env and Gag Immunogens," Journal Virology, Vol. 79, No. 13, 2005, pp. 8024-8031.
36. T. Vanniasinkam and H. C. J. Ertl, "Adenoviral Gene Delivery for HIV-1 Vaccination," Current Gene Therapy, Vol. 5, No. 2, 2005, pp. 203-212.
37. T. Kojaoghlanian, P. Flomenberg and M. S. Horwitz, "The Impact of Adenovirus Infection on the Immunocompromised Host," Review of Medical Virology, Vol. 13, No. 3, 2003, pp. 155-171.
38. Y. Zhu, P. Rota, L. Wyatt, A. Tamin, S. Rozenblatt, N. Lerche, B. Moss, W. Bellini and M. McChesney, "Evaluation of Recombinant Vaccinia Virus: Measles Vaccines in Infant Rhesus Macaques with Preexisting Measles Antibody," Virology, Vol. 276, No. 1, 2000, pp. 202-213.
39. T. Vanniasinkam, S. T. Reddy and H. C. Ertl, "DNA Immunization Using a Non-Viral Promoter," Virology, Vol. 344, No. 2, 2006, pp. 412-420.
40. H. P. Zhao, J. F. Sun, N. Li, Y. Sun, Y. Wang and H. J. Qiu, "Prime-Boost Immunization Using Alphavirus Replicon and Adenovirus Vectored Vaccines Induces Enhanced Immune Responses against Classical Swine Fever Virus in Mice," Veterinary Immunology Immunopathology, Vol. 131, No. 3-4, 2009, pp. 158-166.
41. P. Holst, C. Ørskov, A. R. Thomsen and J. P. Christensen, "Quality of the Transgene-Specific CD8+ T Cell Response Induced by Adenoviral Vector Immunization Is Critically Influenced by Virus Dose and Route of Vaccination," Journal of Immunology, Vol. 184, No. 8, 2010, pp. 4431-4439.
42. A. Patel, Y. Zhang, M. Croyle, K. Tran, M. Gray, J. Strong, H. Feldmann, J. M. Wilson and G. P. Kobinger, "Mucosal Delivery of Adenovirus-Based Vaccine Protects against Ebola Virus Infection in Mice," Journal of Infectious Diseases, Vol. 196, Suppl. 2, 2007, pp. 413- 420.
43. T. J. Bull, S. C. Gilbert, S. Sridhar, R. Linedale, N. Dierkes, K. Sidi-Boumedine and J. Hermon-Taylor, "A Novel Multi-Antigen Virally Vectored Vaccine against Mycobacterium avium subspecies paratuberculosis," PLoS ONE, Vol. 2, No. 11, 2007, p. e1229.

44. R. P. Sekaly, "The Failed HIV Merck Vaccine Study: A Step Back or a Launching Point for Future Vaccine Development," Journal of Experimental Medicine, Vol. 205, No. 1, 2008, pp. 7-12.
45. K. McCoy, N. Tatsis, B. Korioth-Schmitz, M. O. Lasaro, S. E. Hensley, S. Lin, Y. Li, W. Giles-Davis, A. Cun, D. Zhou, Z. Xiang, N. L. Letvin and H. C. J. Ertl, "Effect of Preexisting Immunity to Adenovirus Human Serotype 5 Antigens on the Immune Responses of Nonhuman Primates to Vaccine Regimens Based on Humanor Chimpanzee-Derived Adenovirus Vectors," Journal of Virology, Vol. 81, No. 12, 2007, pp. 6594-6604.
46. A. Pandey, N. Singh, S. V. Vemula, L. Couetil, J. M. Katz, R. Donis, S. Sambhara and S. K. Mittal, "Impact of Preexisting Adenovirus Vector Immunity on Immunogenicity and Protection Conferred with an AdenovirusBased H5N1 Influenza Vaccine," PLoS ONE, Vol. 7, No. 3, 2012, p. e33428.
47. J. S. Richardson, M. C. Abou, K. N. Tran, A. Kumar, B. M. Sahai and G. P. Kobinger, "Impact of Systemic or Mucosal Immunity to Adenovirus on Ad-Based Ebola Virus Vaccine Efficacy in Guinea Pigs," Journal of Infectious Diseases, Vol. 204, Suppl. 3, 2011, pp. S1032- S1042.
48. J. F. Engelhardt, L. Litzky and J. M. Wilson, "Prolonged Transgene Expression in Cotton Rat Lung with Recombinant Adenoviruses Defective in E2a," Human Gene Therapy, Vol. 5, No. 10, 1994, pp. 1217-1229.
49. A. E. Morelli, A. T. Larregina, R. W. Ganster, A. F. Zahorchak, J. M. Plowey, T. Takayama, A. J. Logar, P. D. Robbins, L. D. Falo and A. W. Thomson, "Recombinant Adenovirus Induces Maturation of Dendritic Cells via an NF-Kappab-Dependent Pathway," Journal of Virology, Vol. 74, No. 20, 2000, pp. 9617-9628.
50. L. Trotman, N. Mosberger, M. Fornerod, R. Stidwill and U. Greber, "Import of Adenovirus DNA Involves the Nuclear Pore Complex Receptor CAN/Nup214 and Histone H1," Nature Cell Biology, Vol. 3, No. 12 2001, pp. 1092- 1100.
51. S. Uematsu and S. Akira, "Toll-Like Receptors and Type I Interferons," Journal of Biological Chemistry, Vol. 282, No. 21, 2007, pp. 15319-15323.
52. P. Verdino, D. A. Witherden, W. L. Havran and I. A. Wilson, "The Molecular Interaction of CAR and JAML Recruits the Central Cell Signal Transducer PI3K," Science, Vol. 329, No. 5996, 2010, pp. 1210-1214.
53. N. Bessis, F. GarciaCozar, M. C. Boissier, "Immune Responses to Gene Therapy Vectors: Influence on Vector Function and Effector Mechanisms," Gene Therapy, Vol. 11, Suppl. 1, 2004, pp. S10-S17.
54. S. Worgall, G. Wolff, E. Falck-Pedersen and R. Crystal, "Innate Immune Mechanisms Dominate Elimination of Adenoviral Vectors Following in Vivo Administration," Human Gene Therapy, Vol. 8, No. 1, 1997, pp. 37-44.
55. J. M. Hammond, E. S. Jansen, C. J. Morrissy, W. V. Goff, G. C. Meehan, M. M. Williamson, C. Lenghaus, K. W. Sproat, M. E. Andrew, B. E. H. Coupar and M. A. Johnson, "A Prime-Boost Vaccination Strategy Using Naked DNA Followed by Recombinant Porcine Adenovirus Protects Pigs from Classical Swine Fever," Veterinary Microbiology, Vol. 80, No. 2, 2001, pp. 101-119.
56. S. J. Kim, H. K. Kim, Y. W. Han, A. G. Aleyas, J. A. George, H. A. Yoon, D. J. Yoo, K. Kim and S. K. Eo, "Multiple Alternating Immunizations with DNA Vaccine and Replication Incompetent Adenovirus Expressing gB of Pseudorabies Virus Protect Animals against Lethal Virus Challenge," Journal of Microbiology and Biotechnology, Vol. 18, No. 7, 2008, pp. 1326-1334.
57. S. Gogev, K. de Fays, M. F. Versali, S. Gautier and E. Thiry, "Glycol Chitosan Improves the Efficacy of Intranasally Administrated Replication Defective Human Adenovirus Type 5 Expressing Glycoprotein D of Bovine Herpesvirus," Vaccine, Vol. 22, No. 15-16, 2004, pp. 1946-1953.
58. S. D. Perkins, A. J. Williams, L. M. O'Brien, T. R. Laws and R. J. Phillpotts, "CpG Used as an Adjuvant for an Adenovirus-Based Venezuelan Equine Encephalitis Virus Vaccine Increases the Immune Response to the Vector, but not to the Transgene Product," Viral Immunology, Vol. 21, No. 4, 2008, pp. 451-457.

59. L. Bracci, I. Canini, M. Venditti, M. Spada, S. Puzelli, I. Donatelli, F. Belardelli and E. Proietti, "Type I IFN as a Vaccine Adjuvant for Both Systemic and Mucosal Vaccination against Influenza Virus," Vaccine, Vol. 24, Suppl. 2, 2006, pp. 56-57.
60. H. M. Vordermeier, M. A. Chambers, B. M. Buddle, J. M. Pollock and R. G. Hewinson, "Progress in the Development of Vaccines and Diagnostic Reagents to Control Tuberculosis in Cattle," The Veterinary Journal, Vol. 171, No. 2, 2006, pp. 229-244.
61. H. M. Vordermeier, B. Villarreal-Ramos, P. J. Cockle, M. McAulay, S. G. Rhodes, T. Thacker, S. C. Gilbert, H. McShane, A. V. Hill, Z. Xing and R. G. Hewinson, "Viral Booster Vaccines Improve Mycobacterium Bovis BCGInduced Protection against Bovine Tuberculosis," Infection and Immunity, Vol. 77, No. 8, 2009, pp. 3364-3373.
62. Z. Xing, C. T. McFarland, J. M. Sallenave, A. Izzo, J. Wang and D. N. McMurray, "Intranasal Mucosal Boosting with an Adenovirus-Vectored Vaccine Markedly Enhances the Protection of BCG-Primed Guinea Pigs against Pulmonary Tuberculosis," PLoS ONE, Vol. 4, No. 6, 2009, p. e5856.
63. [Y. Sun, N. Li, H. Y. Li, M. Li and H. J. Qiu, "Enhanced Immunity against Classical Swine Fever in Pigs Induced by Prime-Boost Immunization Using an Alphavirus Replicon-Vectored DNA Vaccine and a Recombinant Adenovirus," Veterinary Immunology and Immunopathology, Vol. 137, No. 1-2, 2010, pp. 20-27.
64. J. Schneider, S. C. Gilbert, C. M. Hannan, P. Dégano, E. Prieur, E. G. Sheu, M. Plebanski and A. V. Hill, "Induction of $CD8^+$ T Cells Using Heterologous Prime-Boost Immunisation Strategies," Immunological Reviews, Vol. 170, No. 1, 1999, pp. 29-38.
65. A. de Mare, A. J. Lambeck, J. Regts, G. M.van Dam, H. W. Nijman, H. Snippe, J. Wilschut and T. Daemen, "Viral Vector-Based Prime-Boost Immunization Regimens: A Possible Involvement of T-Cell Competition," Gene Therapy, Vol. 15, No. 6, 2008, pp. 393-403.
66. B. Asefa, N. Korokhov and F. Lemiale, "Heterologous HIV-Based Lentiviral/ Adenoviral Vectors Immunizations Result in Enhanced HIV-Specific Immunity," Vaccine, Vol. 28, No. 20, 2010, pp. 3617-3624.
67. E. K. Roediger, K. Kugathasan, X. Zhang, B. D. Lichty and Z. Xing, "Heterologous Boosting of Recombinant Adenoviral Prime Immunization with a Novel Vesicular Stomatitis Virus-Vectored Tuberculosis Vaccine," Molecular Therapy, Vol. 16, No. 6, 2008, pp. 1161-1169.
68. F. H. J. Top, B. A. Dudding, P. K. Russel and E. L. Buescher, "Control of Respiratory Disease in Recruits with Types 4 and 7 Adenovirus Vaccines," American Journal of Epidemiology, Vol. 94, No. 2, 1971, pp. 142- 146.
69. B. Zeshan, L. Zhang, J. Bai, X. Wang, J. Xu and P. Jiang, "Immunogenicity and Protective Efficacy of a Replication-Defective Infectious Bronchitis Virus Vaccine Using an Adenovirus Vector and Administered in ovo," Journal of Virology Methods, Vol. 166, No. 1-2, 2010, pp. 54-59.
70. B. Barefoot, N. J. Thornburg, D. H. Barouch, J. S. Yu, C. Sample, R. E. Johnston, H. X. Liao, T. B. Kepler, B. F. Haynes and E. Ramsburg, "Comparison of Multiple Vaccine Vectors in a Single Heterologous Prime-Boost Trial," Vaccine, Vol. 26, No. 48, 2008, pp. 6108-6118.
71. R. C. Carlisle, S. S. Briggs, A. B. Hale, N. K. Green, K. D. Fisher, T. Etrych, K. Ulbrich, V. Mautner and L. W. Seymour, "Use of Synthetic Vectors for Neutralizing Antibody Resistant Delivery of Replicating Adenovirus DNA," Gene Therapy, Vol. 13, No. 22, 2006, pp. 1579- 1586.
72. J. Millar, D. Dissanayake, T. C. Yang, N. Grinshtein, C. Evelegh, Y. Wan and J. Bramson, "The Magnitude of the $CD8^+$ T Cell Response Produced by Recombinant Virus Vectors Is a Function of Both the Antigen and the Vector," Cellular Immunology, Vol. 250, No. 1-2, 2007, pp. 55-67.
73. L. Fischer, J. P. Tronel, C. Pardo-David, P. Tanner, G. Colombet, J. Minke and J. C. Audonnet, "Vaccination of Puppies Born to Immune Dams with a Canine Adenovirus-Based Vaccine Protects against a Canine Distemper Virus Challenge," Vaccine, Vol. 20, No. 29-30, 2002, pp. 3485-3497.

74. Z. Q. Xiang, G. P. Gao, A. Reyes-Sandoval, Y. Li, J. M. Wilson and H. C. Ertl, "Oral Vaccination of Mice with Adenoviral Vectors Is not Impaired by Preexisting Immunity to the Vaccine Carrier," Journal of Virology, Vol. 77, No. 20, 2003, pp. 10780-10789.
75. R. D. Wesley and K. M. Lager, "Overcoming Maternal Antibody Interference by Vaccination with Human Adenovirus 5 Recombinant Viruses Expressing the Hemagglutinin and the Nucleoprotein of Swine Influenza Virus," Veterinary Microbiology, Vol. 118, No. 1-2, 2006, pp. 67-75.
76. W. Chen, M. Liu, Y. Jiao, W. Yan, X. Wei, J. Chen, L. Fei, Y. Liu, X. Zuo, F. Yang, Y. Lu and Z. Zheng, "Adenovirus-Mediated RNA Interference against Foot-andMouth Disease Virus Infection Both in Vitro and in Vivo," Journal of Virology, Vol. 80, No. 7, 2006, pp. 3559-3566.
77. Y. Li, K. Reddy, S. M. Reid, W. J. Cox, I. H. Brown, P. Britton, V. Nair and M. Iqbal, "Recombinant Herpesvirus of Turkeys as a Vector-Based Vaccine against Highly Pathogenic H7N1 Avian Influenza and Marek's Disease," Vaccine, Vol. 29, No. 46, 2011, pp. 8257-8266.
78. R. Hu, S. Zhang, A. R. Fooks, H. Yuan, Y. Liu, H. Li, C. Tu, X. Xia and Y. Xiao, "Prevention of Rabies Virus Infection in Dogs by a Recombinant Canine Adenovirus Type-2 Encoding the Rabies Virus Glycoprotein," Microbes and Infection, Vol. 8, No. 4, 2006, pp. 1090-1097.
79. P. Reddy, N. Idamakanti, C. Pyne, A. Zakhartchouk, D. L. Godson, Z. Papp, M. E. Baca-Estrada, L. A. Babiuk, G. K. Mutwiri and S. K. Tikoo, "The Immunogenicity and Efficacy of Replication-Defective and Replication-Competent Bovine Adenovirus-3 Expressing Bovine Herpesvirus-1 Glycoprotein gD in Cattle," Veterinary Immunology and Immunopathology, Vol. 76, No. 3-4, 2000, pp. 257- 268.
80. T. Tuboly and E. Nagy, "Construction and Characterization of Recombinant Porcine Adenovirus Serotype 5 Expressing the Transmissible Gastroenteritis Virus Spike Gene," Journal of General Virology, Vol. 82, No. 1, 2001, pp. 183-190.
81. J. S. Rothel, D. B. Boyle, G. W. Both, A. D. Pye, J. G. Waterkeyn, P. R. Wood and M. W. Lightowlers, "Sequential Nucleic Acid and Recombinant Adenovirus Vaccination Induces Host-Protective Immune Responses against Taenia Ovis Infection in Sheep," Parasite Immunology,Vol. 19, No. 5, 1997, pp. 221-227.
82. A. Francois, C. Chevalier, B. Delmas, N. Eterradossi, D. Toquin, G. Rivallan and P. Langlois, "Avian Adenovirus CELO Recombinants Expressing VP2 of Infectious Bursal Disease Virus Induce Protection against Bursal Disease in Chickens," Vaccine, Vol. 22, No. 17-18, 2004, pp. 2351-2360.
83. J. Domenech, J. Lubroth and K. Sumption, "Immune Protection in Animals: The Examples of Rinderpest and Foot-and-Mouth Disease," Journal of Comparative Pathology, Vol. 142, Suppl. 1, 2010, pp. S120-S124.
84. J. G. Heldens, J. R. Patel, N. Chanter, G. J. Ten Thij, M. Gravendijck, V. E. Schijns, A. Langen and T. P. Schetters, "Veterinary Vaccine Development from an Industrial Perspective," The Veterinary Journal, Vol. 178, No. 1, 2008, pp. 7-20.
85. P. H. Flore, "Commercialization of Veterinary Viral Vaccines," Animal Health Research Reviews, Vol. 5, No. 2, 2004, pp. 239-242.
86. A. K. Shen, "The U.S. Vaccine and Immunization Enterprise: Working to Sustain and Foster Vaccine Innovation," Human Vaccines, Vol. 5, No. 10, 2009, pp. 649- 653.
87. D. A. Brake, M. McIlhaney, T. Miller, K. Christianson, A. Keene, G. Lohnas, C. Purcell, J. Neilan, C. Schutta, J. Barrera, T. Burrage, D. E. Brough and B. T. Butman, "Human Adenovirus-Vectored Foot-and-Mouth Disease Vaccines: Establishment of a Vaccine Product Profile Through in Vitro Testing," Developments in Biologicals (Basel), Vol. 134, 2012, pp. 123-133.
88. M. J. Grubman, M. P. Moraes, C. Schutta, J. Barrera, J. Neilan, D. Ettyreddy, B. T. Butman, D. E. Brough and D. A. Brake, "Adenovirus Serotype 5-Vectored Foot-andMouth Disease Subunit Vaccines: The First Decade," Future Virology, Vol. 5, No. 1, 2010, pp. 51-64.

89. M. Bublot, N. Pritchard, D. E. Swayne, P. Selleck, K. Karaca, D. L. Suarez, J. C. Audonnet and T. R. Mickle, "Development and Use of Fowlpox Vectored Vaccines for Avian Influenza," Annals of the New York Academy of Sciences, Vol. 1081, No. 1, 2006, pp. 193-201.
90. J. Coon, M. Johnston, J. Minke, R. Pruett, S. Thompson and D. Carithers, "Analysis of Recombinant Canarypox Vectored West Nile Virus Vaccine Stability Post-Reconstitution," International Journal of Applied Research in Veterinary Medicine, Vol. 4, No. 2, 2006, pp. 115-120.
91. H. Poulet, S. Brunet, C. Boularand, A. L. Guiot, V. Leroy, J. Tartaglia, J. Minke, J. C. Audonnet and P. Desmettre, "Efficacy of a Canarypox Virus-Vectored Vaccine Against Feline Leukaemia," Veterinary Record, Vol. 153, No. 5, 2003, pp. 141-145.
92. United States Department of Agriculture, "Veterinary Biological Products-Licensees and Permittees," 2011.
93. J. Esparza, "Progress in the Development of an Adenovirus 26 Vector Platform for HIV Vaccines," Expert Reviews in Vaccines, Vol. 12, No. 5, 2013, pp. 477-480.
94. Z. C. Hartman, D. M. Appledorn and A. Amalfitano, "Adenovirus Vector Induced Innate Immune Responses: Impact upon Efficacy and Toxicity in Gene Therapy and Vaccine applications," Virus Research, Vol. 132, No. 1-2, 2008, pp. 1-14.
95. L. S. Wyatt, P. L. Earl, W. Xiao, J. L. Americo, C. A. Cotter, J. Vogt and B. Moss, "Elucidating and Minimizing the Loss by Recombinant Vaccinia Virus of Human Immunodeficiency Virus Gene Expression Resulting from Spontaneous Mutations and Positive Selection," Journal of Virology, Vol. 83, No. 14, 2009, pp. 7176-7184.

CHAPTER 20

Evidence-Based Use of Antibiotics in Veal Calves with Diarrhea

Michael Hässig, Susanne Kretschmar

Section for Ambulatory Field Clinic and Herd Health, Department of Farm Animals, University of Zurich, Zurich, Switzerland

ABSTRACT

Diarrhea is the leading cause of mortality in beef and dairy calves during the first week of life and results in substantial financial loss [1] . Diarrhea is a multifactorial disease and can be infectious or non-infectious. However, in the majority of calves, infectious organisms, especially Cryptosporidium parvum, rotavirus, coronavirus, and E. coli, are the primary cause [2] . The aim of this study was to generate a decision tree, based on prevalence, diagnostic testing and treatment and to estimate associated costs or risk. For each of the four main pathogens, two principal approaches are outlined and compared. The first approach relies on a detailed diagnostic workup and allows for specific etiological treatment. The second approach relies on the trial-and-error method, which involves the use of a first-choice antibiotic, followed by a second- and third-choice antibiotic if the previous ones failed to resolve the disease. In Switzerland, the prevalence of diarrheic calves infected with E. coli is approximately 1% suggesting that the use of antimicrobials for the treatment of scouring calves, in the absence of a diagnostic workup, is not always justified. However, for all four major pathogens, the trial-and-error method affords cheaper treatment compared with treat- ment based on an etiological diagnosis. This creates a quandary in view of the current worldwide efforts to reduce the use of antibiotics in animal agriculture.

Keywords: Bovine, Calf, Antibiotic, Antimicrobial Susceptibility Testing, Decision Tree Analysis, Diarrhea

1. INTRODUCTION

Diarrhea is the leading cause of death in beef and dairy calves in the first week of life [1] and has a major economic impact on animal production. Calf losses, reduced weight gains, increased workload, and the cost of treatment and preventive measures result in substantial financial expenditures [3] .

Diarrhea is a multifactorial disease and may be caused by viruses, bacteria or parasites. Infectious diarrhea is often a sequel to failure of passive transfer because of inadequate colostrum quality or quantity. Crowding, immune status, and environmental and management factors also play important roles in the pathogenesis of diarrhea [5] . The four most prevalent infectious microorganisms found in calves with diarrhea are Cryptosporidium parvum, rotavirus, coronavirus, and enterotoxic E. coli. These agents are responsible for 75% to 95% of all cases of diarrhea worldwide in newborn calves [2] . Mixed infections and consecutive infections by different infectious agents are common [2] [4] .

The etiological diagnosis in a calf with diarrhea usually requires laboratory testing and typically cannot be made based on the clinical signs and the type of diarrhea alone. The establishment of an etiological diagnosis is crucial, especially in herd problems, to initiate specific treatment and metaphylactic and prophylactic measures [5] . Selection of an effective treatment for scouring calves can be difficult without an etiological diagnosis.

The use of antibiotics for the treatment of diarrhea in calves is controversial. A recent review article on antimicrobial decision making [1] referred to studies that favored the use of antibiotics for the treatment of calves with diarrhea [6] as well as studies that discouraged the use of antibiotics because of contraindication or lack of efficacy [7] . Several studies have shown that antibiotics can reduce the mortality rate and shorten the duration of diarrhea [8] . The two main indications for the use of antibiotics for the treatment of scouring calves are the prevention of secondary bacteremia and the reduction in the number of coliform bacteria in the small intestine [8] . This review investigated the overall cost of veterinary services, diagnostic tests, treatment, reduced weight gain, and mortality in calves with diarrhea. The goal was to create a decision tree to show how calf diarrhea can be addressed diagnostically and therapeutically. Special emphasis was given to comparison of the trial-and-error method and treatment based on an etiological diagnosis after the identification of the infectious agent and evaluation of potential resistance to therapeutic drugs. The different risks that were compared using this decision tree were defined as cost multiplied by probability. The prevalence of the four major diarrheal pathogens in calves in Switzerland was multiplied by veterinary costs. A further goal was to examine whether the cost of antimicrobial susceptibility testing was justified; the trial-and-error method minimized the cost of diagnostic testing but also minimized the likelihood of successful treatments, whereas antimicrobial susceptibility testing increased the cost of diagnostic testing but minimized the number of futile antibiotic treatments.

2. ANIMALS, MATERIALS AND METHODS

The operational approach needed no clearance by an animal protection ethics committee, since all animals were treated when clinically ill. All raw data were extracted from a Filemaker®-database (FileMaker, Inc. 5201 Patrick Henry Drive, Santa Clara, CA) where all herds transferred to the Section for Ambulatory Field Clinic and Herd Health, Department of Farm Animals, University of Zurich were registered from 1990 until 2013. The data were transferred to Stata® for further statistical evaluation such as means and counts for specific selections (Stata statistical software: release 12.0. Statcorp LP, College Station, TX, USA). The decision tree was analyzed according to Altman [9] . The decision tree was divided into four parts with one overview for convenience (Figures 1-5). Four trees have been drawn that show only one branch in detail. The division was made according to the four main diarrhea pathogens. Each field was marked with a combination of a letter and a number. The boxes with the rounded corners represent a free choice. In addition, they contain costs (K). The boxes with the pointed corners show the possible consequences of a given decision. They include the probability (P) linked to a particular decision and the costs thereby incurred (K). The information relating to probability and cost is given in Tables 1-6. The risks expressed as probability multiplied by cost (RK) are listed. The boxes in Figures 1-5, labeled "B2 non-infectious cause" and those containing information about other infectious agents are shown for the sake of comprehensiveness and are not dealt with further because they do not relate to the comparison of targeted diagnostic procedures and and the trial-and-error method.

The sources for the decision tree were visited between January 2014 and March 2015 in the following three databases: (http://www.ncbi.nlm.nih.gov/), ScienceDirect (http://www.sciencedirect.com/) and cabdirect (http:// www.cabdirect.org/). The following search terms were used in different combinations: calf/ calves diarrhea/scours, cryptosporidiosis, rotavirus, coronavirus, E. coli, treatment, fluids, antimicrobials, prevalence Switzerland. The weighted average of the listed prevalence was calculated for the calculation of the prevalences in the decision tree. The drugs were selected according to the treatment guidelines used at our clinic (Cryptosporidium parvum: Halofuginone (Halocur®, MSD Animal Health GmbH, 2 ml/10 kg BW on 7 days, 100 µg/kg BW on 7 days), rotavirus and coronaviruses: symptomatic therapy with rehydration, E. coli: antibiotics). The order of antibiotics was chosen according to the treatment guidelines used at our clinic [4] : 1st choice: Ampicillin (Albipen® LA, MSD Animal Health GmbH, Switzerland, Ampicillinum anhydricum 100 mg, E312 0.0875 mg; Aluminii monostearas, Oleum cocos ad susp. pro 1 ml), 2nd choice: trimethoprim-sulfonamide (Borgal® 24%, MSD Animal Health GmbH, Sulfadoxinum 200 mg, Trimethoprimum 40 mg, Natrii hydroxidum, Glyceroli formalum, Aqua ad

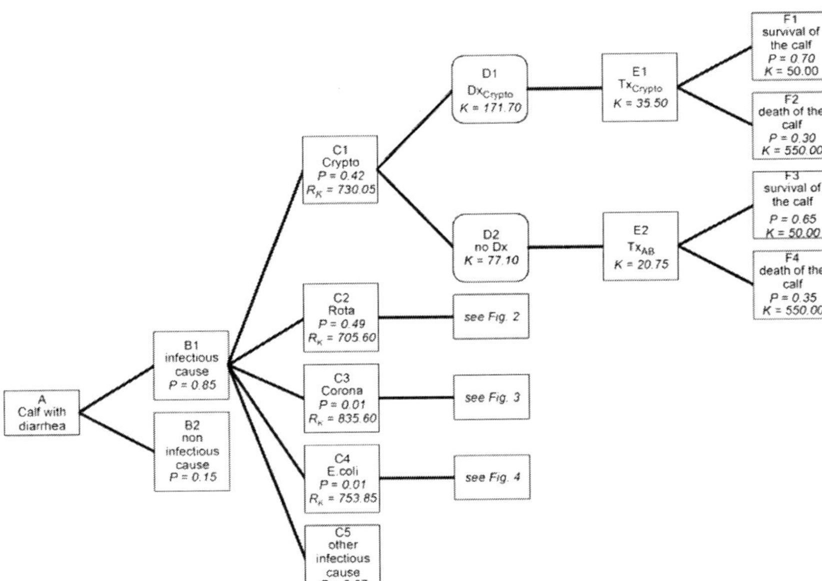

Figure 1. Decision tree for treatment outcome in diarrheic calves infected with Cryptosporidium parvum. Crypto: Cryptosporidium parvum; Rota: rotavirus; Corona: coronavirus; Dx: diagnosis; Tx: therapy; AB: antibiotics; P: probability; K: costs; R_K: risks as probability multiplied by cost.

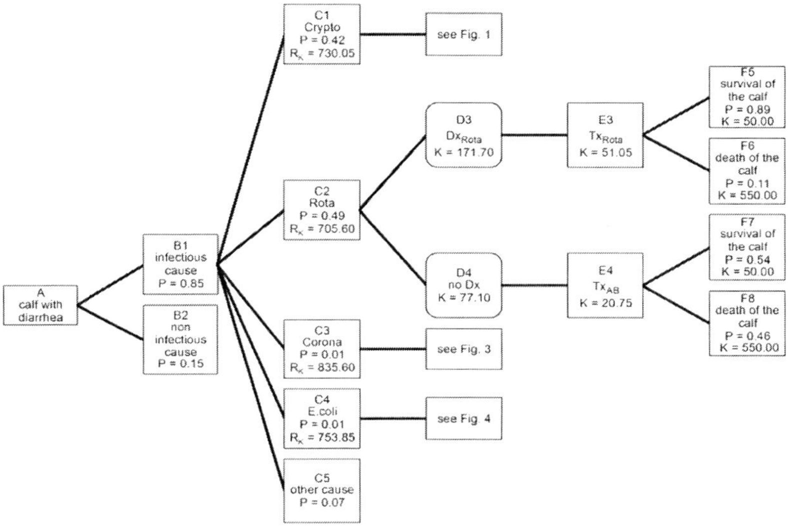

Figure 2. Decision tree for treatment outcome in diarrheic calves infected with rotavirus. Crypto: Cryptosporidium parvum; Rota: rotavirus; Corona: coronavirus; Dx: diagnosis; Tx: therapy; AB: antibiotics; P: probability; K: costs; R_K: risks as probability multiplied by cost.

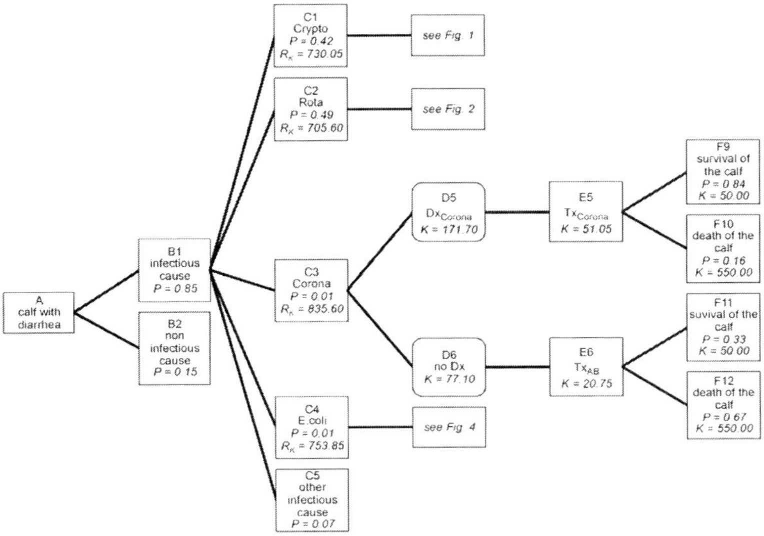

Figure 3. Decision tree for treatment outcome in diarrheic calves infected with coronavirus. Crypto: Cryptosporidium parvum; Rota: rotavirus; Corona: coronavirus; Dx: diagnosis; Tx: therapy; AB: antibiotics; P: probability; K: costs; R_K: risks as probability multiplied by cost.

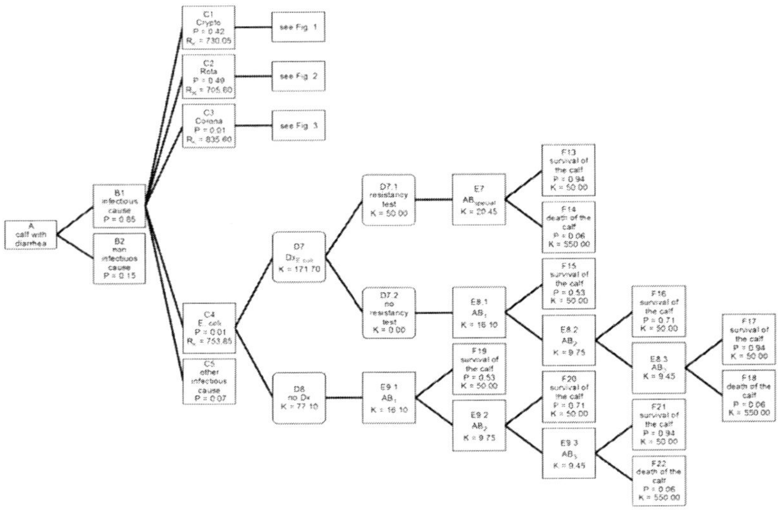

Figure 4. Decision tree for treatment outcome in diarrheic calves infected with E. coli. Crypto: Cryptosporidium parvum; Rota: rotavirus; Corona: coronavirus; Dx: diagnosis; Tx: therapy; AB: antibiotics; ABN: N = choice, i.e. first, second or third choice according to materials and methods; AB special: Choice of AB according to bacteriological examination and resistance test. P: probability; K: costs; RK: risks as probability multiplied by cost.

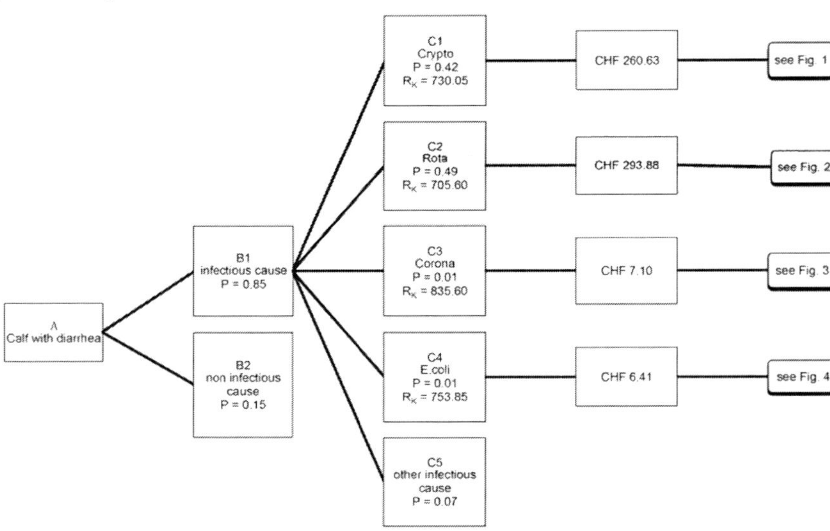

Figure 5. Decision tree for outcome of diarrhea in calves. Crypto: cryptosporidiosis; Rota: rotavirus; Corona: coronavirus. P: probability; RK: risks as probability multiplied by the cost.

iniectabilia pro 1 ml), 3rd choice: enrofloxacin (Baytril® 10%, Provet AG, Switzerland, Enrofloxacinum 100 mg, Kalii hydroxidum, Alcohol butylicus, Aqua ad iniectabilia qs ad 1 ml). Treatment success was defined as resolution of diarrhea in a sick calf or survival of a treated calf. Treatment failure was defined as continuation of diarrhea despite treatment, death of the calf, or the need to euthanize the calf on humane grounds. Information regarding the spontaneous cure rate in calves with diarrhea was gained from studies that used placebo-treated control groups of experimentally or naturally infected calves (references are given in Tables 3-5). Spontaneous cure was defined as resolution of diarrhea in the absence of treatment or survival of a diarrheic calf with resolution of diarrhea. Prevalence of different infectious agents is given in Table 1. Treatment success and spontaneous cure rates are shown in Table 2 for diarrheic calves with cryptosporidiosis, in Table 3 for calves infected with rotavirus, in Table 4 for calves infected with coronavirus, and in Table 5 for calves with E. coli diarrhea. The dosages and prices used for calculation in Tables 6-8 are based on the fee schedule of our clinic as of June 2015. For calculations, the amount of 0.00 CHF (Swiss Francs) was used for a surviving calf, 500.00 CHF was used for a dead calf, and 5.00 CHF was used for the cost of treatment, diagnostic testing, and to account for the loss in weight gain per day [23] .

3. RESULTS

The prevalence of important pathogens causing diarrhea in neonatal calves in Switzerland is shown in Table 1 and the prevalence of treatment success for different pathogens is shown in Tables 2-5. Prevalence is used to express probabilities. The cost of treatment is given in Tables 6-8 and the steps in the decision tree explain how the cost was calculated. Individual risks are additive and generate the total risk contained in a main branch of the decision tree [23] . Table 9 shows the formulae for the risk calculations (RK) for the fields C1 to C4 in Figures 1-5. For clarity, the branches for cryptosporidiosis (Figure 1), rotavirus (Figure 2), coronavirus (Figure 3) and E. coli (Figure 4) are shown separately. Figure 5 provides a synopsis of Figures 1-4 and details the costs associated with the different branches. An infectious pathogen was isolated in 85% of scouring calves in a Swiss study [3] , [5] ; rotavirus and Cryptosporidium parvum each accounted for 40% to 50% of all cases and coronavirus and E. coli each accounted for approximately 1% of cases. As can be seen in the decision trees (Figures 1-4), the costs are higher for all four infectious agents when diagnostic testing was done compared with the trial-and-error method. The risk, defined by probability of occurrence multiplied by cost, is equivalent to CHF 700.00 to 840.00 for all four infectious agents; it is highest for coronavirus (CHF

835.60) and lowest for rotavirus (CHF 705.60, Figure 5). The branches B2 (non-infectious causes) and C5 (other infectious causes) are also shown in Figures 1-5 but detailed calculations were not done.

Table 1. Prevalence of major pathogens causing diarrhea in calves in Switzerland.

Definition	Etiology	Prevalence	Source	N
Prevalence	General infectious influenza etiology	91.2%	[5]	147
		>78.0%	[3]	46
Herd prevalence of major pathogens	*Cryptosporidium parvum*	41.7%	[5]	147
	Rotavirus	52.1%		
	Coronaviruses	2.1%		
	E. coli K99	2.1%		
Prevalence of major pathogens in diarrheic calves	*Cryptosporidium parvum*	43.0%	[3]	46
	Rotavirus	46.0%		
	Coronaviruses	0.0%		
	E. coli K99	0.0%		

Table 2. Therapeutic success and spontaneous cure rate in diarrheic calves infected with Cryptosporidium parvum.

Tx: treatment.

	Prevalence	Definition	Source	n
Tx success	65.0%	Treatment with halofuginone, no diarrhea on day 7 (dosage: 100 µg/kg/day)	[10]	30
	46.0%	Treatment with halofuginone, no diarrhea and no shedding of oocysts on day 7 (dosage: 100 µg/kg/day)	[11]	50
	100.0%	Treatment with halofuginone, no diarrhea (60 µg/kg/day and 120 µg/kg/day)	[12]	10
Spontaneous cure rate	75.0%	Placebo-treated calves, no diarrhea on day 20	[13]	24
	65.0%	Untreated calves, no diarrhea on day 10	[10]	30
	55.0%	Placebo-treated calves, no diarrhea on day 14	[11]	47

Tx: treatment.

Table 3. Therapeutic success and spontaneous cure rate in diarrheic calves infected with rotavirus.

	Prevalence	Definition	Source	N
Tx success	100.0%	Clinically healthy after treatment with OES	[14]	11
	79.0%	Clinically healthy after intravenous rehydration	[15]	19
	83.0% 92.0%	Survival rate after treatment with OES/bicarbonate Survival rate after treatment with OES/acetate	[16]	12
Spontaneous cure rate	75.0%	Survival rate 30 days after experimental infection with rotavirus	[17]	12
	<33.0%	Survival rate after treatment with OES without alkalizing component	[16]	12

Tx: treatment; OES: oral electrolyte solution

Table 4. Therapeutic success and spontaneous cure rate in diarrheic calves infected with coronaviruses.

	Prevalence	Definition	Source	N
Tx success	80.0%	Clinically healthy after treatment with OES	[14]	15
	79.0%	Clinically healthy after intravenous rehydration	[15]	19
	83.0% 92.0%	Survival rate after treatment with OES/bicarbonate Survival rate after treatment with OES/acetate	[16]	12
Spontaneous cure rate	<33.0%	Survival rate after treatment with OES without alkalizing component	[16]	12

Tx: treatment; OES: oral electrolyte solution.

Table 5. Therapeutic success and spontaneous cure rate in diarrheic calves infected with E. coli. The therapeutic success is based on results of susceptibility testing of isolated fecal E. coli strains to various antibiotics.

	Prevalence	Definition	Source	N
Tx success	25.0% 73.3% 96.5%	Resistance against. Ampicillin (1st choice) Trimethoprim-Sulphonamide (2nd choice) Enrofloxacin (3rd choice)	[18]	NA
	72.7%	Resistance against. Ampicillin (1st choice)	[19]	176
	72.6% 86.3%	Resistance against. Ampicillin (1st choice) Enrofloxacin (3rd choice)	[20]	95
	40.6% 68.8% 100.0%	Resistance against. Ampicillin (1st choice) Trimethoprim-Sulphonamide (2nd choice) Enrofloxacin (3rd choice)	[21]	251
Spontaneous cure rate	22.2%	Mortality in calves infected with *E. coli* B44	[6]	9
	0.0%	Mortality in calves infected with *E. coli* B44	[22]	4
	18.0%	Mortality in calves infected with *E. coli* B44	[8]	11

Tx: Treatment, NA: not available.

4. DISCUSSION

Calf diarrhea is a complex disease caused by a variety of infectious microorganisms and non-infectious factors. Infectious agents include viruses, parasites and bacteria [2] . The decision trees shown in this study are limited to a few variations of cost and risk analysis in calves with diarrhea. By virtue of the public health axiom 'intention to treat' [24] , only those branches of the decision tree are shown that involve treatment. The premise was that calves with diarrhea presented for veterinary examination were either treated or euthanized and that non-treat- ment was not an option for humane reasons.

Some pathogens including Cryptosporidium parvum, rotavirus, and coronavirus have intermittent shedding patterns [5] and therefore, negative results must be interpreted accordingly regardless of the sensitivity and specificity of the tests used. Furthermore, diarrhea pathogens also are commonly isolated from clinically healthy calves [5] .

Two studies of the prevalence of diarrhea pathogens in calves in Switzerland found an infectious cause in 85% of affected calves; 93% of the infectious agents belonged to the four main pathogens Cryptosporidium parvum, rotavirus, coronavirus, and E. coli [3] [5] . This is consistent with the observation that 75% - 95% of diarrheic calves worldwide are affected by these pathogens [2] . Other potential diarrhea pathogens mentioned in the literature include vero-cytotoxic and necrotoxic E. coli, bovine torovirus, calicivirus, norovirus, Giardia spp., Salmonella spp., Clostridium perfringens type B and C, Eimeria spp., and Campylobacter jejuni [2] but these were not considered for calculation in this study.

Table 6. Treatment costs for diarrheic calves infected Cryptosporidia based on the fee schedule at our clinic as of June 2015 and [23] .

Decision§	Diagnostics	Treatment	Costs (CHF) 50 kg BW calf
D1	Dx cryptosporidiosis. Veterinary visit and laboratory fecal examination (without susceptibility testing)	Visit	30.00
		professional fee 15 min. clinical examination	28.50
			18.60
		fecal sampling	4.60
		PU. VU. BU	90.00
		Total	**171.10**
D2	No Dx. Veterinary visit. no laboratory examination	Visit	30.00
		professional fee 15 min. clinical examination	28.50
			18.60
		Total	**77.10**
E1	Tx cryptosporidiosis	Halofuginone	24.50
		Vitamin injection	11.00
		Total	**35.50**
E2	Tx antibiotics	Trimethoprim sulfadoxine	3.75
		Injection i/v or i/m	6. 00
		Vitamin injection	11.00
		Total	**20.75**
F1. F3	Calf survived	Daily loss over 10 days	50.00
		Total	**50.00**
F2. F4	Loss of calf	Daily loss over 10 days	50.00
		Loss of calf	500.00
		Total	**550.00**

Dx: diagnosis; Tx: treatment; PU: parasitological examination; VU: viral examination; BU: bacteriological examination; CHF: Swiss francs; BW: body weight; i/v: intravenous; i/m: intramuscular; s/c: subcutaneous. §: decision according to decision tree in Figures 1-4.

Table 7. Treatment costs for diarrheic calves infected with rotavirus and coronavirus based on the fee schedule at our clinic as of June 2015 and [23]
.

Decision[§]	Diagnostics	Treatment	Costs (CHF) 50 kg BW calf
D3, D5	Dx rotavirus, Dx coronavirus. Veterinary visit and laboratory fecal examination (without susceptibility testing)	Visit	30.00
		professional fee 15min	28.50
		clinical examination	18.60
		fecal sampling	4.60
		PU, VU, BU	90.00
		Total	**171.10**
D4, D6	No Dx. Veterinary visit, no laboratory examination	Visit	30.00
		professional fee 15min	28.50
		clinical examination	18.60
		Total	**77.10**
E3, E5	Tx rotavirus, Tx coronavirus	Placing catheter	20.70
		i/v rehydration	15.10
		Rehydration p/o for 2 days	4.25
		Vitamin injection	11.00
		Total	**51.05**
E4, E6	Tx antibiotics	Trimethoprim sulfadoxine	3.75
		Injection i/v or i/m	6.00
		Vitamin injection	11.00
		Total	**20.75**
F5, F7 F9, F11	Calf survived	Daily loss over 10 days	50.00
		Total	**50.00**
F6, F8 F10, F12	Loss of calf	Daily loss over 10 days	50.00
		Loss of calf	500.00
		Total	**550.00**

The prevalences presented in this study confirm that viral and parasitic pathogens are much more commonly involved in calf diarrhea than bacterial pathogens suggesting that in the majority of cases the use of antibiotics is not justified. A Swedish study reported that 30% of calves with diarrhea in the first three months of life were treated with antibiotics [7] , and a similar substantial over-treatment of diarrheic calves with antimicrobial drugs was reported in another study [25] . The use of antibiotics for the treatment of calves with diarrhea is controversial [1] and several authors have been critical of this practice [7] [25] [26] . Calves treated prophylactically for calf diarrhea during the first two weeks of life with neomycin or tetracycline in the milk had lower weight gain, lower feed intake, and more days with diarrhea than calves not receiving antibiotics in the milk. However, the use of antimicrobials other than in-milk antimicrobials used in calves with diarrhea, fever, anorexia and depression was beneficial [27] . The use of antibiotics is recommended in diarrheic calves with systemic signs of disease including anorexia, dehydration, lethargy, fever, or when the diarrheic stool contains blood or sloughed mucous membranes. This is to prevent bacteremia following bacterial overgrowth in the intestines and to reduce the duration of the disease and the mortality rate [28] .

Table 8. Treatment costs for diarrheic calves infected with E. coli based on the fee schedule at our clinic as of June 2015 and [23] .

Decision[§]	Diagnostics	Treatment	Costs (CHF) 50 kg BW calf
D7	Dx *E. coli*. Veterinary visit and laboratory fecal examination (without susceptibility testing)	Visit	30.00
		professional fee 15 min	28.50
		clinical examination	18.60
		fecal sampling	4.60
		PU, VU, BU	90.00
		Total	**171.10**
D7.1	Susceptibility testing	Susceptibility testing	50.00
		Total	**50.00**
D7.2	No susceptibility testing	Susceptibility testing	0.00
		Total	**0.00**
D8	No Dx. Veterinary visit, no laboratory examination	Visit	30.00
		professional fee 15 min	28.50
		clinical	18.60
		Total	**77.10**
E7	effective AB	Enrofloxacin	3.45
		injection i/v or i/m	6.00
		Vitamin injection	11.00
		Total	**20.45**
E8.1, E9.1	AB_1	Ampicillin	5.10
		Vitamin injection	11.00
		Total	**16.10**
E8.2, E9.2	AB_2	Trimethoprim sulfadoxine	3.75
		Injection i/v or i/m	6.00
		Total	**9.75**
E8.3, E9.3	AB_3	Enrofloxacin	3.45
		injection i/v or i/m	6.00
		Total	**9.45**
F13, F15, F16, F17, F19, F20, F21	Calf survived	Daily loss over 10 days	50.00
		Total	**50.00**
F14, F18, F22	Loss of calf	Daily loss over 10 days	50.00
		loss of calf	500.00
		Total	**550.00**

AB: antibiotic; Dx: diagnosis; Tx: treatment; PU: parasitological examination; VU: viral examination; BU: bacteriological examination; CHF: Swiss francs; BW: body weight; i/v: intravenous; i/m: intramuscular; s/c: subcutaneous; §: decision according to decision tree in Figures 1-4.

Table 9. Calculation of the risk (RK) for each of the 4 main infectious agents according to Figures 1-4. For probabilities (P) see Figures 1-4, for costs CN to FNN see Tables 6-8.

Decision		Calculation
C1	$R_K =$	(P * F1) + (P * F2) + E1 + D1 + (P * F3) + (P * F4) + E2 + D2
C2	$R_K =$	(P * F5) + (P * F6) + E3 + D3 + (P * F7) + (P * F8) + E4 + D4
C3	$R_K =$	(P * F9) + (P * F10) + E5 + D5 + (P * F11) + (P * F12) + E6 + D6
C4	$R_K =$	(P * F13) + (P * F14) + E7 + D7.1 + (P * F17) + (P * F18) + E8.3 + (P * F16) + E8.2 + (P * F15) + E8.1 + D7.2 + D7 + (P * F21) +, (P * F22) + E9.3 + (P * F20) + E9.2 + (P * F19) + E9.1 + D8

There are three main management tools aimed at the reduction of antimicrobial use in young calves [1] . Maternal vaccination against enterotoxic E. coli, rotavirus and coronavirus in the last trimester increases specific colostral immunity against these pathogens. A sufficient amount of high-quality colostrum fed to newborn calves within a few hours after birth prevents failure of passive transfer. Finally, attention to good hygiene and other appropriate management factors reduces the pathogen load in the environment.

Regardless of the etiology of the diarrhea, fluid therapy and replacement of electrolyte deficits are a crucial part of treatment of diarrheic calves [26] . The inclusion of an alkalinizing agent in the electrolyte mixture seems to be beneficial; between 83% and 92% of diarrheic calves infected with rotavirus or coronavirus recovered after treatment with electrolyte solutions that contained bicarbonate or acetate compared with 33% of calves that did not receive an alkalinizing agent [16] .

The decision trees show that empirical treatment of diarrheic calves infected with Cryptosporidium parvum, rotavirus or coronavirus in the absence of a diagnostic workup is significantly cheaper than specific treatment according to a laboratory diagnosis. This means that from a strictly economic point of view, the indiscriminate use of antibiotics compares favorably with targeted treatment based on an etiological diagnosis. This difference is caused by the high cost of diagnostic testing. Calves infected with rotavirus or coronavirus receiving fluid therapy had a significantly higher survival rate than the same calves treated with antibiotics alone, whereas this difference was less pronounced for calves with cryptosporidiosis treated with fluids, antibiotics, or halofuginone alone.

With respect to the branch of the decision tree for E. coli, the costs are highest when bacteriological culture and antimicrobial susceptibility testing are carried out. The trial-and-error methods provided cheaper treatment even in cases in which three different antibiotics were used consecutively.

In contrast to calves with pneumonia, in which diagnostic laboratory testing is advantageous from an economic standpoint [18] [23] , the same does not seem to be true for calves with diarrhea. Nevertheless, the producer should always be given the option for additional diagnostic testing, particularly when a herd problem exists or when antimicrobial misuse is suspected. Once an etiological diagnosis has been made, specific prophylactic measures can be instituted to reduce further cases of diarrhea and associated costs. This will pay off in the long term and offset the cost of the initial testing; however, this aspect was not taken into account in these calculations.

The risk analysis calculations generated costs between CHF 700.00 and 840.00 for diarrheic calves infected by any of the four pathogens. The costs are highest for calves infected with coronavirus, which accounts for only about 1% of diarrheic calves. The high cost associated with coronavirus diarrhea is attributable mainly to to the low spontaneous cure rate and loss of the calf [29] .

5. CONCLUSION

There is a need for standard operating procedures for the management of diarrhea in young calves when the goal is to prevent unnecessary use of antibiotics in diarrheic calves. It is expected that the implementation of standard operating procedures for evidenced-based treatment of calves with diarrhea will increase therapeutic costs. This is also likely to increase the prices in the entire chain of animal food products. However, if the use of antimicrobial drugs must be reduced, consumers must tolerate paying higher food prices.

CITE THIS PAPER

Michael Hässig, Susanne Kretschmar (2016) Evidence-Based Use of Antibiotics in Veal Calves with Diarrhea. *Open Journal of Veterinary Medicine*,**06**,28-39.

REFERENCES

1. Smith, G. (2015) Antimicrobial Decision Making for Enteric Diseases of Cattle. Veterinary Clinics of North America: Food Animal Practice, 31, 47-60.
2. Radostits, O.M., Gay, C. C., Hinchcliff, K.W. and Constable, P.D. (2007) Veterinary Medicine. 10 Issues. Saunders, Elsevier.
3. Luginbühl, A., Reitt, K., Metzler, A., Kollbrunner, M., Corboz, L. and Deplazes, P. (2005) Field Study about Prevalence and Diagnostics of Diarrhea Causing Agents in the New-Born Calf in a Swiss Veterinary Practice Area. Schweizer Archiv für Tierheilkunde, 147, 245-252.
4. Braun, U. (2013) Durchfall beim Kalb in den ersten Lebenswochen. Script Innere Medizin Rind, Departement für Nutztiere, Universit&aumlt Zürich, Zürich.
5. Uhde, F.L., Kaufmann, T., Sager, H., Albini, S., Zanoni, R., Schelling, E. and Meylan, M. (2008) Prevalence of Four Enteropathogens in the Faeces of Young Diarrhoeic Dairy Calves in Switzerland. Veterinary Record, 163, 362-366.
6. Lofstedt, J., Miller, L., Duizer, G. and Daley, J. (1996) Comparison of Efficacy of Sulbactam: Ampicillin to Ampicillin and Saline for Treatment of Experimentally

Induced Escherichia coli Diarrhea in Neonatal Calves. Canadian Journal of Veterinary Research, 60, 210-215.
7. Ortman, K. and Svensson, C. (2004) Use of Antimicrobial Drugs in Swedish Dairy Calves and Replacement Heifers. Veterinary Record, 154, 136-140.
8. Valente, C., Fruganti, G., Tesei, B., Ciorba, A., Cardaras, P., Floris, A. and Bordoni, E. (1988) Vaccination of Pregnant Cows with K99 Antigen of Enterotoxigenic Escherichia coli and Protection by Colostrum in Newborn Calves. Comparative Immunology, Microbiology & Infectious Diseases, 11, 189-198.
9. H&aumlssig, M., Eugster, S. and Lewis, F.I. (2015) Evidence-Based Use of Antibiotics in Meat Calves. Open Journal of Veterinary Medicine, 5, 68-72.
10. Bonita, R., Beaglehole, R. and Kjellstr&oumlm, T. (2013) Einführung in die Epidemiologie. WHO Press, Colorado.
11. Bj&oumlrkman, C., Svensson, C., Christensson, B. and de Verdier, K. (2003) Cryptosporidium parvum and Giardia intestinalis in Calf Diarrhoea in Sweden. Acta Veterinaria Scandinavica, 44, 145-152.
12. Grove-White, D. (2007) Practical Intravenous Fluid Therapy in the Diarrhoeic Calf. In Practice, 29, 404-408.
13. Berge, A.C., Moore, D.A., Besser, T.E. and Sischo, W.M. (2009) Targeting Therapy to Minimize Antimicrobial Use in Pre-Weaned Calves: Effects on Health, Growth, and Treatment Costs. Journal of Dairy Science, 92, 4707-4714.
14. Constable, P.D. (2009) Treatment of Calf Diarrhea: Antimicrobial and Ancillary Treatments. Veterinary Clinics of North America: Food Animal Practice, 25, 101-120.
15. Heiniger, D., van den Borne, B.H.P., Lechner, I., Tschopp, A., Strabel, D., Steiner, A. and Meier, H. (2014) Kosten-Nutzen-Analyse einer Intervention zur Verbesserung der Eutergesundheit in Schweizer Milchviehbetrieben. Schweizer Archiv für Tierheilkunde, 156, 473-481.
16. Pereira, R.V.V., Santos, T.M.A., Bicalho, M.L., Caixeta, L.S., Machado, V.S. and Bicalho, R.C. (2011) Antimicrobial Resistance and Prevalence of Virulence Factor Genes in Fecal Escherichia coli of Holstein Calves Fed Milk with and without Antimicrobials. Journal of Dairy Science, 94, 4556-4565.
17. De Verdier, K., Nyman, A., Greko, C. and Bengtsson, B. (2012) Antimicrobial Resistance and Virulence Factors in Escherichia coli from Swedish Dairy Calves. Acta Veterinaria Scandinavica, 54, 2.
18. ARCH-Vet (2013) Bericht über den Vertrieb von Antibiotika in der Veterin&aumlrmedizin und das Antibiotikaresistenzmonitoring bei Nutztieren in der Schweiz. Eidgen&oumlssisches Departement des Innern, Bundesamt für Lebensmittelsicherheit und Veterin&aumlrwesen. http://www. anresis.ch/ index.php/resistenzdaten-veterinaermedizin.html
19. Eugster, S. (2013) Antibiotikaresistenz in K&aumllbermastbetrieben. Master's Thesis, Vetsuisse Faculty, University of Zurich, Zürich.
20. Castrucci, G., Ferrari, M., Frigeri, F., Traldi, V. and Angelillo, V. (1994) A Study on Neonatal Calf Diarrhea Induced by Rotavirus. Comparative Immunology, Microbiology & Infectious Diseases, 17, 321-331
21. Naylor, J.M., Petrie, L., Rodriguez, M.I. and Skilnick, P. (1990) A Comparison of Three Oral Electrolyte Solutions in the Treatment of Diarrheic Calves. Canadian Veterinary Journal, 31, 753-760.
22. Bouda, J., Doubek, J., Medina-Cruz, M., Paasch, M.L., Candanosa, A.E., Dvorák, R. and Soska, V. (1997) Pathophysiology of Severe Diarrhoea and Suggested Intravenous Fluid Therapy in Calves of Different Ages under Field Conditions. Acta Veterinaria Brno, 66, 87-94.
23. Glover, A.D., Puschner, B., Rossow, H.A., Lehenbauer, T.W., Champagne, J.D., Blanchard, P.C. and Aly, S.S. (2013) A Double-Blind Block Randomized Clinical Trial on the Effect of Zinc as a Treatment for Diarrhea in Neonatal Holstein Calves under Natural Challenge Conditions. Preventive Veterinary Medicine, 112, 338-347.

24. Keidel, J. and Daugschies, A. (2013) Integration of Halofuginone Lactate Treatment and Disinfection with P-Chloro-M-Cresol to Control Natural Cryptosporidiosis in Calves. Veterinary Parasitology, 196, 321-326.
25. Naciri, M., Mancassola, R., Yvore, P. and Peeters, J.E. (1993) The Effect of Halofuginone Lactate on Experimental Cryptosporidium parvum Infections in Calves. Veterinary Parasitology, 45, 199-207.
26. Joachim, A., Krull, T., Schwarzkopf, J. and Daugschies, A. (2003) Prevalence and Control of Bovine Cryptosporidiosis in German Dairy Herds. Veterinary Parasitology, 112, 277-288.
27. Lallemond, M., Villeneuve, A., Belda, J. and Dubreuil, P. (2006) Field Study of the Efficacy of Halofuginone and Decoquinate in the Treatment of Cryptosporidiosis in Veal Calves. Veterinary Record, 159, 672-676.
28. Altman, D.G. (1994) Practical Statistics für Medical Research. Chapman & Hall, London.
29. Jacks, T.M., Schleim, K.D., Miller, B.M., Shungu, D.L., Weinberg, E. and Gadebusch, H.H. (1983) Quaternary Heterocyclylamino Beta-Lactams. V. L-640,876 Treatment of Induced Enterotoxigenic Colibacillosis (Scours) in Calves and Piglets. The Journal of Antibiotics, 36, 70-75.

Index